Green Chemistry Metrics

Green Chemistry Metrics: Measuring and Monitoring Sustainable Processes

Edited by

Alexei Lapkin
Centre for Sustainable Chemical Technologies
Department of Chemical Engineering
University of Bath
Bath BA2 7AY
UK

David J.C. Constable
GlaxoSmithKline
One Franklin Plaza
Mail Stop: FP1870
Philadelphia, PA 19101
USA

WILEY
A John Wiley and Sons, Ltd, Publication

This edition first published 2008

Reprinted with corrections December 2009

Blackwell Publishing was acquired by John Wiley & Sons in February 2007. Blackwell's publishing programme has been merged with Wiley's global Scientific, Technical, and Medical business to form Wiley-Blackwell.

Registered office
John Wiley & Sons Ltd, The Atrium, Southern Gate, Chichester, West Sussex, PO19 8SQ, United Kingdom

Editorial office
9600 Garsington Road, Oxford, OX4 2DQ, United Kingdom
2121 State Avenue, Ames, Iowa 50014-8300, USA

For details of our global editorial offices, for customer services and for information about how to apply for permission to reuse the copyright material in this book please see our website at www.wiley.com/wiley-blackwell.

Library of Congress Cataloging-in-Publication Data
Green chemistry metrics : measuring and monitoring sustainable processes / edited by Alexei Lapkin, David Constable.
p. cm.
Includes bibliographical references and index.
ISBN 978-1-4051-5968-5 (hardback : alk. paper)
1. Environmental chemistry–Industrial applications. 2. Chemistry, Analytic–Quantitative. I. Lapkin, Alexei. II. Constable, David, 1958-
TP155.2.E58G746 2008
660.028'6–dc22

2008012704

A catalogue record for this book is available from the British Library.

Set in 10/12 pt Minion by Newgen Imaging Systems (P) Ltd, Chennai, India

1 2009

Contents

Contents

Preface

Following the establishment of the 12 Principles of Green Chemistry by Anastas and Warner,[1] there has been a steady growth in our understanding of what Green Chemistry means. Happily, this increased understanding has spurred many notable efforts to implement chemical processes and technologies that are incrementally moving society towards more sustainable practices and products. However, the Green Chemistry movement has always struggled to convince many people in the chemical community that Green Chemistry is different from the historical pollution prevention initiatives of the 1980s that Green Chemistry grew out of. The idea of changing the way chemistry is done, the development of new chemistries, the establishment of chemical processes that are significantly more efficient using non-petrochemical and renewable feedstocks has been a challenging notion for a world that has been surrounded by the products of petroleum for more than a century.

The writing of this book was undertaken because it was intended to be the first work that solely focuses on chemistry, and what appropriate metrics for Green Chemistry might be. We hope the book provides an up-to-date and authoritative text on the current development of environmental concepts in chemical technologies from clean and green to sustainable development. We also think it provides up-to-date information on the problems of metrics: fundamental aspects of metrics, practical realisations and real-world case study examples. The concepts and approaches of metrics are related to the fundamental problems in chemistry and the main focus is on the use of metrics to promote the development and implementation of green chemistry and technology solutions.

Different approaches to metrics have been described by noted authors for various parts of the chemical enterprise and examples of their applications in practice have been provided. The main focus has been on chemical processes but the breadth and interconnections chemistry has with other technology sectors (e.g., energy, electronics, healthcare, food, consumer products) means that many aspects of the book will be relevant to those related sectors as well.

The functional approach to products and processes, sometimes known as 'product design', or more specifically for this book 'green product design', has also been used as a basis for discussion of the different metrics approaches that stimulate R&D and implementation of new concepts.

This book has been written for a reasonably prepared reader who is comfortable with chemistry, chemical technologies, and the processes used to make modern materials of commercial interest. However, there are two introductory chapters that start fairly gently and will help to prepare the relatively novice reader. The book is generally intended for university, graduate and post-graduate level training. Some undergraduate and especially master level courses should be able to draw from the book as should advanced practitioners in Industry and Government.

References

1. Anastas, P.T. & Warner, J.C. (1998) *Green Chemistry: Theory and Practice.* Oxford University Press, New York.

Contributors

Chapter 1

Professor Martin A. Abraham, P.E., received his B.S. in chemical engineering in May 1982 from Rensselaer Polytechnic Institute and his Ph.D. in May 1987 from the University of Delaware. Dr Abraham joined Youngstown State University as Professor of Chemical Engineering and Founding Dean of the College of Science, Technology, Engineering and Mathematics in July 2007, after serving as Professor and Dean of the College of Graduate Studies at the University of Toledo. Dr Abraham has over 60 refereed publications and over 100 technical presentations in the area of green engineering and sustainability. He was the chair for the multidisciplinary conference on 'Green Engineering: Defining the Principles' and the editor for the book *Sustainability Science and Engineering: 1. Defining Principles*, the first book in the new Elsevier series on Sustainability, for which he also serves as a co-editor. He also serves as editor for the American Institute of Chemical Engineer's quarterly, *Environmental Progress*, and is the chair of AIChE's Sustainable Engineering Forum. He is a member of the American Chemical Society and Sustainability subcommittee chair for their Committee on Environmental Improvement.

Professor James Clark is a graduate of Kings College, London. Following periods in Canada and the UK he was appointed to the new chair in Industrial and Applied Chemistry at York in 1994. James held an EPSRC-RAEng Clean Technology Fellowship between 1996 and 2001. He is now Director of the Green Chemistry Centre, which brings together research, industrial collaboration, education and worldwide networking. James is founding director of the Green Chemistry Network and was founding scientific editor of the RSC journal *Green Chemistry*. He has published extensively in green chemistry related topics as well as giving Presidential, Plenary, Keynote and numerous other invited lectures on green chemistry around the world. He has won several honours including the RSC John Jeyes medal, the SCI Environment medal and the diploma of honour from the University of Patras as well as leading the team that won Clean Technology awards from the EU and the Royal Society of Arts.

Dr Neil Winterton is a graduate of University College London, gaining his Ph.D. in the study of reaction mechanisms of organotransition metal carbonyl compounds with M. D. Johnson. He is currently University Research Fellow at the University of Liverpool, UK. In 1999, he career-shifted to his present position after 25 years working for ICI's (then) halogenated products businesses on the R&D for their solvent, fire-fighting and plasticizer applications. It was during this time he developed his interests in sustainable technology, being a founder member (and later Deputy Chairman) of the Industrial Advisory Board of the QUESTOR Centre at the Queen's University of Belfast. His current research focuses on seeking to understand (and thereby, possibly, circumvent) those characteristics of ionic liquids that prevent their wider use in chemicals processing. While at ICI he played a critical role in the conception of QUILL, the Queen's University Ionic Liquid Laboratory. Dr Winterton is a Fellow of the Royal Society of Chemistry and European Editor of *Clean Technology and Environmental Policy*.

Chapter 2

Dr Nicholas D. Anastas is an environmental toxicologist with degrees in Environmental, Earth and Ocean Sciences as well as in pharmacology. His doctoral work focused on molecular design for safer chemicals using Green Chemistry. He works for the Massachusetts Department of Environmental Protection focusing on the identification, effects and treatment of pharmaceuticals, personal care products and endocrine disrupting chemicals in the environment. A major focus area of this effort is to engage chemical manufacturers to implement the principles of green chemistry and green chemical design.

Chapter 3

Dr Richard Miller trained as a chemist and spent several years as an academic specializing in measurement science before joining industry as a research manager with Unilever. There he became increasingly involved with renewable feedstocks and mild chemistry, becoming Vice President R&D for Unichema International, an oleochemical subsidiary of Unilever. He moved with the company to ICI becoming Director of Knowledge and Sustainability for one of the group companies. Since leaving ICI to form his own consultancy business he has focused on sustainability and green chemistry, with a particular emphasis on innovation for sustainability. He has worked with many clients, both in the private and public sector, helping them to develop practical routes to sustainability.

Chapter 4

Dr John Andraos earned a B.Sc. and Ph.D. in 1987 and 1992 from the University of Toronto (Canada) in physical organic chemistry and then was a post-doctoral fellow and a Junior Research Fellow at the University of Queensland (Australia). Since his appointment as Lecturer and Course Director at York University in 1999 he has taught and developed courses in organic chemistry. In 2002 he launched the first industrial and 'green' chemistry course in the history of the Department of Chemistry at York. His current research is broadly defined as reaction optimization and discovery including: application of reaction metrics for analysis of organic reactions and total syntheses of organic molecules; optimization of recycling and reagent retrieval protocols; discovery of new multi-component reactions by structural combinatorial techniques; unified mathematical analysis of green metrics; synthesis tree analysis; molecular and topological complexity and connectivity; route selection, and strategy; and quantification, plasticity and optimization of dynamic kinetic systems relevant to stereoselective syntheses of pharmaceuticals.

Currently he is undertaking an ambitious project which is the construction of a database of synthesis plans fully quantified by green metrics analysis for pharmaceuticals, important natural products, dyestuffs, agrichemicals and molecules of theoretical interest. He has given invited addresses on green chemistry research and education at the University of Western Ontario, the National Academy of Sciences in Washington, D.C., the 15th Organic Process Research and Development Conference in San Diego, 2007, BioVerdant in San Diego, Merc & Co. in Rahway, New Jersey, and Apotex Pharmachem in Brantford. He is the author of more than 40 scientific papers in refereed journals and the founder of CareerChem (http://www.careerchem.com/MainFrame.html).

Chapter 5

Dr Bruno Bühler studied biotechnology at the Swiss Federal Institute of Technology (ETH), Zürich, Switzerland, and received his Ph.D. in 2003 at the Institute of Biotechnology at ETH Zürich. After a postdoctoral period, he became group leader of the Applied Biocatalysis group at the Chair of Chemical Biotechnology at the Technical University of Dortmund, Germany, in October 2004. His research focuses on redox biocatalysis with special emphasis on oxyfunctionalization reactions and involves all steps of bioprocess development from gene to product. Beside biocatalyst development and engineering (enzyme engineering, metabolic engineering), this also includes integrated process development (downstream processing and scaling) and economical and ecological process evaluation issues. In this respect, Dr Bühler is involved in several projects with the Deutsche Bundesstiftung Umwelt (DBU) and the network for sustainable chemical and biotechnological production processes ChemBioTec.

Professor Erick M. Carreira was born in Havana, Cuba, in 1963. He received his B.S. degree from the University of Illinois at Urbanas Champaign working with Scott Denmark and his Ph.D. from Harvard University (1990) working under the direction of David A. Evans. After postdoctoral research at the California Institute of Technology with Peter Dervan, he joined the faculty at California Institute of Technology as an assistant professor (1992). He was promoted to associate professor in 1996, and subsequently to professor (1998). He is currently Professor of Chemistry at the ETH-Zürich. He is the recipient of the American Chemical Society Award in Pure Chemistry, Nobel Laureate Signature Award, Fresenius Award, a David and Lucile Packard Foundation Fellowship in Science, Alfred P. Sloan Fellowship, Camille and Henry Dreyfus Teacher Scholar Award, Merck Young Investigator Award, Eli Lilly Young Investigator Award, Pfizer Research Award, National Science Foundation CAREER Award, Arnold and Mabel Beckman Young Investigator Award, and a Camille and Henry Dreyfus New Faculty Award. He is also the recipient of the Associated Students of the California Institute of Technology Annual Award in Teaching and a Richard M. Badger Award in Teaching.

His research program focuses on the asymmetric synthesis of biologically active, stereochemically complex, natural products. Target molecules are selected which pose unique challenges in asymmetric bond construction. A complex multistep synthesis endeavour provides a goal-oriented setting within which to engage in reaction innovation and design. Drawing from the areas of organometallic chemistry, coordination chemistry, and molecular recognition, Carreira's group is developing catalytic and stoichiometric reagents for asymmetric stereocontrol.

Dr Marco Eissen received his Ph.D. in organic chemistry in 2001 from the University of Oldenburg, Germany, under the direction of Professor Jürgen O. Metzger. He was a postdoctoral fellow at the Swiss Federal Institute of Technology Zürich, Switzerland, under the direction of Professor Konrad Hungerbühler and at the Saarland University with Professor Elmar Heinzle, Germany. In 2003 Marco Eissen began teaching chemistry and mathematics in Göttingen and since 2005 in Ganderkesee, Germany. The development of assessment methodologies in early stages of synthesis design is the main focus of his work. Since 2008 Marco Eissen cooperates with IFU Hamburg, Germany, which is a software company developing solutions in the fields of material flow management, life cycle assessment (LCA) and environmental management.

Dr Christian Fischer earned a M.Sc. and Ph.D. in 1999 and 2004 from the Swiss Federal Institute of Technology Zürich in organic chemistry under the direction of Professor Erick M. Carreira. His research focused on transition-metal catalysed additions of alkynes to imines and the development of novel chiral ligands for late transition metals. In 2004 he joined the group of Professor Gregory C. Fu at MIT as a DAAD postdoctoral fellow where he worked on catalytic asymmetric cross-couplings and organocatalytic Heck reactions. Since 2006 Dr Fischer is a Senior Research Chemist in the Department of Drug Design and Optimization at Merck Research Laboratories Boston.

Dr Georg Geisler is a product safety expert and modeller working with RCC Ltd, a Contract Research Organisation based in Basel, Switzerland. In this function, he conducts environmental risk assessments of pesticides, biocides and other chemicals, as well as safety assessments for pesticide residues in the food chain. In 2003, Georg Geisler earned his Ph.D. on environmental life-cycle assessment of pesticides at ETH Zürich. In 1999, he had received a Diploma in environmental chemistry at the Friedrich-Schiller University, Jena, Germany.

Professor Konrad Hungerbühler studied Chemical Engineering at ETH-Zürich and carried out his Ph.D. work under the guidance of Professor J.R. Bourne (1979). He immediately joined Ciba-Geigy where his industrial career started within a central process engineering group directed by Professor W. Regenass. He joined Ciba Grenzach to establish a process engineering group within the Chemicals Division. In 1989 he became head of a research and development group of dye stuff and tannery products. Konrad Hungerbühler appointed to be Professor at the ETH Zürich in 1994. His research topics cover the environmental and safety oriented assessment of chemical products and processes.

Professor Andreas Schmid graduated as biologist (microbiology and biochemistry) at the University of Stuttgart (Germany) in 1992 and recieved his Ph.D. with a topic on environmental microbiology at the University of Stuttgart in February 1997. After joining the Laboratory of Waste Air Purification (Department of Civil Engineering, University

Stuttgart, 1995 to 1996) and an internship at the Swiss Federal Institute for Environmental Science and Technology (EAWAG, Dübendorf, Switzerland), he joined the Institute of Biotechnology (ETH Zürich) as a PostDoc. From 1998 to 2004 he was head of the Technical Enzymology Laboratory at the same Institute. He recieved his venia legendi for biotechnology and microbiology in 2004 from ETH Zürich. He then moved to the University of Dortmund, where he is heading the Chair of Chemical Biotechnology since October 2004 and since January 2005 the 'Single Cell Analysis' Laboratory at the Institute for Analytical Sciences (ISAS, a member of the Leibniz Association). Professor Schmid has an intense collaboration with the Deutsche Bundesstiftung Umwelt (DBU) and leads the network for sustainable chemical and biotechnological production processes ChemBioTec. For more information, please refer to http://www.bci.uni-dortmund.de and http://www.chembiotec.de.

Chapter 6

Dr David J.C. Constable is currently the Director of EHS Product Stewardship Team, Corporate Environment, Health and Safety, at GlaxoSmithKline. In this role David is responsible for championing EHS Support for New Product Development and Supply and shares responsibility for developing programs, systems, tools and methodologies that integrate Sustainability, Life Cycle Inventory/Assessment, Green Chemistry and Green Technology activities into existing R&D and Global Manufacturing Supply business processes. David joined SmithKline Beecham's Environmental Research Laboratory in 1991 to assist in the development of a variety of fate, effects and treatability data for drug substances in SB's portfolio and has held a variety of positions in Corporate Environment, Health and Safety since then. Prior to joining SB, David worked for ICI Americas as a Group Leader in the SHEA Analytical Services group where he developed a variety of non-routine air sampling methodologies. David obtained a Ph.D. in Analytical Chemistry from the University of Connecticut in Storrs, CT, USA and a B.S. in Environmental Sciences, Air and Water Pollution from Slippery Rock University, Slippery Rock, PA, USA.

Dr Concepcion 'Conchita' Jimenez-Gonzalez is Manager of Sustainable Processing and New Product Support at the Corporate Environment, Health and Safety department of GlaxoSmithKline. She has been an active contributor in the fields of green engineering, sustainability, Life Cycle Inventory and Assessment (LCI/A), green chemistry and the integration of EHS and chemical engineering concepts to design more efficient, greener, safer processes. She holds a Ph.D. degree in Chemical Engineering from North Carolina State University; an M.S. in Environmental Engineering from ITESM; Monterrey, México and a B.S. in Chemical and Industrial Engineering at the Chihuahua Institute of Technology, México. Following the Spanish tradition, she is also known as Conchita.

Dr Alexei Lapkin is a Senior Lecturer in Chemical Engineering at the University of Bath, UK, where he is also a Deputy Director of the Centre for Sustainable Chemical Technologies. He obtained his Master in Chemistry from the Novosibirsk State University, Russia, and Ph.D. in Chemical Engineering from the University of Bath. His research is focused on process intensification, bioprocessing, design and applications of functional materials.

Chapter 7

Dr Dana Kralisch studied environmental chemistry at the Friedrich-Schiller-University (FSU), Jena, Germany. After two years of gaining practical experience in making environmental analyses at the Agency for Agriculture of the Federal State of Thuringia, she started to work as a research assistant at the Institute of Technical and Environmental Chemistry at the FSU. In 2006, she obtained her Ph.D. from the School of Chemical and Earth Sciences of the FSU. In her work she concentrated on the integration of sustainability criteria into chemical process development. Since 2007 she is leading the green process engineering and evaluation research group at the Institute of Technical and Environmental Chemistry.

Chapter 8

Dr Lauren Heine is the Principal for the Lauren Heine Group LLC in Bellingham, Washington and a Senior Advisor with Clean Production Action. As such, she guides the development of technical tools and approaches that help organizations integrate Green Chemistry and Green Engineering into their product and process design and development activities. In her previous role as Director of Applied Science at GreenBlue, Lauren was the driving force and guiding hand behind the development of CleanGredients™. Prior to that, she was Director of Green Chemistry and Engineering at the Portland, OR-based, Zero Waste Alliance (ZWA) and a Fellow with the American Association for the Advancement of Science in the Green Chemistry Program of the Industrial Chemicals Branch of the U.S. EPA in Washington, D.C. She also taught Organic Chemistry labs at Bowdoin College in Brunswick, ME, where she helped to develop the Microscale Organic Lab program. Lauren earned her doctorate in Civil and Environmental Engineering from Duke University.

Teresa McGrath is a chemist with NSF International, a non-profit public safety company based in Ann Arbor Michigan. Before working for NSF Teresa McGrath earned her Masters degree in Clean Chemical Technology at the University of York. She worked for two years at the U.S. EPA in the Design for the Environment (DfE) Branch of the Office of Pollution Prevention and Toxics (OPPT). She now is the Lead Chemist in the new Green Chemistry Program at NSF International.

Chapter 1
Introduction

Martin A. Abraham, James Clark and Neil Winterton

'Green chemistry', 'green engineering' and 'sustainability' – these terms are often used interchangeably to describe the concept of making processes and products that have reduced environmental impact. However, if one examines these concepts more deeply, it becomes clear that there are significant differences in their philosophy, which impacts the applicability of the methodologies and techniques in the development of an environmentally conscious society (see Figure 1.1).

Green chemistry is most appropriately defined by the twelve principles of Anastas and Warner.[1] These principles speak to issues associated with chemical reactions and chemical products, and describe synthesis concepts critical to making chemicals in a purposefully more environmentally benign way. Thus, specific ideas such as the use of benign reaction solvents, reducing the level of derivatization, and the concept of atom economy, or incorporating all of your raw materials into your product are focal points of green chemistry. Although not specifically included within the principles, disciplines such as toxicology and thermodynamics play an important role in green chemistry. In many respects, green chemistry can be considered as the scientific underpinning of environmentally preferable manufacturing.

Green engineering, on the other hand, seeks to apply fundamental engineering concepts for the improvement of manufacturing processes. According to the definition promoted

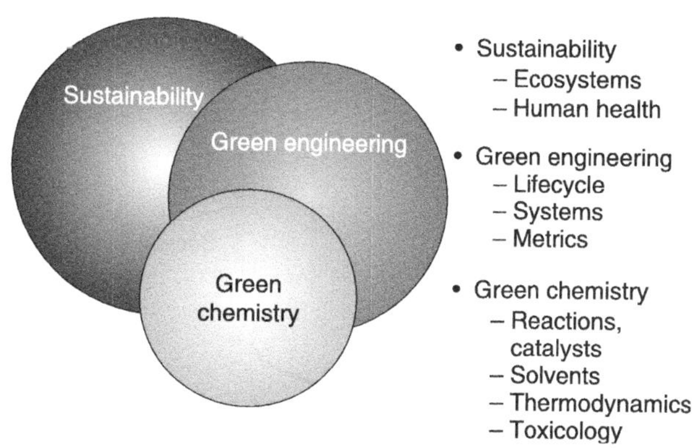

Figure 1.1 Description of the relationships between green chemistry, green engineering and sustainability.

by the Environmental Protection Agency (EPA), green engineering is 'the design, commercialization and use of processes and products that are feasible and economical while minimizing: generation of pollution at the source, and risk to human health and the environment'.[2] The green engineer uses the tools of recycle, process intensity, and design optimization to maximize the efficiency of the process, and reduce its burdens on the environment. Green engineering evaluates the manufacturing process as a system and seeks to optimize its design, and in the truest sense, incorporates the concepts of life-cycle analysis and environmental economics into an appropriate evaluation of the overall environmental impact. Because green engineering is based on optimization, it requires the development of a set of metrics that appropriately evaluate the environmental parameters that we seek to control.

Finally, *sustainable design* looks out even more broadly at *systems* issues to try to understand the relationships between the manufacturing system and the ecosystem, including issues surrounding environmental impacts on human health. Sustainability focuses on the triple bottom line, the integration of ecological integrity, societal responsibility, and economic viability. Sustainability takes the broadest level systems approach, looking at the planet as the system of interest. But in order to optimize design at this scale, new ways of measuring human impacts on the environment will be required (see Figure 1.2).

The chemical industry faces as tough a challenge as any other. The twentieth century saw enormous growth in chemicals manufacturing which fed the parallel economic growth in the developed world. However, this growth has come at a cost. Inefficient processes leading to unacceptable levels of pollution, hazardous operations resulting in a number of well publicized disasters, and inadequate product testing causing often irrational public concerns over product safety, all leading to an exponential growth in chemicals legislation. The industry now needs to achieve environmentally acceptable and economically viable manufacturing in the toughest ever legislative framework while meeting the increasing demands of a growing population. Sustainable chemical production for the new century can only be realized through a reassessment of the entire chemical product life-cycle from resources, to manufacturing and production, through to product use and ultimate fate (Figure 1.1).

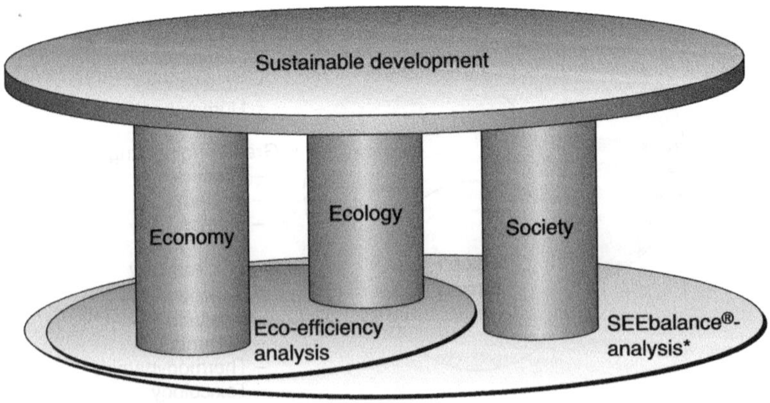

Figure 1.2 The three pillars of sustainable development. After BASF [3].

A useful example of sustainable design comes from BASF, and their development of the eco-efficiency tool.[3] This tool seeks to integrate the combined aspects of each of the three pillars in an attempt to quantify the most sustainable products and illustrates some of the concerns associated with evaluating sustainable products. For example, the economic analysis includes a total cost of ownership that goes beyond the purchase price of a product to incorporate the cost of operation, the cost of environmental health and safety, and the cost of labour. Thus, even though a product may have a lower purchase price, it may be more expensive to use and thus costlier over the total life cycle of the product.

The ecological impacts are evaluated on a life cycle basis, looking at elements such as energy consumption, emissions of global-warming and ozone-depleting gases, water-borne toxins and ecotoxicity, risk potential, land use, consumption of raw materials and similar measures. These ecological parameters can then be combined in a systematic way in order to understand the total environmental impact of the product, on the basis of user benefit. Eco-efficiency analysis combines the ecological analysis with the total cost of ownership to provide an overall measure of benefit relative to environmental impact.

Sustainability clearly must be addressed on a global, systems basis. For example, the fuel efficiency of the US automobile has increased by approximately 60% since the early 1970s. However, the consumption of transportation fuels has increased from about 16 quads to about 25 quads (and the emissions from this fuel consumption has increased by a similar amount) over that time. Why? First, the number of vehicles has steadily increased over that time, from about 300 million in 1975 to over 600 million today. In addition, we are driving more often by ourselves, and we have tended to drive larger vehicles, totally offsetting the advantages gained by the improved fuel economy. Only now, with the cost of gasoline approaching record high levels is vehicle fuel economy once again becoming important.

In some respects, our success in increasing the fuel efficiency of our vehicles has allowed the consumer to use their vehicles in a less environmentally appropriate way. This ‘Catch-22’ continues to demand greater attention to the overall transportation system as we look at alternative technologies. Had gasoline prices continued to rise through the 1970s, it is likely that the expansion of vehicle usage may not have occurred, and the challenges of global climate change might not be so severe today. While it is not possible to undo the mistakes of the past, it is critical that we have techniques available to predict the impacts of similar unintended consequences as we role out new economies for the future.

In the area of renewable resources, we are already seeing such consequences, with the growth in the price of corn (also known as maize) and its impact on the cost of foods throughout our society. Because corn is used both for ethanol production and animal feed, one needs to carefully evaluate the societal benefits of diverting corn for fuel. These analyses must include measures that consider the impact of increased land use for crop growth (with its potential impacts on biodiversity), the impact of increased fertilizer use on water quality, increased cost of foodstuffs and its impact on less affluent societies, and similarly broad issues in the design of metrics that incorporate the fullest systems analysis, allowing us to optimize these processes on a global scale.

The most profound challenge to sustainable systems design will be to reconcile the meeting of the essential needs of a growing population with the minimization of its impact on the natural world. Central to the question of sustainable development is an assessment of the degree to which the natural resources of the planet are sufficient and in a condition, now and in the future, to supply the needs of the human population, to assimilate wastes that human-kind produces, whilst not affecting the natural diversity. Such a question can be cast graphically (if simplistically) into the form of an equation, that relates the Earth's carrying capacity (or environmental burden) to the product of the Earth's total human population, P, some notional measure of average individual prosperity, wealth or economic activity, W_P, and a factor, B_W, to estimate the aggregated burden on the planet arising from such units of wealth or economic activity. This is the so-called 'master' equation.[4]

$$\text{Carrying Capacity} = P \times W_P \times B_W$$

The world population (P) continues to increase, though there is evidence that the rate of increase may be declining. The aspirations for the populations of developing countries to improve their economic circumstances and political pressure to eliminate excessive disparities of wealth in developed countries are such that average prosperity or economic activity (W_P) is also unlikely to show any marked reduction in the foreseeable future. If, as some contend, the world is close to, even beyond, its carrying capacity, then to prevent any further increases (or even to reduce it) it will be necessary to offset increases in P and W_P by proportionate decreases in B_W.

B_W, the burden on the planet per unit of economic activity, can be reduced in two ways:

by more efficient use of raw materials (and the use of less energy in so doing) ('dematerialization')

and replacement of feedstocks, energy sources and raw materials whose use is environmentally damaging with alternatives that are less so and by treating all wastes as resources ('transmaterialization').

Dematerialization and transmaterialization represent major technological challenges concerned with the whole processing chain from raw material extraction, intermediate production, transformation and processing, assembly (and then use and, finally, recovery and disposal) of all the products we employ. Each separate stage involves combinations of chemical or physical manipulation as well as effects or consequences, some beneficial, others deleterious, arising from chemical or physical phenomena. Chemistry and chemical engineering, along with other key disciplines, thus play a central role in responding effectively to such challenges. Critical will be the need to develop less waste-producing and more efficient chemical technologies, the essential focus of green chemistry and green chemical engineering.

The alternative, that is, doing without technology altogether, is not an option (except, possibly, as an individual lifestyle choice). In the UK, for instance, it has been estimated that a wholly agrarian, pre-industrial, economy could support only 15 M of the current 60 M population. As Norbert Wiener, the father of cybernetics, said over 50 years ago: 'we live by the grace of invention'.[6] This is even more true today. The challenge, now, is to make the right 'invention' using insights from cleaner chemical processing, and developing and applying practical measures of the net benefits that arise from proposed new developments.

As a representative example of our current technological dependency, we can consider the manufacture of the polyester, poly(ethylene terephthalate) (PET) and its use in the production of synthetic fibre for clothing, and to examine the extent to which it might be possible to replace PET fibre by an equivalent quantity of the natural fibre, wool. Some

Box 1.1 Land-use intensification from chemicals production. Reproduced from Winterton [5], copyright 2003, with kind permission from Springer Science and Business Media.

Factor	Dimension	Comment
PET production (as fibre)	500 000 $t\,y^{-1}$	World-scale plant
Area occupied by PET production	ca. 4 ha	Assume area needed for infrastructure (raw material/energy provision, processing) for chemicals production and sheep-rearing are similar
Average weight of sheep fleece	3.8 kg	Round up to 5 kg
Number of sheep to give weight of wool equivalent to PET fibre	ca. 100 000 000	Assume technical and performance equivalence of PET fibre and wool
Number of sheep sustained by best pasture	ca. 25 per ha	Assume all pasture able to sustain at this level
Area of pasture required	4 000 000 ha	40 000 km^2 (land area equivalent to the size of Holland)
Land-use intensification factor	10^6	To within ±1 order or magnitude

Other factors: Disposal of 15–20 $M\,y^{-1}$ sheep carcasses
Waste produced from meat and wool processing
Impact of climatic, seasonal and disease factors
Hazards associated with sheep shearing

simple sums (see Box 1.1), based on reasonable assumptions, show the impracticality of seeking to supply all of the world's PET needs by farming sheep for wool. In addition, and more subtly, the very high intensification of land usage (ca. $10^{6\pm1}$) associated with the chemical process compared with sheep-rearing releases land from intensive agriculture, enabling some of it to be saved for wilderness conservation and the protection of biodiversity.

The production of polyesters such as PET is, currently, almost totally based on the chemistry of fossil-resource-derived intermediates. PET is a condensation polymer, formed by an esterification (or transesterification) of a diacid (or diester) and a diol. DuPont have developed a process in which the diol employed, propane-1,3-diol, is obtained from a renewable feedstock, corn or maize, by a fermentation process. This hybrid of fossil and renewable-sourced intermediates will probably be typical of the initial stages of moving the production of chemicals from a total dependency on fossil feedstocks to one that maximizes the use of renewables. This transition will require a prior full appreciation of the likely impact of a product or process on the environment (through the methods of Life-Cycle Assessment and the input from the environmental, epidemiological and ecological sciences) in addition to the development of efficient processes from a closer integration of chemistry, biochemistry, process engineering, product design and economics to enable cost–benefit comparisons to be made between technologies to assist the identification of the optimum process or product. Even a cursory understanding of current chemicals manufacture reveals how closely integrated are the operations of chemicals production, with the by-product streams of one process being used as feeds to another. Such integration leads to a more effective use of primary raw materials and the parallel with ecological systems (in which waste from one population can become the food of another) has been recognized and has inspired a purposeful examination of the wider application of such concepts across all of manufacturing, known as industrial ecology.[7] Because of the greater inter-dependence of the component parts of a highly integrated system, critics suggest that securing radical change (such as the move to renewable from fossil feedstocks) becomes more difficult, locking in sub-optimum components.

To provide a perspective for the future, particularly to get an idea of the scale of challenges to be met, it is worth taking stock of the immediate past. For instance, our ability to feed ourselves over the last one hundred years (accepting the imperfect global distribution of such food) was dependent, to a significant degree, on the development of the Haber process for the production of ammonia from nitrogen and hydrogen, a development that occurred just at the time that the world supply of naturally 'fixed' nitrogen was running out. In addition, the meeting of society's need for food, clean water, sanitation, health care, shelter and clothing has been achieved, in significant measure, by the application of the chemical sciences, engineering and technology, a conclusion not invalidated by any concerns about the excesses of a consumption-obsessed society.

The increase in efficient use of resources associated with technological development and the economies of scale are self-evident, driven by competitive pressures and a motivation to enhance profit margins by cutting costs. As disposal of waste involves cost, and greater regulatory and social pressures add to these costs, seeking economic benefit through finding uses for waste has driven the formation of whole new industries. Indeed, the dyestuffs and pharmaceuticals sectors began in just such a manner.

The degree to which industrialized societies have 'dematerialized' can be seen from Figure 1.3, in which the amount of material consumed (expressed as carbon from fossil feedstocks used) per unit of added value (using gross domestic product as a measure of economic activity) has declined steadily over the past 30 years of industrial development.

Interestingly, such trends provide (from the slope of these plots) a rough estimate of the factor by which industrial efficiency improved in long-industrialized countries. In the USA, for example, this has changed by a factor of 2–2.5 over the past 30–50 years. Estimates of between 4 and 10 have been made of the factor by which such material and energy efficiency must be improved.[8] This is *more than twice* the historic rate in the *most* productive economies. Fortunately, as Figure 1.3 also shows, some of the more recently developed countries, such as South Korea, have factors of 2–2.5 over 15 years, possibly a consequence of an increased pace of technology transfer and not having to displace earlier, less efficient, technologies. Encouraging the emerging economies to leapfrog traditional approaches to chemicals manufacture may bring about a more rapid introduction of more sustainable technology.

The new levels which the efficiency of energy and material usage must reach, and the critical importance of using technologies that impact less on the environment, highlight the need for novel approaches, in addition to those responsible for past improvements. Many of the concepts for the cleaner and more sustainable technologies of the future are encompassed within the aims of green chemistry and green chemical engineering. Such concepts, while necessary, are not sufficient. Indeed, only if chemists and chemical engineers can develop and apply these concepts in a broader environmental, industrial, economic and sociological context, are they likely to have the quickest and greatest impact.

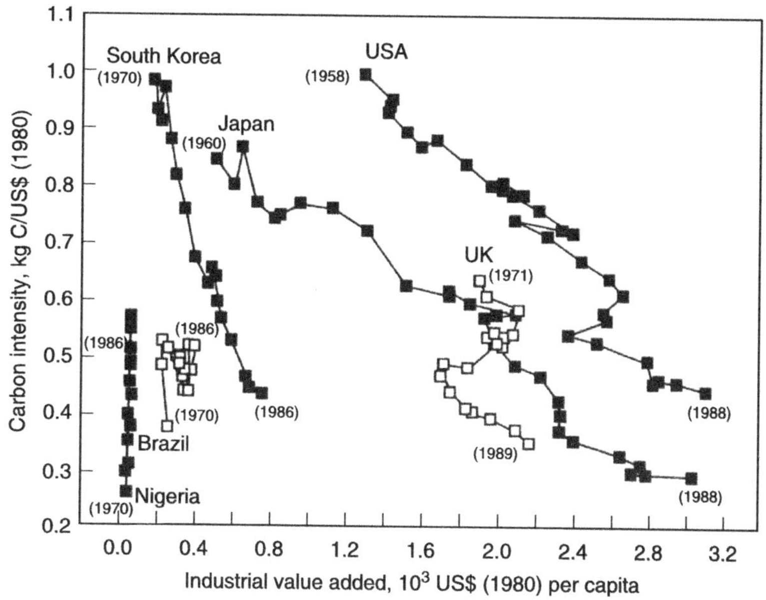

Figure 1.3 Relative decline in resources consumed to create a unit of economic value. Reproduced from Grübler [9] with permission.

A foreseeable problem associated with improving material efficiency is the consequence of increased demand associated with the lowering of costs. Meeting such demand may fail to bring about any absolute reduction in waste, even though the waste produced per ton of product has been reduced.

This raises two fundamentally important questions. First, the importance of substituting existing feedstocks and resources by those whose extraction, processing, use and disposal present less of a threat to the environment. Such 'transmaterialization' is usually associated with the use of renewable feedstocks. The second, more profound, question asks where waste comes from and why it cannot be eliminated altogether.

Addressing the second question first leads to a critical constraint when thinking about new, more sustainable, technological developments, that is, the universal applicability of the laws of thermodynamics to all physical, chemical and biological processes. A central and inescapable fact is the inevitability of waste formation. One statement of the second law of thermodynamics says that heat cannot be converted completely into work. Or, in other words, the energy output of work is *always* less than the energy transformed to accomplish it. A consequence of this is that, *even in principle*, it is impossible for any real process to proceed without the generation of some sort of waste.

It is possible, artificially, to so draw the boundaries of a process to give the appearance that waste is not produced. However, a simple examination of the inputs and outputs (and their origin) for any process will show where such wastes are produced. A topical example of this misunderstanding is represented by the mistaken belief that electric vehicles (whether battery- of fuel-cell driven) are non-polluting compared to those powered by the internal combustion engine (Figure 1.4). For any process, therefore, there needs to be a full

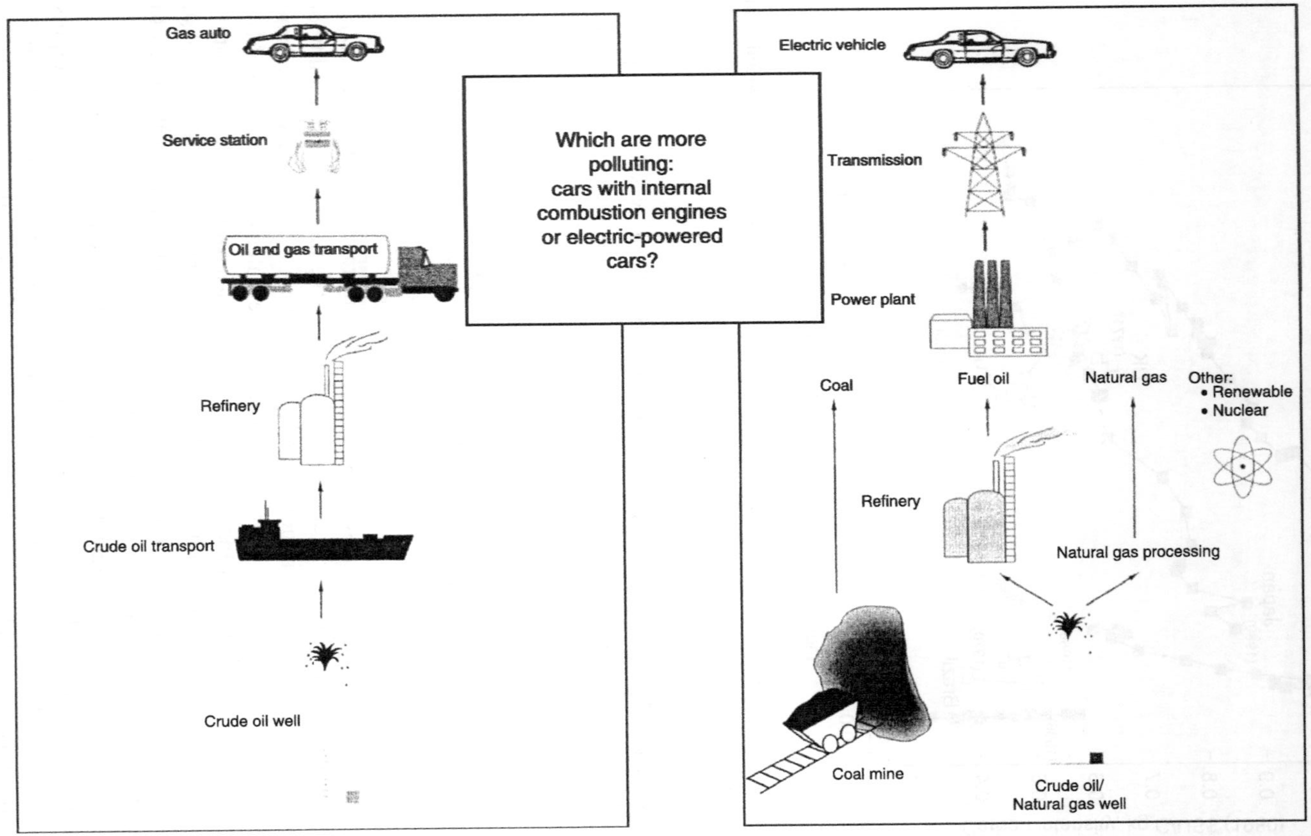

Figure 1.4 Analysing the primary sources of energy required to propel vehicles using internal combustion engines or electricity. Reproduced from Keoleian [10] with permission.

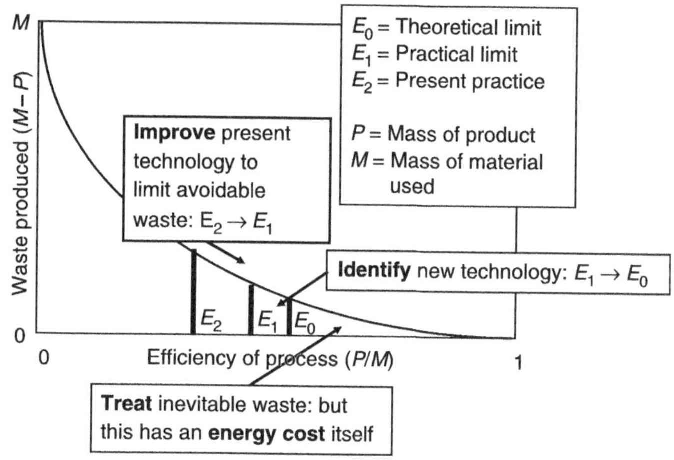

Figure 1.5 Plot of waste produced versus process efficiency. After Jackson *et al.* [11].

inventory made of all feedstocks and energy sources used as an input (back to the primary sources of raw materials, including those used for energy generation) and all outputs, up to and including those emitted at the various stages of production and in use *and* to provide for ultimate disposal according to appropriate ethical and regulatory standards.

Figure 1.5 shows a plot (based on reference [11]) of waste produced in a process (expressed by the difference between the total mass of materials used in production (M) less the mass of useful product (P)) versus process efficiency (expressed by the ratio P/M). The area bounded by the curve to the right of the vertical lines may be seen as representing waste associated with the process, that to the left representing material converted into the useful product (in the limit of M). Perfect efficiency (the thermodynamically impossible 'zero-waste' process) lies at $P/M = 1$ with $M - P = 0$, that is, when all material used in a process finds its way into the product. A real process might lie at E_2. This may then have evolved into, or may have been displaced by, a new, more efficient, process, E_1, in which the area representing waste is reduced. The practical limit, given by E_0, occurs when waste produced is minimized to that inevitably produced taking account of thermodynamic limitations. Past process improvements (through competitive pressure or the need to reduce costs) will have driven the move from E_2 to E_1. Such developments might arise from technological improvements, whether step-change or incremental, or from changes in the raw materials used in the process. A good example is represented by innovation in the catalysed manufacture of ammonia from dinitrogen and dihydrogen, the Haber process, in which dramatic improvements arise from changes in the feedstock used as a source of dihydrogen. Future innovation (perhaps driven, additionally, by customer preferences, by government regulation or global agreement) will continue to drive further cost reductions, thereby moving efficiencies towards the thermodynamic limit.

However, some waste will be inevitable. While some waste represents minimal risk to the environment when emitted (such as the release to the marine environment of depleted brine in chlor-alkali production), a fully sustainable process will (other things being equal) require all harmful emissions to be treated to ensure zero impact on the environment. This might require, for instance, all hydrocarbon emissions from chemicals manufacture to be converted into carbon dioxide and water. In addition, there might be a need to recover and sequester carbon dioxide itself if the release of the latter is on a scale such as to materially add to the anthropogenic component of climate change. So, waste treatment represents a process that itself is waste-producing. This is not the end of the story. It is no longer considered acceptable that a producer ignores the consequences of the use of a product once it has been purchased by a user or consumer. This will often be driven by customer preference, as choices based, for instance, on fuel or energy efficiency, may well decide which product is purchased (and which producer benefits). In addition to the consumption of resources and the impact on the environment during use, similar considerations will attend the final disposal of the product at the end of its useful life. Such factors need to be fully accounted for. An analysis of the overall environmental impact associated with the use of a paper, plastic or ceramic cup (Box 1.2) illustrates the complexities of such a comparison for an everyday object of apparent simplicity. Many relevant factors need to be taken into account and often the data are not available or do not have the precision or the provenance that can allow reliable comparisons. The analysis, based on that of Hocking,[12] shows that the value of the parameter used for the comparison (x, the number of disposable cups needed to exceed the aggregated energy used for x uses of ceramic cup washed after use) is very dependent on precisely how the comparison is made. This has its origins in the different materials employed (and how they are obtained), the technology of cup manufacture, the make-up of

Box 1.2 A comparison of the environmental impact of reusable and disposable cups. After Hocking [12].

TO REUSE OR TO THROW AWAY?
'Which is more environmentally friendly: a disposable cup or a reusable cup?'
(With acknowledgements to M. B. Hocking [12])

If we clean and reuse a reusable ceramic cup x times, we can calculate the value of x (the '<u>break-even point</u>') at which the energy used becomes the same for:

(a) ceramic cup manufacture and its reuse after washing x times
(b) manufacture of x disposable cups (e.g., made with paper or expanded polystyrene).

Factors which affect such a calculation include:

- Raw material energy use per g to produce: *minerals and process them to ceramic*
  oil/naphtha and process styrene to polystyrene
  forest/wood/paper
- Sum of waste produced during manufacture per g
- Sum of all energy consumed during fabrication per cup
- Weight of ceramic and polystyrene cups
- Electrical energy consumed per cup in washing reusable cup
- Water utilisation for each wash per cup
- Efficiency of electrical production (country dependent)
- Factors linked to final disposal of cup: *garbage collection/sorting*
  recycle as feedstock
  net energy produced on incineration (of paper or polystyrene cup)
  impact as solid waste

Box 1.2 Continued.

To calculate the break-even point for *x*, we need values for:

a_1	Energy (kJ/g) of production of ceramic (+ cup fabrication)
a_2	Weight (g) of ceramic cup
b_1	Energy (kJ/g) of production of polystyrene (+ cup fabrication)
b_2	Weight (g) of polystyrene cup (same capacity)
C	Energy required for one wash
G	Net energy recoverable from ceramic cup on ultimate disposal
H	Net energy recoverable from polystyrene cup on ultimate disposal

Energy for production of 1 ceramic cup, $A = a_1 \times a_2$
For x washes, total energy use for ceramic cup: $A + C\,x$
Energy for production of 1 polystyrene cup, $B = b_1 \times b_2$
Equivalent energy for x polystyrene cups, thrown away after single use: $B\,x$

Break-even point occurs when $A + C.x = B\,x$ or $x = A/(B - C)$

Or, taking account of energy recovery on ultimate disposal: $x' = (A - G)/((B - H) - C)$

	Energy requirement (kJ/g)	Weight of 8–9 fl oz cup (g)	Energy requirement (kJ/cup)
Expanded polystyrene	**104.3**	**1.9**	**198**
Paper	**66.2**	**8.3**	**549**
Ceramic	**48.2**	**198.6**	**14088**

Pairs of cups compared	***x***
Ceramic/paper	**39**
Ceramic/polystyrene	**1006**

Note 1: Other scenarios might take account of washing of polystyrene cups or repeat use of cups before washing.

Note 2: Consider other important use-attributes: convenience, domestic or institutional use, heat retention, aesthetics, safety (e.g., insulation, breakage/leak resistance), hygiene (efficiency of washing process).

the energy production ('energy mix') in the country concerned, how the cups are disposed of. This neglects all the non-technical factors relating to ease of use, aesthetics and hygiene. In this particular case, the calculation is simple (even if the raw data are more difficult to obtain[12]). The calculation in Box 1.2 shows that to be less energy-consuming than a polystyrene cup, a reusable cup must be reused >1000 times, whereas only 39 reuses are needed if the comparison is with a paper cup. A good exercise is to think through why this is the case. When it comes to the integrated production of chemicals the problem becomes even more complex. Simply finding a common metric on which to base comparisons between technological choices represents a major challenge. This is dealt with in more detail in a later chapter.

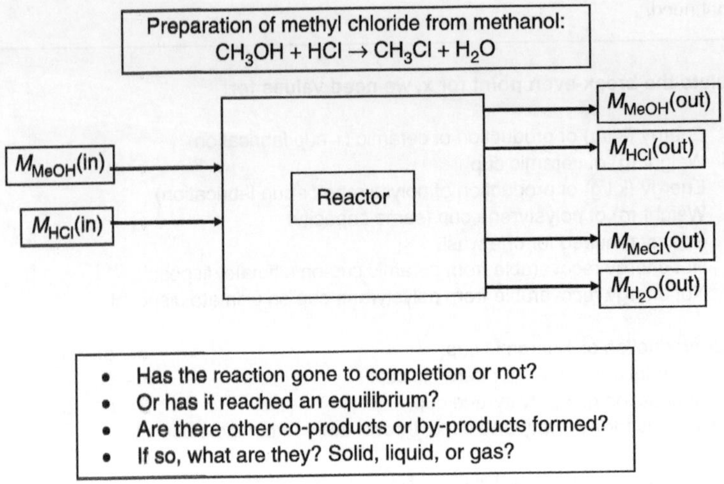

Figure 1.6 Simplified representation of inputs and outputs of a reaction to produce methyl chloride from methanol.

The starting point, therefore, for any evaluation of the degree to which any process may be considered sustainable requires a full inventory of all the inputs and outputs. For chemicals production, this necessarily requires the focus to be on the engineered process, not simply on the chemical transformation that occurs in the reactor. While necessary, the latter is not sufficient.

The notions of mass balance, selectivity and conversion (see Figure 1.6) need to be added to the idea of yield when chemists consider a chemical transformation, particularly if this consideration takes place in the context of pollution prevention or green chemistry.[13]

A chemist seeking to prepare chloromethane from methanol might only be concerned with how much of the desired product is formed. However, yield is an inadequate metric for estimating overall process efficiency. Yield focuses only on a single (usually, the desired) product, expressing the proportion of the stoichiometric amount that is actually isolated, assuming a reaction of known and perfect stoichiometry. A more realistic representation (Figure 1.7) would include the formation of co- and by-products, allow for equilibrium limitations, and include the recovery of unconverted starting materials. Even such an approach is incomplete, as a reaction is usually carried out in a solvent or other medium in the presence of a catalyst or other reaction auxiliaries. In addition, product separation, recovery and isolation are further processes that can involve the use of additional materials, such as solvents, acids or alkali for neutralization, desiccants for drying and so on. Chemists usually also tend to ignore the utilities and services employed in bringing the reaction about (see Figure 1.8). These must be accounted for when engineering and operating a chemical process on the large scale as they can represent a significant cost that must be recouped. These include electricity for heating and agitation, water for cooling and extraction, inert gases for reaction blanketing. (There are, of course, other process- and engineering-related factors that are not included in this introductory analysis, such as the materials employed in the fabrication of the unit operations employed in the overall process.)

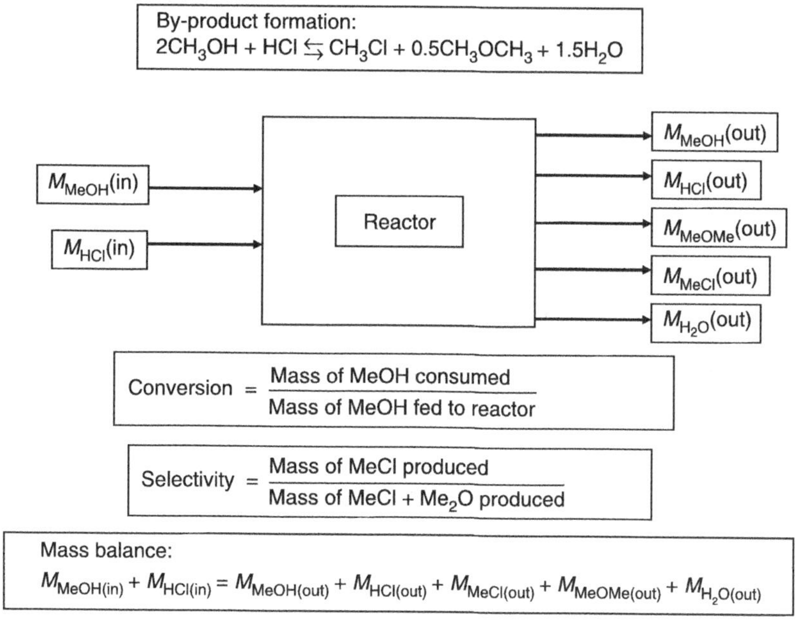

Figure 1.7 Conversion, selectivity and mass balance for a reaction to produce methyl chloride from methanol.

Having established that waste is an inevitable product of any real process, one consequence of increased efficiency, and the associated reduced cost, can be greater demand and turnover. The benefits of technological improvements will be offset (and possibly completely eliminated) by greater product volume. Other strategies of addressing the problem of waste thus need to be developed. We can now address the first question posed above.

One key approach is to limit the consumption of fossil resources, that is, those long stored in the Earth and which cannot quickly be replaced, particularly where their consumption adds wastes that are now known to be contributing to climate change. We need to find new ways of generating the material, energy and food that a growing world population needs, doing so while limiting environmental damage, particularly that which has the most serious consequences for human-kind. The approach is called 'transmaterialization', in which fossil-derived or 'harmful' products are replaced by those judged to be 'renewable'.

An *ideal* renewable resource will be one that can be replenished over a relatively short timescale or is essentially limitless in supply. The latter will include solar radiation, geothermal energy, oxygen, carbon dioxide and water. Nor should production or consumption of these resources contribute to the net atmospheric burden of carbon dioxide. Advantage can be taken of the fixation of atmospheric carbon dioxide into plant material by the process of photosynthesis.

The production of chemicals is characterized by a degree of integration and interdependence that arises from need to find an outlet for (and add value to) products that previously may have been seen as wastes. Bearing in mind this integration of existing chemical technology, in moving it from its current heavy dependence on fossil feedstocks, such as, coal, oil and natural gas, to a greater use of renewable resources represents a series of major

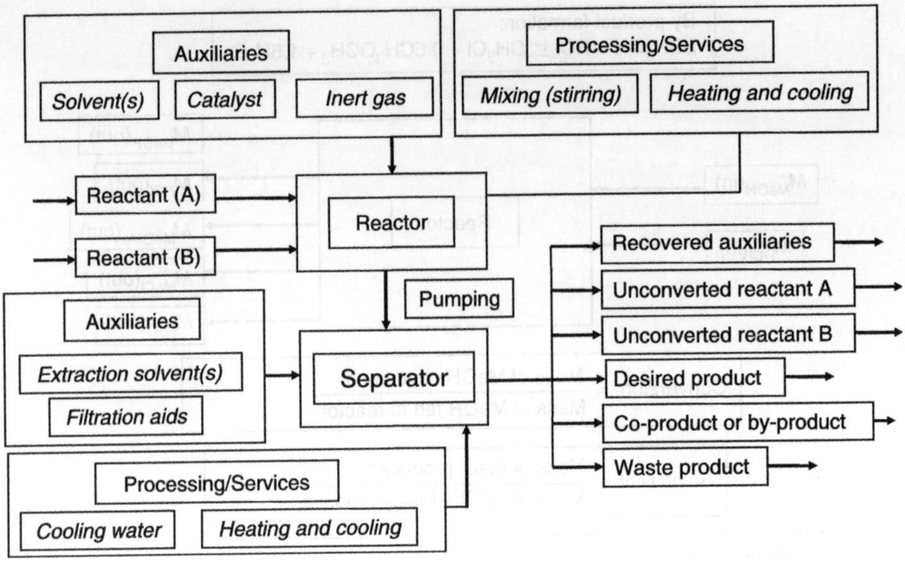

Figure 1.8 A simplified representation of a chemical process including separations and provision of services and utilities.

challenges, some technological, others economic, organizational and political. Of these, possibly the most straightforward to meet is the technological challenge. The others require global agreements where the trade-offs between competing interests often lead to outcomes that fall short of what is truly needed. Such agreements are also often slow to achieve, with patchy observance.

Bearing in mind that most chemicals manufacture arises from companies that are subject to market mechanisms modified, to varying degrees, by regulatory constraint, it is unlikely that there will be any single approach to technological development that will dominate.

Approaches to the fundamental need to shift from fossil to renewable feedstocks for chemicals production will range from modifications to, and developments of, traditional chemical, engineering and biotechnological methods (that may be implemented on a relatively short timescale, say, 10–15 years) to much more radical processes (such as direct capture of solar energy, through artificial photosynthesis), requiring longer time to implement (say 15–30 years).

Therefore, combinations of methods will be needed that, over time, will bring about the desired change. These may include the early substitution of one fossil-derived product or one intermediate by one that is renewable-derived. Ultimately, over time, whole new supply chains using only renewable feedstocks will arise.

Early identification of specific technical challenges may result from considerations of the greater use of existing plant-derived chemical intermediates or products to replace those that are fossil-derived. An example might be the wider use of sugar-derived ethanol as a chemicals feedstock. Why for instance, are such renewable-derived materials not more widely used? Is it because the necessary investment (and the associated 'learning') has not reduced unit costs? Or are there technical or engineering problems that are proving intractable? These might include problems of ensuring a secure, stably priced, source of feedstock or of increasing the productivity of a biotechnological process (such as a fermentation) or finding a cost-effective means of separating a product from a dilute aqueous medium.

As assured sources of plant-derived intermediates become increasingly available, they are then more likely to be used for further downstream derivatization to provide new intermediates and products.

Two broad approaches can be envisaged that might exploit the availability of new high-tonnage plant- or biomass-derived feedstocks. One might involve conversion into some of the relatively small number of basic building blocks that might then feed the existing (largely hydrocarbon-chemistry-based) chemical infrastructure. This would have the advantage that the need for wholly new investment might be limited and could be introduced with relatively limited disruption and its availability could influence investment decisions taking place in the burgeoning economies of China, India and Brazil.

The conversion of biomass into synthesis gas to feed basic chemicals production (linked, in all likelihood, with energy production) would fall into this category. The engineering challenges associated with the processing of highly complex and variable raw material should not be underestimated, however; nor should the required infrastructural changes linking raw-material supply with processing centres.

Alternatively, an entirely new downstream process and product chain, using renewable raw materials, can be conceived (the 'biorefinery'). The chemistry will be more focused on that of oxohydrocarbons (particularly carbohydrates) rather than hydrocarbons. Understanding the materials chemistry of biomass and related products would need to be enhanced. However, work has already been undertaken to identify the top sugar-derived intermediates (Figure 1.9) on which down-stream chemical processing might be derived.

1.1 Ecosystems and the biorefinery concept

Ecosystem services can be considered as being the goods and services provided by coupled ecological and social systems. They provide our material needs and quality of life, and we all depend on the sustainable use of ecosystems services. The Millennium Ecosystems Assessment has brought the continual capacity for coupled ecological–social systems to deliver eco-systems services into public attention.[15] It states that the ability of many systems to deliver valuable services has been compromised by over-exploitation of several resources and environmental degradation.

The extraordinarily rapid development of biofuels is a striking example of this. Greener subsidies, tax-breaks and other industry incentives are accelerating the emergence of biodiesel now and bioethanol and biobutanol in the near future as topics of great concern. Liquid biofuel production in the EU in 2004 amounted to 2040 Mtoe or about 0.3% of the market, an increase of some 30% from 2003. About 80% of this production was biodiesel, mostly produced from rapeseed with the balance being bioethanol produced from wheat and to a small extent sugar beet in Spain, France and Sweden. Volumes are even larger in other parts of the world, notably Brazil, and in most regions, the growth in biofuel production and the government incentives for growth are seemingly unstoppable. But is such large-scale production of fuels from plants, trees and agricultural products really sustainable? The first critical issue is over resources. The biomass potential in the EU based on a recent briefing of the European Environment Agency is summarized in Table 1.1.

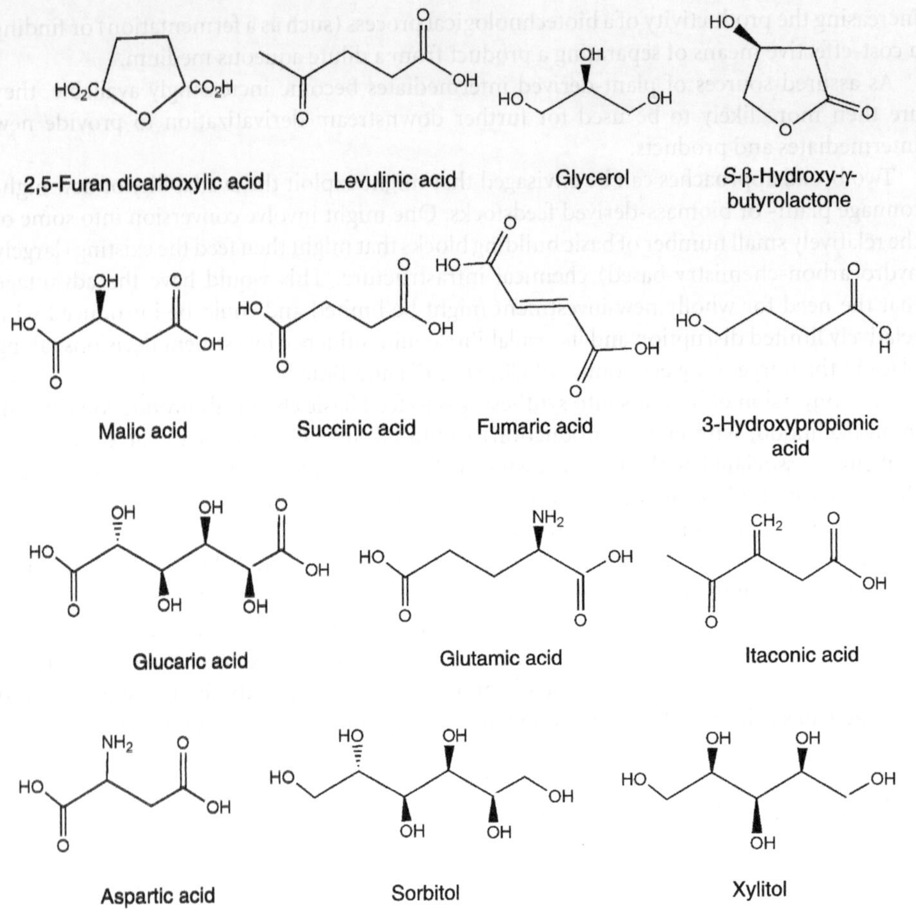

Figure 1.9 Top sugar-derived potential key intermediates identified by the US National Renewable Energy Laboratory. After Werpy *et al.* [14].

These values show considerable growth and overall volume potential (the value for 2003 was 69 Mtoe). They suggest that the EU can continue to plan for a sustainable increasing fraction of its fuel (and possibly chemical and polymer) needs being met from biomass. But can the EU environment cope with such large-scale biomass exploitation?

Biofuels rely on organic feedstocks such as plant oil, food wastes and trees but their larger scale and rapid exploitation to meet government targets is stressing large areas of land and associated systems such as water, food production and recreation. A truly sustainable future for biofuels and other eco-system exploitation for industrial value requires a better understanding and more quantitative assessment of a number of critical issues:

- How external forces change the quality and availability of ecosystem services and how activities in one region affect the ecosystem elsewhere.
- The linkage between livelihoods and human well-being and ecosystem services.
- The management of ecosystem services especially when under pressure.

Table 1.1 Biomass potential in the EU.[16]

	Biomass potential (Mtoe)		
	2010	2020	2031
Forest products	43	39–45	39–72
Organic wastes	100	100	102
Energy crops	43–46	76–94	102–142
Total	186	215–239	243–376

Within the constraints of these factors it is becoming increasingly likely that we will move progressively from petro-refineries to bio-refineries for the production of a large proportion of our organic chemical and material polymer needs and a significant proportion of our energy needs. The chemical and energy industries are still very much dependant on crude oil, natural gas and coal. In 2006 the annual imports of crude oil into the EU was ca. 3500 million barrels of which 10% was used as a chemical feedstock (84 million metric tonnes of oil equivalent) for various industrial chemical industries. An additional almost equivalent amount (ca. 82 MMt of oil equivalent) was used as an energy source to drive these chemical processes (heating, pressure, etc.). In total, the EU chemical industry is by far the most energy intensive industry sector. At the other end of the lifecycle, the chemical industry is also a massive source of greenhouse gases with OECD estimates showing total direct CO_2 emissions value of 10^9 Mt or 4% of global CO_2 emission worldwide. For the EU, a switch from fossil to biomass as a source of organic carbon and energy for chemical manufacturing would provide economic and security benefits by reducing its dependence on imported oil, and environmental benefit by reducing its emission of greenhouse gases. Further economic and social benefits would accrue from providing new income to its agriculturally intense and often isolated tree-rich regions.

Chemicals, including many pharmaceuticals, have been isolated from plant extracts for many years. Ethanol (a petrochemical product in the UK) may be produced by fermentation from sugars derived from plant materials, a process operated on the large scale in Brazil. A major source of chemicals would be 'biomass'. Can sufficient be grown to meet the world's needs? And considering a life-cycle perspective, can sufficient be grown in a sustainable manner; in other words, without the use of depleting petrochemicals for energy, fertilization, and other mechanical systems.

Starch and cellulose are potentially important renewable resources for chemical production. Glucose (a component of starch) is relatively easy to obtain from plant material and is used to synthesize existing chemicals. While this is so, the production of such 'renewable' materials, a full life-cycle assessment of the requirements for their production suggest that much fossil-sourced energy and material would still be employed in the growing, harvesting and processing of biomass.

In addition, other questions arise: is sufficient land area available to grow crops in sufficient quantity? Can the needs of such agro-industrial processes for water be met? To what extent will encroachments on wilderness limit our ability to protect species biodiversity? Bearing in mind the seasonal, climatological and disease effects on biomass production, how can the supply chain for materials, food and energy respond to possible intermittency

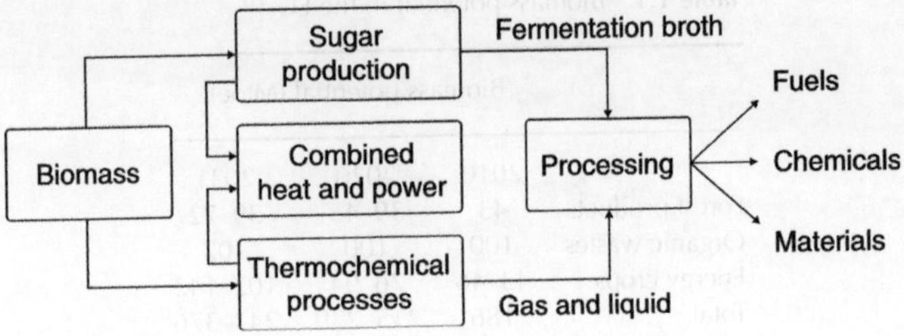

Figure 1.10 The biorefinery concept.

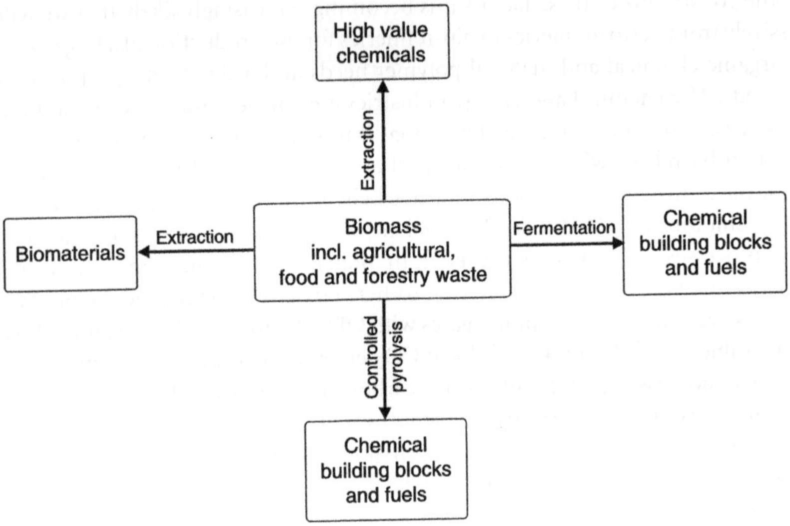

Figure 1.11 Proposed model for second-generation biofuels.

of supply? Can all our requirements for materials be obtained from renewable sources? Are there other bottlenecks that might arise from a limitation of supply of a key intermediate (or material, such as a catalyst) if the volume of renewables demand increases substantially?

Biorefineries will take renewable matter derived from trees, grasses, plants and crops and convert them into non-food as well as food products (Figure 1.10). The biorefinery is the term that is used to describe the facilities that will carry out these conversions. In today's limited number of biorefineries, one raw material, for example, tree biomass, is converted into simple chemicals and energy products, for example, bioethanol, cellulose and ethyl acetate. In second-generation biorefineries the technologies will be more elaborate and will be designed to maximize carbon utilization including extraction of surface chemicals (waxes), fractionation to produce higher value products, and use of energy efficient processes such as controlled pyrolysis to provide valuable chemical building blocks (Figure 1.11). It is very important that these future biorefineries largely use forestry, agricultural and other wastes as feedstock. At present, biorefineries producing non-food products are perceived as being in competition with food production. For economical as well as social reasons this 'fuel

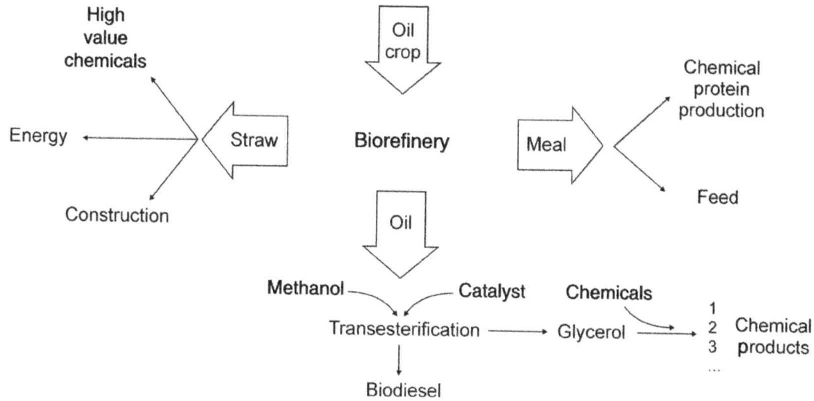

Figure 1.12 A future biorefinery model based on a biochemical production plant.

versus food' debate should be avoided. Given the rapid establishment of biofuels as major commodity products and with growing markets, one of the most likely and least controversial biorefinery models would be those with a biofuel being the major product and all wastes from the feedstock processing through to the biofuel production being utilized to manufacture chemical, material and other energy products. A good example of this is a biodiesel plant as a nucleus for a future biorefinery, with other products including those derived from initial wax extraction from the straw, from proteins, from the meal and from the glycerol co-product of the biodiesel process (Figure 1.12).

1.2 Green and sustainable supply chains

How can we ensure that future biorefineries are the starting point for green and sustainable supply chains for products with genuinely lower environmental footprints? A renewable resource as the primary building block is a good start, but does not guarantee this. It is essential that we both employ green chemical technologies throughout the processing used to convert that raw material into the final product, and that we have effective ways of measuring the overall 'greenness'. At the first stage, valuable surface chemicals (fatty acids, steroids, polycosanols, hydrocarbons) can be extracted using supercritical CO_2. The bulk biopolymers, cellulose and lignin may be separated using ultrasonic technologies.[17] The bulk materials can be broken down into small building blocks ('platform') chemicals or into energy products (bioethanol and biobutanol) by fermentation processes and/or by controlled pyrolysis.

There is some similarity between the cracking of petroleum and the cracking of biomass. However, biomass is more complex chemically both in terms of structural types and functional groups. In petrochemistry, hydrocarbons are fractionated and they are then functionalized by oxidation, halogenation, nitration and other chemical processes so as to add value. The commodity chemicals are then built up into more complex molecules using such popular synthetic methods as Friedel Craft reactions, Michael and aldol condensations, and Heck and Suzuki couplings. The speciality products of these reactions are then further elaborated into formulations for use in everyday applications ranging from personal care

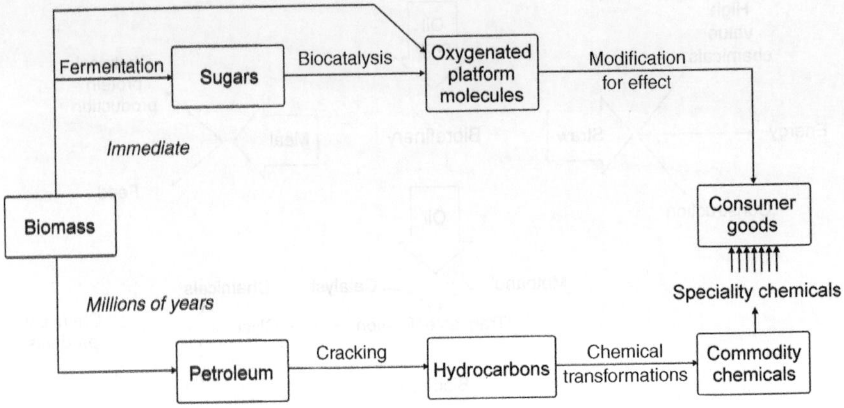

Figure 1.13 Traditional petrochemical and proposed new biomass sources to consumer goods.

products to advanced materials, coatings and dyes, agrochemicals and pharmaceuticals. The separation of biomass is more difficult but as stated there are technologies available for bulk and partial conversion into valuable chemicals, materials and fuels. The closest comparison to the petrochemical process described above is biomass platform molecules which are those relatively simple chemical compounds that we can now obtain in substantial quantities from various types of biomass. In contrast to the relatively simple hydrophobic hydrocarbon petrochemical platform molecules (ethene, propene, benzene, etc.), biomass platform molecules are oxygenated, polyfunctional and hydrophilic. The top biomass platform molecules have been identified and include compounds such as lactic acid, glycerol, succinic acid and ethanol. In fermentation processes, their compounds are produced as dilute, aqueous and impure broths – in striking contrast to those from petroleum. It is very important that we do not expend large amount of energy in separating these broths and purifying individual compounds. Instead, we must learn to do more chemistry in dilute aqueous mixtures using robust water-tolerant catalysis, especially porous solid catalysis that can partition the organic molecules from aqueous broth into the catalyst domain. Furthermore, and more ambitiously, we should rethink the traditional concept of separate-purify-functionalize-formulate, which is very elaborate, multi-step, energy intensive and often inefficient. Since nature gives us multi-component systems in biomass, it would seem wise to develop methods that can convert biomass multi-compound systems into multi-component formulations with the properties necessary to achieve the effects we desire, such as cleaning power and surface protection (Figure 1.13).

By using green chemistry methods throughout the processing of biomass building blocks and by producing environmentally compatible products including consumer formulations we can create green and sustainable supply chains for future chemical products.

1.3 Green chemical technologies

To achieve sustainable production through biorefineries, engineers can look to specific chemical processes that draw on green technological alternatives. Whereas the principles of green chemistry focus on the development of reactions that are 100% atom economic, green engineering principles speak towards optimizing the amount of product formed. As a result, separations and system efficiencies are important in green design.[18] Proper solvent selection, process intensification, and the design of optimal reactor/catalyst systems thus play key roles in the development of green chemical technologies.

1.3.1 Solvent usage

Many processes require the use of solvents, which historically are organic materials. Organic solvents have traditionally been used for easier processing, for extraction of the desired product, for coatings, and for liquid-phase reactions. Many of these organic solvents have a variety of inherent properties that make them, for example, flammable, carcinogenic, toxic, and/or volatile to name just a few. As a result, the development and selection of appropriate alternatives can be used to decrease environmental, health and safety impacts of processes. Numerous alternatives are becoming available that may reduce the dependence on organic solvents, including the use of an organic solvent that is less hazardous than the one originally included within the process, the use of non-organic solvents such as water, supercritical CO_2, or ionic liquids, replacement of the solvent with a species present in the reaction scheme, and solventless processing. While each of these options has certain technical limits, all can reduce the environmental impact associated with chemical manufacturing.

Solvent selection should be based on reactant and product solubilities, the particular type of chemistry used, the process requirements, looking at the specific desired performance parameters, the cost of the solvent, and its reactivity with other materials present in the process.[19] Then, the design engineer examines other constraints on solvent selection, such as EHS or regulatory requirements. Finally, a list of potential solvents that meet the process constraints can be compiled, and the performance of the trial solvents can be analysed using simulation tools. Based on the performance goals and process requirements, the best solvent can be selected from the group of solvents that meet the established constraints, as described conceptually in Figure 1.14. The US EPA has developed a green chemistry expert system, solvent selection module that can be used as an aid in the selection of environmentally preferred solvents.

Recent research has shown that water may be used as a solvent for some industrial processes; for example, the oxidation of alcohol using a soluble catalyst.[20] Water, which in most cases is inexpensive, comparatively non-toxic, and non-flammable, also may reduce the production of solvent-related CO_2 from use and/or manufacture or reduce the risk of VOC emissions when the solvent is removed. These reactions not only take advantage of the environmental benefits of using water, but also allow selective chemistry that would not be possible in traditional organic solvents. Water may be an especially appropriate solvent for reactions of biomolecules. However, when considering the total process life cycle, the use of water as a solvent may not always turn out to be preferred. Water has a high boiling point, so there may be significant energy consumption required for heating or separations. While water is a viable solvent for certain reactions, many industrially important chemicals have only a limited solubility in water, so the use of surfactants or other agents may be required, limiting its environmental, health, safety and cost advantages.

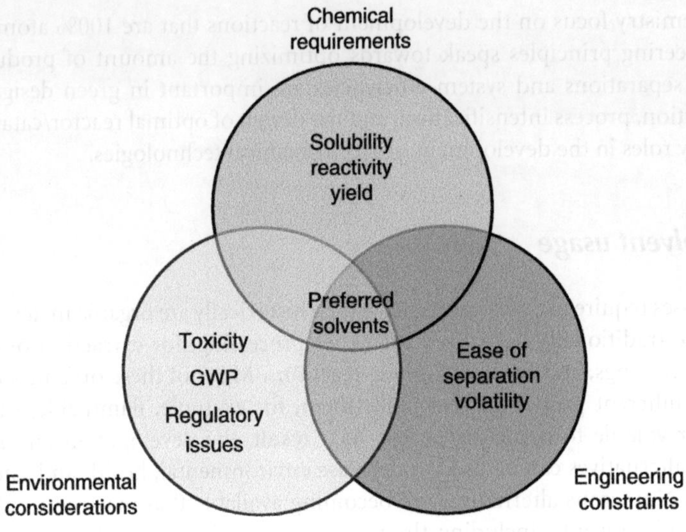

Figure 1.14 Conceptual basis for environmentally preferable solvent selection.

Supercritical carbon dioxide ($scCO_2$) is also being investigated as a solvent, particularly for specialty chemicals and pharmaceuticals. Although CO_2 is relatively inexpensive, the costs of compression generally make $scCO_2$ a relatively high cost solvent, and thus its use is generally limited to high value-added products or perhaps to particular specialized extractions of natural products. On the other hand, CO_2 is environmentally benign, and it can be quickly and completely separated from the materials of interest through a pressure reduction that vaporizes the solvent, with no residue. Commercially, $scCO_2$ is now used regularly in food processing because of its desirable properties, and it has also shown promise as a cleaning solvent in the semiconductor industry.[21]

Finally, there has been substantial recent interest in the use of ionic liquid solvents.[22] These materials are non-volatile salts that are liquids at room temperature and frequently, across a broad temperature range. Because of their ionic nature, these salts have little or no vapour pressure, and thus produce no vapour emissions. Chemically, they are quite interesting as, for example, they can promote certain types of reactions catalytically in addition to their solvent capabilities. On the other hand, these materials are highly expensive, many of them have substantial or as-yet unknown toxicity, and their recyclability and/or regeneration is unknown, so their overall environmental benefits may be somewhat reduced. Recent research suggests that ionic liquids have unique solvent properties that can be used as an advantage to control the reaction chemistry and to dissolve materials that may not otherwise be soluble. Of particular note is the observation that cellulose can be dissolved in selected ionic liquids, the subject of a recent Presidential Green Chemistry Challenge Award,[23] and that chemistry can be done on the dissolved cellulose for its conversion into more valuable products. Substantial barriers to using ionic liquids remain for many reactions before ionic liquids are used as a matter of routine. For example, product isolation without the use of organic solvents may not be possible, and the corrosion characteristics for some are unknown and could be substantial. In addition, the complexity of the cations and anions suggests that the life cycle environmental impacts from their production may be considerable in some cases and may still be based in petrochemical feedstocks.

1.3.1.1 Process intensification

The combination of multiple unit operations into a single compact device is the concept underlying process intensification. The development of these devices and techniques holds significant promise for substantial improvements in production processes, reductions in the size of production equipment, lower investment costs, and lower energy use and waste production.[24] Process intensification is a green strategy in that it allows certain chemical and engineering process combinations that would not otherwise be possible, and it offers opportunities to decrease the use of materials in the process. It further offers the ability to more precisely control the addition of reactant and the removal of product, can be used to overcome unfavourable chemical equilibrium, and minimize the generation of unwanted by-products. Although in principle any combination of unit operations could be considered, practical aspects generally include combined reaction and separation, reaction and heat exchange, or heat exchange and separation.

Some of the advantages or process intensification includes:

1 Better product quality.
2 The ability to use just-in-time chemical manufacturing because of ultra-short residence times in microscale reactors.
3 Distributed (rather than centralized) manufacture may become economically viable.
4 Lower by-product formation that reduces downstream purification costs.
5 Smaller inventories lead to improved intrinsic safety.
6 Better control of process irreversibilities which can lead to lower energy consumption.

Reactive distillation is one of the classic techniques of process intensification. This combination of reaction and distillation was first developed by Eastman Kodak under the 1984 patent in which methyl acetate was produced from methanol and acetic acid.[25] One of the key elements of the design is to use the acetic acid as both a reactant and an extraction solvent within the system, thereby breaking the azeotrope that exists within the system. Likewise, the addition of the catalyst to the system allowed sufficient residence time such that high yields could be obtained, making the process commercially viable. Other examples in which reactive distillation may enhance selectivity include those of serial reactions, in which the intermediate is the desired product, and the reaction and separation rates can be systematically controlled to optimize the yield of the desired intermediate.[18]

In a related process, enhanced reaction yields can be obtained through the use of membrane reactors, in which the desired product is selectively transported through a membrane, or in which an undesired inhibitor is selectively removed. Application has been demonstrated for use of membrane reactors in combination with homogeneous catalysts, wherein the catalyst is contained within the reactor and the products selectively pass through the membrane.[26] Such a system is more environmentally appropriate since the catalyst can be recovered and reused, thereby increasing the number of turnovers achieved while maintaining the selectivity of the homogeneous catalyst. However, significant catalyst modification is often required, and the reactor must be designed specifically to allow the products to diffuse through the membrane. Fouling of the membrane reactor impacts its performance and the lifetime of the operation.

The use of process intensification for the development of fuel processors for hydrogen generation through reforming (for use in fuel cell technologies) is an area of current research interest. Applications exist for the separation of hydrogen using membrane technologies, combined sulphur removal with reaction, and the control of heat exchange via microscale reactor techniques.[27]

In one example, a multichannel reactor architecture has been derived in which the heat from combustion of the fuel is used to drive the endothermic reforming reaction. In this case, fuel and air are introduced from one side of the reactor while fuel and water are introduced on the opposite side. The reactor is constructed of metal foil on which the catalyst is placed; the catalyst must be active both for combustion and reforming. The ability to transfer heat efficiently across the foil allows for an extremely compact design that is among the smallest and most lightweight systems available for hydrogen production from liquid fuels. In addition, because the feed to the channels can be cycled between reforming and combustion, catalyst deactivation by coking can be minimized and the catalyst's life extended.

Summary

The transition to a sustainable society clearly requires scientists and engineers to develop the necessary technologies to make better (more efficient) use of the resources that are available to us. Equally important are the appreciation and acceptance by wider society, nationally, regionally and globally, of the costs, constraints and changes needed to effect their timely implementation. In the short term, incremental gains that more efficiently use our non-renewable resources will extend the viable lifetimes of existing technologies to cushion some of the more severe consequences of the transition. Use of waste products as raw materials ('cradle-to-cradle') will provide one means by which the use of non-renewables can be extended in a more sustainable fashion. Mid-term, inclusion of an increasing proportion of renewable components as feedstocks for traditional processes will further ease the transition by developing the infrastructure needed for wholly renewable supply chains. In the long term, wholly new processes based exclusively on renewable materials will be developed. However, when evaluating the sustainability (or otherwise) of all these processes, account must be taken of the consequences of increased and more intense land usage needed to generate high yields of renewables, including the likely need (using today's technologies) for non-renewable resources for their production. Such challenges are not beyond the capabilities of today's scientists and engineers, but require a new way of thinking about process and product design, in which the constraints include the global impacts of our technology on society and the environment. Success in confronting and overcoming these constraints requires a trialogue between experts, the public and governments, the success of which requires, on an unprecedented scale, informed, pragmatic, enlightened and disinterested decision-making: as others have said: we require 'a deeper kind of prudence'[28] based on a 'capacity to worry intelligently'.[29]

References

1. Anastas, P.T. & Warner, J.C. (1998) *Green Chemistry: Theory and Practice.* Oxford University Press, New York.
2. http://www.epa.gov/oppt/greenengineering/, accessed on 6-1-07.
3. Shonnard, D.R., Kicherer, A., Saling, P. (2003) Industrial Applications Using BASF Eco-Efficiency Analysis: Perspectives on Green Engineering Principles. *Environmental Science and Technology,* **37**, 5340–5348.
4. See, for example, Ehrlich, P.R., Holdren, J.P. (1971) Impact of Population Growth. *Science,* **273,** 1212.
5. Winterton, N. (2003) Sense and Sustainability: The Role of Chemistry, Green or Otherwise. *Clean Technology and Environmental Policy,* **5**, 8.
6. Wiener, N. (1993) *Invention: The Care and Feeding of Ideas.* MIT Press, Cambridge, MA, p. 3.
7. Graedel, T.E. & Allenby, B.R. (1995) *Industrial Ecology.* Prentice-Hall, Englewood Cliffs, NJ.
8. Robèrt, K.-H., Schmidt-Bleek, B., Aloisi de Larderel, J. *et al.* (2002) Strategic Sustainable Development – Selection, Design and Synergies of Applied Tools. *Journal of Cleaner Production,* **10**, 197.
9. Grübler, A. (1995) Industrialization as a Historical Phenomenon. IIASA (International Institute for Applied Systems Analysis). Working Paper WP-95-029.
10. Keoleian, G.A. (1999) Centre for Sustainable Systems (formerly the National Pollution Prevention Center), University of Michigan.
11. Jackson, T., Costanza, R., Overcash, M., Rees, W. (1993) The 'Biophysical' Economy – Aspects of the Interaction between Economy and Environment. See http://www.amazon.co.uk/gp/reader/0873718844/ref=sib_rdr_cr?ie=UTF8&p=S008&j=0#reader-page
12. Hocking, M.B. (1994) Reusable and Disposable Cups – an Energy-based Evaluation. *Environmental Management,* **18**, 889–899.
13. Winterton, N. (2001) Twelve More Green Chemistry Principles. *Green Chemistry,* **3**, G73–G75.
14. Werpy, T. & Peterson, G. (eds) (2004) Top Value-Added Chemicals from Biomass, Vol. 1. *Results of Screening of Potential Candidates from Sugars and Synthesis Gas.* Pacific Northwest National Laboratory and National Renewable Energy Laboratory, US, August 2004.
15. (2007) *Ecosystems and Human Well-Being.* Island Press, Washington. http://www.millennium-assessment.org/en/index.aspx
16. *Biofuels in the European Union, a Vision for 2030 and Beyond.* Final report of the Biofuels Research Advisory Council. European Commission, EUR 22066, 2006. Available online: http://ec.europa.eu/research/energy/pdf/biofuels_vision_2030_en.pdf
17. Virtanen, J., Keinänen, P., Tilli, M. Improved Methods and Devices for the Fabrication of Cellulose. WO 2006/053939.
18. Malone, M.F., Huss, R.S., Doherty, M.F. (2003) Green Chemical Engineering Aspects of Reactive Distillation. *Environmental Science and Technology,* **37**, 5325–5329.
19. Gani, R., Jimenez-Gonzalez, C., Constable, D.J.C. (2005) Method for Selection of Solvents for Promotion of Organic Reactions. *Computers and Chemical Engineering,* **29**, 1661–1676.
20. ten Brink, G.-J., Arends, I.W.C.E., Sheldon, R.A. (2000) Green, Catalytic Oxidation of Alcohols in Water. *Science,* **287**, 1636–1639.
21. Blowers, P. & Titus, M. (2004) Use of Life-Cycle Inventory as a Screening Tool for Environmental Performance: Supercritical Carbon Dioxide in the Semiconductor Industry. *Environmental Progress,* **23**, 284–290.
22. Rogers, R.D. & Seddon, K.R. (2003) Ionic Liquids – Solvents of the Future? *Science,* **302,** 792–793.
23. Rogers, R.D. (2005) A Platform Strategy Using Ionic Liquids to Dissolve and Process Cellulose for Advanced New Materials. The University of Alabama.

24. Tsouris, C. & Porcelli, J.V. (2003) Process Intensification: Has its Time Finally Come? *Chemical Engineering Progress*, **99**, 50–55.
25. Agreda, V.H. & Partin, L.R. (1984) U.S. Patent 4435595, Issued March 6, 1984.
26. Dijkstra, H.P, van Klink, G.P.M, van Koten, G. (2002) The Use of Ultra- and Nanofiltration Techniques in Homogeneous Catalyst Recycling. *Acceleration Chemistry Research*, **35**, 798–810.
27. Qi, A., Peppley, B., Karan, K. (2007) Integrated Fuel Processors for Fuel Cell Application: A Review. *Fuel Processing Technology*, **88**, 3–22.
28. Freudenburg, W.R. (2001) Risky Thinking: Facts, Values and Blind Spots in Societal Decisions about Risks. *Reliability Engineering System Safety*, **72**, 125.
29. Kates, R.W. (2001) Environment and Development – Sustainability Science. *Science*, **292**, 641.

Chapter 2

Incentives for Using Green Chemistry and the Presentation of an Approach for Green Chemical Design

Nicholas D. Anastas

2.1 Introduction

Green Chemistry is the utilization of a set of principles that reduces or eliminates the use or generation of hazardous substances in the design, manufacture and application of chemical processes.[1] This non-regulatory approach for pollution prevention occupies a pivotal position in the management of hazardous substances and has been quite successful as evidenced, for example, in the United States by the Presidential Green Chemistry Challenge Awards.[2]

2.2 History of chemical hazard assessment

Assessing the hazards associated with chemicals involves qualitatively and quantitatively examining and evaluating information regarding the adverse outcomes associated with exposures to hazardous chemicals. Physical hazards (e.g., corrosivity, explosivity, flammability), toxicity (e.g., cancer, neurological effects, death) and global hazards (stratospheric ozone depletion and global warming) are all part of a comprehensive treatment of hazard.

2.2.1 Hazard and risk: The risk paradigm

Before continuing, it is necessary to distinguish between the concepts of hazard and risk because they are often used interchangeably which is inappropriate and potentially confusing. Hazard is defined as an inherent capacity of an agent to result in harm, whereas risk is defined as the probability that harm may result under the appropriate set of circumstances. This relationship can be expressed with the following formula:

$$\text{Risk} = \text{Intrinsic hazard} \times \text{exposure} \quad (2.1)$$

This relationship clearly shows that risk is a function of both the intrinsic hazard of a chemical and the exposure to that chemical which includes the frequency, duration and

Table 2.1 Hazard categories and selected examples.

Hazard category	Selected examples
Physical	Corrosivity Explosivity Extremes of pH Flammability Radioactivity Strong oxidizers or reducers
Human toxicity	Acute effects Carcinogenicity Genotoxicity/mutagenicity Developmental Teratogenicity/mutagenicity Neurotoxicity Endocrine disruption
Ecological	Acute and chronic aquatic toxicity Adverse reproductive effects on wildlife Phytoxicity
Environmental	Persistence Bioaccumulation
Global	Global climate change Stratospheric ozone depletion

amount of a defined agent. Reducing intrinsic hazard, exposure, or both results in risk reduction and allows talented researchers to focus their efforts on several key areas. Hazard has been divided into physical, toxicological and global categories for this discussion (Table 2.1).

2.2.2 Traditional approaches to minimizing risk

Historically, risk reduction strategies have been focused on the exposure side of the risk equation and in many cases have been very successful. However, there are inherent flaws in focusing primarily on controlling exposure that will be identified here.

Chemicals are an essential part of modern commerce and cannot be eliminated without catastrophic consequences to public health, the environment and the economy. Limiting exposure to the most hazardous chemicals has been the focus of national and international groups for many years. The USEPA relies on the development and implementation of regulations such as the Clean Water Act, Toxics Uses Reduction Act (TURA) and the Clean Air Act (CAA) as regulatory pollution prevention tools.

Reducing the intrinsic hazard of chemicals is the most effective and most fundamental of the risk reduction options available. Intrinsic risk reduction is based on the principle that the structure of a chemical drives hazard and molecular intentional, informed manipulation will result in the design of safer chemicals. Wastewater treatment plants employ large filters packed with granular activated carbon to remove polluted water. Synthetic chemists are uniquely positioned to reduce hazard at the design phase which is the beginning of the chemical life cycle.

2.3 Existing motivators for implementing green chemistry

In the United States, several federal programmes and focused regulations that incorporate elements of green chemistry have been put into place to reduce risks. The Pollution Prevention Act of 1990 has clearly stated that pollution prevention at the source is the highest and most desired level of environmental protection.[3]

European countries have taken a more aggressive approach in implementing regulations that require hazard reduction for the manufacture, use and exposure to chemicals. The Registration, Evaluation, Authorisation of Chemicals (REACH) legislation has been developed as a practical way to consolidate the numerous existing environmental regulations into one wide-reaching and comprehensive piece of legislation.[4] The current approach is often confusing, inconsistently applied and ineffective at improving chemicals safety. This new ambitious effort hopes to provide a more comprehensive and systematic approach to evaluating and regulating hazardous chemicals and chemical substances by improving and unifying the science supporting hazard ranking. Ideally this process will be transparent and will anticipate potential hazardous chemicals before they are able to reach general commerce and cause damage to human health and the environment.

The fact that both new and existing chemicals will be subjected to the provisions of REACH means that more rigorous examination protocols will likely result in severe restrictions or even outright bans on some chemicals that have been used for many years with little or no restrictions. A search for less hazardous alternatives makes good economic, environmental and public health sense. A natural tool for developing environmentally benign chemicals is green chemistry, and particularly, green chemical design. The remainder of this chapter is devoted to describing a framework for green chemical design and presenting incentives for its implementation.

Over the past several years, metrics for green chemistry have been proposed.[5,6] Effective mass yield modifies the traditional concept of synthetic yield by considering the use of 'non-benign' reagents used in a reaction.[7] There are no qualitative methods for how to measure whether a reagent is benign or not under this approach. The E-factor is a measure of the total waste per kilogram of product and has been widely accepted in the chemical industry as a useful metric for quantifying waste elimination.[8] However, this approach does not address hazard reduction of the target molecule. This chapter provides an overview of the current drivers for using green chemistry and presents a framework for designing safer chemicals that proposes new answers to meet these challenges. These incentives or drivers will prompt innovative companies to look for the next generation of chemicals that comply with new regulations and also add value and safety.

Trost[9] articulated the concept of atom economy, which characterizes the 'greenness' of a synthetic process by calculating the number of atoms from all of the reactants that make it into the final product. Even though this concept is an important part of the overall green chemistry approach it does not address the hazard associated with the final product.

2.3.1 Design rules for hazard reduction

The penultimate step in designing safer chemicals is the articulation of design rules that provide the synthetic chemist with direction for making choices that will result in reduced hazard. Design rules are ideally derived through a comprehensive and systematic scientific endeavour of evaluating existing hazard data and generating new information to support the articulation of structural features and molecular modifications that will lead to the synthesis of less hazard chemicals. Design rules have been developed to minimize carcinogenicity,[10] the acute toxicity of nitriles[11] and reduces the interaction of substituted hydrocarbons with stratospheric ozone.[12] The derivation of these design rules was possible because of advances in the understanding of the modes and mechanisms of each specific hazard. An examination of the existing relationship between structure and property followed by applied professional judgement resulted in an articulation of a putative mechanism or mode of action for these examples. Structural modifications to a molecule that minimize or prevent the mechanistic pathway from occurring completes the articulation of a complete design rule. A 'perfect' design rule would reflect a complete design rule that has been demonstrated to reduce a physical, toxicological or global hazard through appropriate physical, chemical or toxicological testing.

A library of heuristic design rules, assembled through a rigorous, interdisciplinary approach, is the next logical focus area in the quest to design safer chemicals. The endeavour to create safer chemicals is a dynamic process that challenges synthetic chemists, toxicologists, environmental scientists, and other investigators in related disciplines, to use their collective creative talents in a truly holistic way to reduce the intrinsic hazard associated with molecules resulting in a profound improvement in public health and environmental protection. The mechanism to develop a 'green design toolbox' is a framework that provides documentation of some of the approaches that are currently underway thereby encouraging a systematic adoption and to expand the traditional practices to include an evaluation of comprehensive toxicity and global hazards.

Bridging the gap between toxicology and synthetic chemistry is accomplished by making connections between structure and adverse outcomes and identifying molecular changes that minimize hazard. This will require a new breed of chemists, trained to understand the power and possibilities of synthetic chemistry while possessing enough understanding of toxicology to recognize potential hazards associated with a particular chemical structure.

2.3.2 A framework for designing safer chemicals

A framework to design chemicals that pose minimal hazard is proposed in this chapter (Figure 2.1). The goal of this framework is to provide a comprehensive and systematic approach for articulating design rules that can be used by molecular designers as guidelines for making safer chemicals. The framework is targeted to identify and present an approach that will be most useful to the synthetic chemist because of the unique position this specialized group of chemists occupies to influence hazard. The role of toxicologists, atmospheric

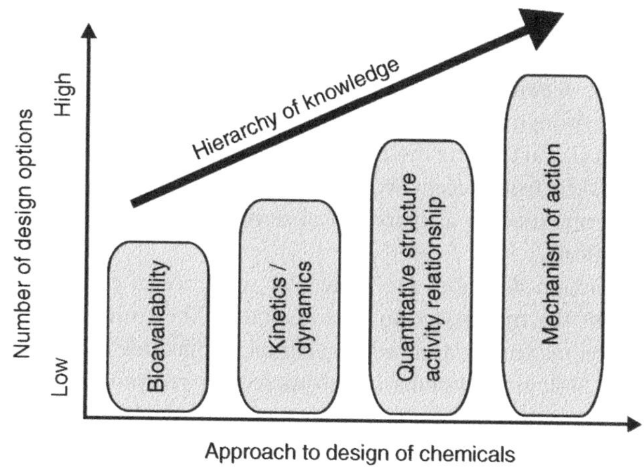

Figure 2.1 Design options.

and environmental scientists, biostatisticians and other allied professionals is critical to the success of this approach.

There are advantages to developing and using a framework for designing safer chemicals including a greater likelihood that a molecule with a more favourable hazard profile will be developed, an increased likelihood that toxicity tests will show minimized hazard, and the financial benefits associated with a more favourable hazard profile including decreased environmental damage and regulatory expenses.

This framework has been divided into four tiers of knowledge to reflect the usefulness of the information as it applies to a synthetic chemist or other individual that is developing design rules. The tiers are not necessarily meant to be used independently, but in a more integrated manner that reflects the weight of evidence that a set of data provides as support for developing a design rule. The maximum utility of the framework is realized when all of the available data are used together to support the development of a design rule based on a total weight of evidence.

2.3.3 *Tier I analysis: Mechanism of hazard*

The most useful level of knowledge for a synthetic chemist is to understand the interactions of a target chemical at the mechanistic level of detail. Over the past twenty years or so, scientists have made great strides in elucidating the mechanisms of physical, global and toxicological hazards. Advances in mechanistic toxicology provide the necessary information to elucidate the molecular events responsible for toxicity. The elucidation of a sequence of molecular events that are associated with hazard provides opportunities for intervention through rational molecular design. In addition to providing information to interpret descriptive toxicity data, and improve risk assessments, mechanistic information can be used as the basis for articulating a green design rule to guide the development of a synthetic pathway and potentially disarm a chemical.

Mechanisms can mean different things to different investigators, in this case synthetic chemists and toxicologists. The use of the terms mechanism and mode of action are often used interchangeably, depending on the background of the investigator. Although these terms are often used synonymously there are subtle but critical differences in their meanings that are important in characterizing their usefulness to molecular designers. For this paper, the definitions are taken from recent work[13,14] that provide important distinctions that differentiate these terms used in a toxicological sense versus its use specifically aimed at designing safer chemicals.

A mechanism of action describes the molecular sequence of events (covalent or non-covalent) that lead to the manifestation of a response. The complete elucidation of the reactions and interactions among and between chemicals, include very complex and varied situations including biological systems (macromolecular receptors, physical phenomena (thermodynamics of explosions) or global systems (ozone depletion). Unfortunately, this level of mechanistic detail is often unavailable but recent advances in molecular toxicology and others hazards, at the molecular level, have provided valuable information that elucidates key steps in a mechanism or mode of action.[15,16]

There are several points along the pathway to hazard that can be influenced through molecular design that are described in more detail later in this chapter. The magnitude and duration of a toxic event can be minimized through influencing the toxicodynamic and toxicokinetic phases associated with the manifestation of toxicity.

Global hazards can be mitigated through targeted molecular design. Uncovering the mechanism of ozone reduction by chlorine, and specifically by chlorofluorocarbons, has led to research chemicals that had minimal impact on stratospheric ozone.

2.3.4 Tier II analysis: Quantitative structure–activity relationships

In cases where there is not a complete understanding of a mechanism of hazard, there is often a body of knowledge on how different structural features within a class of chemicals may correlate with different types and levels of hazard. There is a long history of relating structure and activity or property.[17–20] Quantitative structure–activity relationships (QSAR) extend this concept and use regression analysis to develop numerical relationships between structure and activity.[21,22] These types of models can be very useful to individuals attempting to 'design-out' undesirable adverse consequences. Relationships have also been developed for predicting boiling point, vapour pressure and melting point,[23] atmospheric oxidation of organic chemicals,[24] and soil-sediment coefficients.[25] Understanding how structural features influence hazard is a powerful tool for the synthetic chemist to use in designing safer chemicals.

Computers have greatly enhanced the speed and accuracy of the predictive nature of QSAR,[26,27] however they are only tools to aid investigation and elucidation of the relationship between structure and properties. A clear understanding of the system and the fundamental questions that need to be answered are still a requirement for future research.

Chemicals that possess a common structural feature are called congeners. Some common examples are polychlorinated biphenyls (PCBs), polyaromatic hydrocarbons (PAHs), polybrominated diphenylethers (PBDE) and chlorofluorocarbons (CFC). The common features are modified through the addition or deletion of atoms or slight changes in position that can impart significant differences in hazard profile. PCBs can serve as an example of the concept.

The toxicity of PCBs is determined by the number and position of chlorine atoms on the biphenyl molecule.[28] When the ortho-position contains no or 1 to 2 chlorine atoms, the molecule can assume a planar position and is associated with a higher degree of toxicity. This 'coplanar' subgroup of PCBs is therefore a higher concern and provides a synthetic chemist with guidance on avoiding similar structural features in a new or redesigned molecule.

2.3.5 Tier III analysis: Toxicokinetic/toxicodynamics. The role of ADME in designing safer chemicals

2.3.5.1 Kinetics

An understanding of the role of toxicokinetics and toxicodynamics in the manifestation of hazard is fundamental to designing safer chemicals and can guide early design choices. Toxicokinetics and toxicodynamics use the same principles to study toxicological phenomena as those that are used to study the therapeutic use of chemicals as medicines. Toxicokinetics is concerned with the time course of action of chemicals that involves the disposition of a chemical affected by absorption, distribution, metabolism and excretion commonly referred to by the acronym ADME.

Each of the individual ADME components can be used independently or together to evaluate the influence of molecular modifications on hazard reduction.

2.3.5.2 Absorption

Absorption across biological membranes is often necessary for a chemical to manifest toxicity. In many cases several membranes need to be crossed and the structure of both the chemical and the membrane need to be evaluated in the process. The major routes of absorption are ingestion, inhalation, dermal and, in the case of exposures in aquatic systems, gills. Factors that influence absorption have been reviewed recently.[29] Methods to assess absorption include *in vivo*, *in vitro*, various cellular cultures as well as modelling approaches.[30] Solubility and permeability are barriers to absorption and guidelines have been developed to estimate the likelihood of candidate molecules being absorbed after oral administration.[31]

Molecular structure has been shown to influence absorption.[32] By examining the structural characteristics of drugs that were in use, certain common characteristics of well-absorbed molecules were identified, commonly referred to as the rule of five.[33] Some investigators have used this as a basis for characterizing the drug-likeness of a lead chemical.[34,35] Other factors also come into play including receptor activity, metabolism profile and for CNS-active compounds, an ability to cross the blood–brain barrier.

Molecular designers can exploit these rules and design chemicals that demonstrate characteristics that are likely to lead to decreased absorption and therefore minimize toxicity. Molecules that are permanently charged at physiological pH like the neurotoxin curare or the cholesterol sponge cholestyramine, do not cross the intestinal mucosa and therefore are not harmful by this route of exposure. The approaches medicinal chemists use as part of drug discovery and development to maximize the likelihood of a chemical to reach its site of action and have effect are the same principles that can be used in reverse to minimize or eliminate this scenario.

2.3.5.3 Distribution

Chemicals that enter the circulatory system after being absorbed or that enter the circulatory system through intravenous injection are available throughout the organism. Several factors including lipophilicity, molecular size and molecular charge will influence the degree to which a chemical is distributed, and to which organ the chemical is distributed.

2.3.5.4 Metabolism

Metabolism is a critical component of the disposition of xenobiotics and natural chemicals that can result in toxication of a chemical or a detoxication event.[36] Being able to predict the likely metabolic fate of a chemical provides a chemist with powerful information that can be used to express the molecular changes that will result in a chemical with reduced hazard. The concept of hard, soft and pro-drug development has been described[37,38] and has been used successfully for several decades to predict biotransformation products and their potential activity. Hard drugs are those chemicals that do not undergo metabolism and therefore will not produce metabolic products that are known to be toxic or products with unknown toxic characteristics.[39] Soft drugs are those pharmaceutical agents that are active in the administered state and undergo predictable and controllable metabolism to non-toxic products. This approach is attractive to synthetic chemists because of the opportunity that exists for exploiting the knowledge of toxicokinetics to rationally predict the metabolism of a chemical with an uncharacterized hazard profile. Pesticides are designed to be intentionally toxic to targeted plants and animals. Molecular modifications have been made to enhance the specificity of pesticides to their target and minimize non-target effects.[40]

2.3.5.5 Elimination

Eliminating a compound quickly from an organism can minimize the toxicity through the simple act of removing the compound away from the potential site of action. Chemicals that are hydrophilic are more easily eliminated from living systems so either the parent compound or a metabolite that is quickly eliminated is preferred to a slowly excreted compound.

2.3.5.6 Dynamics

Toxicodynamics is concerned with chemical reactions that occur at the site of biological action whether these reactions are receptor-mediated or involve non-receptor-mediated processes, for example, narcosis. The magnitude of a biological effect is controlled by the binding affinity of a compound for a receptor or other type of biological target, which is directly related to the structural characteristics of both entities. All of the bonding forces familiar to the organic chemist are involved in ligand–receptor interactions (e.g., covalent bonds in alkylation reactions, ionic reactions involved with the acetylcholinesterases, free radical destruction of ozone by halogens released from CFCs).

Influencing the efficacy or potency of chemicals is a strategy used by the pharmaceutical industry as part of the drug discovery process that can be incorporated into designing safer industrial chemicals. Efficacy is the maximal effect, either therapeutic or toxic, that a chemical can achieve. Potency is a measure of the amount of a substance that is needed to attain a given response level. Opioid analgesics are examples of where structural modifications have been used to establish a relationship between structure and activity.[41]

2.3.6 Tier IV analysis: The role of bioavailability in designing safer chemicals

Bioavailability is defined as the portion or fraction of a chemical that is available for biological action and is influenced by several factors including the molecular size and charge of a molecule, structural features of membranes, first pass metabolism, and therefore, bioavailability can be influenced by the molecular structure of a chemical.[42] This situation presents an opportunity for molecular designers to manipulate a chemical's structure to decrease bioavailability and consequently hazard. If the availability of a molecule can be decreased, the amount of chemical at the site of action is decreased which leads to decreased toxicity.

2.3.7 Reverse design for hazard reduction

The new concept of 'reverse design', introduced here, is based on the advancements made by the pharmaceutical industry for designing chemicals that will likely be successful drug candidates as part of the drug discovery and development process. We have described some of the advancements made by the pharmaceutical industry in the understanding of the relationship between molecular structure and biological activity to the point where quantitative treatments are used routinely as part of the discovery and development of new drugs.[43] The search for 'drug-like' candidates has evolved from essentially a trial and error approach where chemicals were tested on biological systems without a quantitative treatment of the relationship of their structure to activity, to more modern approaches of rational drug design. This approach is based on integrating the advancements made by medicinal chemistry and molecular biology with the improvements in computer science to quantitatively develop drugs. Improved three-dimensional structural information on drug receptor targets, coupled with advanced computer software capabilities has provided inspired medicinal chemists and toxicologists with the ability to forecast interactions at the molecular level before a synthetic plan is drawn up. The tools that allow a molecular designer to deliberately optimize the structure of a molecule to influence reactivity can be used in reverse to minimize hazard.

Molecular designers, medicinal chemists and pharmacologists have worked closely together to design molecules that have 'drug-like' characteristics and will therefore have a better chance of making it through the design and development stage. The tools that allow a chemist to optimize the structure of a chemical for deliberate reactivity in living systems can be used in reverse to minimize the likelihood of an adverse event, for example, toxicity, physical or global hazard. By using this 'reverse-design' approach, compounds that are not necessarily designed to influence biological systems are less likely to manifest hazard.

2.4 Metrics for evaluating greener chemicals

Up to this point the discussion has focused on the methods of characterizing hazard at the molecular level but has not yet addressed the underlying question of how hazards are compared so that decisions can be made about safer choices. Hazard is a relative term and therefore the difficult assignment of comparative hazard assessment must be evaluated and quantified if possible. To claim that one chemical is safe (non-toxic) and another is not is meaningless without including a description of how that decision is made.

Progressing toward the goal of designing safer chemicals requires that the relative safety of chemicals can be compared so that a clear decision process can be identified. The metrics to accomplish this evaluation of the relative hazards of chemicals are well established but have not been applied to the field of green chemistry, specifically, designing safer chemicals.

2.5 Physical hazard metrics

Physical hazard metrics are closely associated with hazards that are familiar to most chemists because they often involve spectacular, rapid and destructive events (see Table 2.2).

2.5.1 Toxicological hazard metrics

Toxicity is a relative term and to have any usefulness in designing safer chemicals this concept must be clearly articulated to synthetic chemists. For a chemical to be declared a greener alternative there must be a mechanism for comparing the toxicity of one chemical with another. The literature contains many references to qualitatively and quantitatively evaluating toxicity that can serve as guidance for characterizing the relative reduction in toxicity for a target chemical (see Table 2.3).[44,45]

Table 2.2 Physical hazard metrics.

Class of physical hazard	Assessment metric	Measurement metric
Corrosivity	Corrosivity to metals Corrosivity to tissue	Corrosion rates
Explosivity	Capacity to liberate heat and gases	Structural alerts Heats of formation
Flammability	Ignites and burns quickly	Flash point
pH extremes	pH ranges outside of accepted Normal ranges	$pH < 2$ $pH > 12$
Reducers and oxidizers	Violent reactions with oxygen	

Table 2.3 Toxicological hazard metrics.

Toxic endpoint	Assessment metric	Measurement metric
Lethality Mammalian systems Aquatic vertebrates and invertebrates Plants	Lethal dose$_{50}$ (LD_{50}) Lethal concentration 50 (LC_{50})	Both LD_{50}/LC_{50} values greater than a reference compound
Acute toxicity *In vitro* Mammalian systems Aquatic vertebrates and invertebrates Plants	IC/EC_{50} in appropriate test species	Use of appropriate indicators of acute toxicity, for example, EPA guidance values, reference doses, and so on
Chronic toxicity Mammalian systems Carcinogenicity Neurotoxicity Developmental/reproductive toxicity	Mutagenicity, increased tumours	Cancer slope factors Reference doses, and so on
Aquatic vertebrates and invertebrates Plants	Reproduction and growth	IC_{50}, EC_{50}

Table 2.4 Global hazard metrics.

Physical hazard endpoint	Assessment metric	Measurement metric
Global warming potential (infra-red absorption)	A compound's ability to absorb infra-red radiation	Global warming potential (GWR) (*Preferred GWR less than carbon dioxide*)
Ozone depletion	The ability of a chemical to reach the stratosphere and interact with and destroy ozone	Atmospheric lifetime

2.6 Global hazard metrics

Global warming and ozone depletion are the two primary global hazards associated with chemicals production and use. Chemicals that have structural features capable of absorbing infra-red radiation have the potential to contribute to global warming (see Table 2.4).[46]

Conclusions

A framework for designing safer chemicals using the concepts of green chemistry has been described. The framework is comprehensive because it expands the consideration of hazard from toxicity to include environmental, physical and global hazards. Molecular designers make decisions that will affect human health and the environment as soon as the first caricature of a carbon atom is scratched onto a note pad, displayed on a black board or summoned from a software program onto a computer screen. These choices rely on the experience and training of the chemist to design compounds that fulfill the functional requirements of a target molecule while minimizing the hazard associated with its life cycle; from cradle to cradle. Designing safer chemicals is an extremely complex undertaking requiring a multi- and inter-disciplinary approach of trained scientists that can generate and interpret complex data from a variety of sources.

The drivers for accepting this new approach must include a mechanism to evaluate the relative 'greenness' of new and redesigned chemicals as well as provide clear directions for advancing towards safer chemicals that is economically viable.

References

1. Anastas, P.T. & Warner, J.C. (1998) *Green Chemistry Theory and Practice.* Oxford University Press, Oxford, UK, 135pp.
2. EPA (2007) Office of Pollution Prevention and Toxics. Presidential Green Chemistry Challenge Awards. www.epa.gov/greenchemistry/pubs/pgcc/presgcc.html
3. EPA (1990) Pollution Prevention Act of 1990. 42 USC 13101.
4. REACH (2007) Registration, Evaluations and Authorisation of Chemicals. http://ec.europa.eu/enterprise/reach/overview_en.htm.
5. Constable, D.J.C., Curzons, A.D., Cunningham, V.L. (2002) Metrics to "Green Chemistry" – Which are the Best? *Green Chemistry*, **4**, 521–527.
6. Constable, D.J.C., Dunn, P.J., Hayler, J.D. *et al.* (2007) Key Green Chemistry Research Areas – a Perspective from Pharmaceutical Manufacturers. *Green Chemistry*, **9**, 411–420.
7. Hudlicky, T., Frey, D.A., Koroniak, L., Claeboe, C.D., Brammaer, L.E. (1999) Toward a 'Reagent-Free' Synthesis. *Green Chemistry*, **2**, 57–59.
8. Sheldon, R. (2005) Green Solvents for Sustainable Organic Synthesis-State of the Art. *Green Chemistry*, **7**, 267–278.
9. Trost, B. (1991) The Atom Economy – a Search for Synthetic Efficiency. *Science*, **254**, 1471–1477.
10. Lai, D.Y., Woo, Y.T., Argus, M.F., Arcos, J.C. (1996) Cancer Risk Reduction through Mechanism Based Molecular Design of Chemicals. In: *Designing Safer Chemicals* (eds. S. DeVito & R. Garrett) 254pp. American Chemical Society, Symposium Series 640, pp. 63–72.
11. DeVito, S. (1996) *Designing Safer Nitriles in Designing Safer Chemicals* (eds. S. DeVito & R. Garrett), 254pp. American Chemical Society. Symposium Series 640, pp. 194–223.
12. Manzer, L.E. (1990) The CFC-ozone Issue: Progress on the Alternative to CFCs. *Science*, **249**, 31–35.
13. Borgert, C.J., Quill, T.F., McCarty, L.S., Mason, A.M. (2004) Can Mode of Action Predict Mixture Toxicity for Risk Assessment? *Toxicology and Applied Pharmacology*, **210**, 85–96.
14. Walker, J.D., Jaworska, J., Comer, M.H.I., Schultz, T.W., Deardon, J.C. (2003) Guidelines for Developing and Using Quantitative Structure Activity Relationships. *Environmental Toxicology Chemistry*, **22**(8), 1653.
15. Saraf, S.R., Rogers, W.J., Mannan, M.S. (2003) Prediction of Reactive Hazards Based on Molecular Structure. *Journal of Hazardous Material*, **A98**, 15–29.

16. Russom, C.L., Bradbury, S.R., Broderius, S.R., Hammermeister, D.E., Drummond, R.A. (1997) Predicting Modes of Toxic Action from Chemical Structure: Acute Toxicity in the Fathead Minniw (*Pimephales promelas*). *Environmental Toxicology Chemistry*, **16**(5), 948–967.
17. Debnath, A.K. (2001) Quantitative Structure–Activity Relationship (QSAR) Paradigm – Hansch Era to New Millennium. *Mini Reviews in Medicinal Chemistry*, **1**, 187–195.
18. McKinney, J.D., Richard, A., Waller, C., Newman, M.C., Gerbick, F. (2000) The Practice of Structure Activity Relationships (SAR) in Toxicology. *Toxicological Sciences*, **56**(8), 8–17.
19. Cronin, M.T.D., Deardon, J.C., Duffy, J.C. *et al.* (2002) The Importance of Hydrophobicity and Electrophilicity Descriptors in Mechanistically-Based QSARs for Toxicological Endpoints. *SAR and QSAR in Environmental Research*, **13**(1), 167–176.
20. Selassie, C.D., Mekapati, S.B., Verma, R.P. (2002) QSAR: Then and Now. *Current Topics Medicinal Chemistry*, **2**, 1357–1379.
21. Hansch, C. & Fujita, T. (1964) ρ-σ-π Analysis. A Method for the Correlation of Biological Activity and Chemical Structure. *Journal of the American Chemical Society*, **85**, 1616–1626.
22. Harder, A., Escher, B.I., Schwarzenbach, R.P. (2003) Applicability and Limitation of QSAR for the Toxicity of Electrophilic Chemicals. *Environmental Science and Technology*, **37**, 4955–4961.
23. Deardon, J.C. (2003) Quantitative Structure–Property Relationships for Prediction of Boiling Point, Vapor Pressure and Melting Point. *Environmental Toxicology and Chemistry*, **22**(8), 1696–1709.
24. Meylan, W.M. & Howard, P.H. (2003) A Review of Quantitative Structure–Activity Relationship Methods for the Prediction of Atmospheric Oxidation of Organic Chemicals. *Environmental Toxicology and Chemistry*, **22**(8), 1724–1732.
25. Doucette, W.J. (2003) Quantitative Structure–Activity Relationships for Predicting Soil-Sediment Sorption Coefficients for Organic Chemicals. *Environmental Toxicology and Chemistry*, **22**(8), 1771–1788.
26. Barratt, M.D. & Rodford, R.A. (2001) The Computational Prediction of Toxicity. *Current Opinion in Chemistry and Biology*, **5**, 383–388.
27. Richard, A. (1995) Role of Computational Chemistry in Support of Hazard Identification (ID): Mechanism-Based SARs. *Toxicology Letters*, **79**, 115–122.
28. Safe, S. (1998) Hazard and Risk Assessment of Chemical Mixtures Using the Toxic Equivalency Factor Approach. *Environmental Health Perspectives*, **106**(Suppl. 4), 1051–1058.
29. Hansch, C., Leo, A., Mekapaty, S.B., Karup, A. (2004) QSAR and ADME. *Bioorganic and Medicinal Chemistry*, **12**, 3391–3400.
30. Lipinski, C.A. (2000) Drug-like Properties of Poor Solubility and Poor Permeability. *Journal of Pharmacological and Toxicological Methods*, **44**(1), 235–249.
31. Veith, M., Siegel, M.G., Higgs, R.E. *et al.* (2004) Characteristic Physical Properties and Structural Fragments of Marketed Oral Drugs. *Journal of Medicinal Chemistry*, **47**, 224–232.
32. Veber, D.F., Johnson, S.R., Cheng, H.Y., Smith, B.R., Ward, K.W., Kopple, K.D. (2002) Molecular Properties that Influence the Oral Bioavailability of Drug Candidates. *Journal of Medicinal Chemistry*, **45**, 2671–2623.
33. Lipinski, C.A., Lombardo, F., Dominey, B.W., Feeney, P.J. (1997) Experimental and Computational Approaches to Estimate Solubility and Permeability in Drug Discovery and Development Settings. *Advances in Drug Delivery Review*, **46**, 3–26.
34. Ajay Walters, W.P., Murcko, M.A. (1998) Can we Learn to Distinguish between 'Drug-like' and 'Non-Drug-like' Molecules? *Journal of Medicinal Chemistry*, **41**, 3314–3224.
35. Wenlock, M.C., Austin, R.P., Barton, P., Davis, A.M., Leeson, P.D. (2003) A Comparison of Physicochemical Property Profiles of Development and Marketed Oral Drugs. *Journal of Medicinal Chemistry*, **46**, 1250–1256.
36. Jacoby, W.B. (1980) *The Enzymatic Basis of Detoxication, Volumes I and II. Biochemical Pharmacology and Toxicology Monograph Series.* Academic Press, New York.

37. Buchwald, P. & Bodor, N. (1999) Quantitative Structure Activity Relationships; Steric Effects in the Enzymatic Hydrolysis of Carboxylic Esters. *Journal of Medicinal Chemistry*, **42**, 5160–5168.
38. Bodor, N. & Buchwald, P. (2006) Designing Safer (Soft) Drugs by Avoiding the Formation of Toxic and Oxidative Metabolites. *Molecular Biotechnology*, **26**(2), 123–132.
39. Lin, J., Sahakian, D.C., deMorais, S.M.F., Xu, J.J., Polzer, R.J., Winter, S.M. (2003) The Role of Absorption, Distribution, Metabolism, Excretion and Toxicity in Drug Discovery. *Current Topics in Medicinal Chemistry*, **3**, 1125–1154.
40. Pauli, B.D., Coulson, D.R., Berrill, M. (1999) Sensitivity of Amphibian Embryos and Tadpoles to Mimic 240 LV Insecticide Following Single or Double Exposures. *Environmental Toxicology and Chemistry*, **18**(11), 2538–2544.
41. Hardman, J.G., Limbird, L.E., Gilman, A.G. (2001) *Goodman & Gilman's The Pharmacological Basis of Therapeutics*, 10th ed. McGraw-Hill, New York.
42. Kwan, K.C. (1997) Oral Bioavailability and the First Pass Effect. *Drug Metabolism and Disposition*, **25**(12), 1329–1336.
43. Lin, J., Sahakian, D.C., deMorais, S.M.F., Xu, J.J., Polzer, R.J., Miller, S.M. (2003) The Role of Absorption, Distribution, Metabolism, Excretion and Toxicity in Drug Discovery. *Current Topics in Medicinal Chemistry*, **3**, 1125–1154.
44. EPA (1989) *Risk Assessment Guidance for Superfund. Volume I. Human Health Evaluation Manual.* EPA/540/1-89/002.
45. EPA (1998) Framework for Ecological Risk Assessment. 63 FR 93. EPA/630/R-95/002F.
46. Climate Change (2007) The physical science basis. Summary for Policymakers. Intergovernmental Panel on Climate Change. *Climate Change 2007: The Physical Sciences Basis*, IPCC.

Chapter 3
Green Product Design

Richard Miller

3.1 Introduction

The chemical industry has been very successful in progressively reducing its environmental impact. Figure 3.1 shows that for the European chemical industry over a 15-year period from 1990 to 2004, energy consumption stayed constant and greenhouse gas emissions dropped by over 20% whilst production volume increased by over 50%.[1] The industry has also significantly reduced other impacts including sulphur dioxide, nitrogen oxides and volatile organic compounds (VOCs) to air, and nitrogen and phosphorus to water.[2]

These gains have been driven by external pressures from governments and regulators, reflecting the interests of society, and by a fundamental business case. Regulators have encouraged a reduction in emissions to air, water and land, and the use of materials with a lower intrinsic toxicity. At the same time, process engineers have been working to improve the overall efficiency of the chemical industry by reducing materials use, minimising waste and reducing energy consumption. These two forces have combined over many years to provide a steady continuous improvement in the overall environmental impact of the chemical industry.

Unfortunately, these gains are not seen in all sectors of the industry. Roger Sheldon of Delft University has described an efficiency, or E-factor, for materials use in a chemical process by dividing the amount of waste produced by the amount of product.[3] Table 3.1 shows some typical examples for different sectors of the chemical industry. They range from 0.1 or less for oil refining to more than 100 in pharmaceuticals. Low E-factors tend to be found where there are high production volumes, continuous processes and low margins. The problem of what to do with large volumes of waste, and the economic value of small improvements in process efficiency, have encouraged the development of ever more efficient processes. Regulatory attention has also tended to focus on high volume commodity products where there is a greater scope for reducing emissions. In contrast, high E-factors are found for high-value, high margin, low production volume, batch processes. Looking at the energy efficiency of processes would produce a similar ranking.

Applying the techniques that have worked so well in commodity chemicals to the higher value sectors will reduce environmental impact and improve the economics, but the gains will be smaller. In particular, the economic benefits may not justify the investment in improvement. Further reduction in environmental footprint will require a new way of looking at the problem.

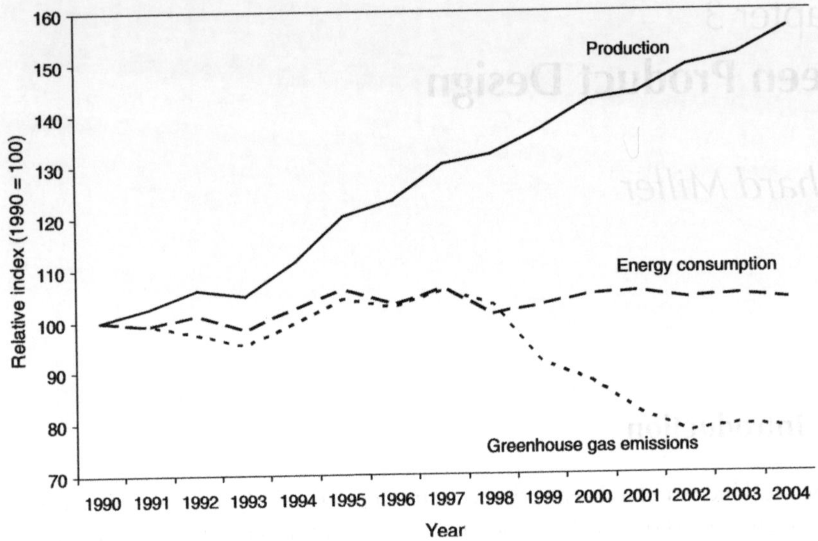

Figure 3.1 Reduction in environmental impact of the European Chemical Industry.[1]

Table 3.1 Typical E-factors for chemical industry sectors.[3]

Sector	Product tonnage	E-factor (kg waste per kg product)
Oil refining	10^6–10^8	<0.1
Bulk chemicals	10^4–10^6	<1 to 5
Fine chemicals	10^2–10^4	5 to >50
Pharmaceuticals	10–10^2	25 to >100

Current thinking focuses on the production process. What goes on inside the plant, and what is under the direct control of the business. It tends to ignore what happens, both upstream and downstream; where and how the raw materials are sourced, and how the products are actually used. What is needed in the chemical industry is a shift to life-cycle thinking.

Understanding the total life-cycle opens up many more areas for innovation. If we understand what our customers and our customer's customers are trying to do, right through to the ultimate end-user, we have many more opportunities to participate in their innovation. Their challenges can become our ideas for innovation.

For example, the packaging industry has many challenges coming from their customers, regulators and the public. Among other things they would like:

- Stronger polymers so they can produce lighter packaging
- Polymers with better barrier properties for oil, water and gases, so that they can preserve foods better

- Polymers with better printability
- Materials to use in tamper-evident packaging
- Better adhesives for assembling complex packages
- Polymers from renewable feedstocks
- Biodegradable polymers
- New ways to recycle complex multi-material packaging systems.

For every other industry one can produce a list of unmet demands for products that will help reduce environmental footprint. We can no longer focus on fixing the problems we think we own. The opportunity is in fixing other people's problems.

3.2 Green product design

Green product design can be defined as designing a product or service to reduce the overall environmental impact, whilst maintaining or improving economic, technical and social performance. It is essential that we do not focus simply on environmental issues. The goal is to be able to produce a product or service which can compete in the marketplace, not simply to win awards for green design. The history of green product design is littered with failed products and services that did not work as well as, or were more expensive than, the existing solutions. Only rarely can an inferior solution succeed; usually where there is some pressing safety issue.

The concept behind green product design has been around since the early days of the environmental movement. It has been given many names; including, eco-design, eco-innovation, sustainable product design, design for environment, and design for sustainability. The label varies from industry to industry as each has struggled to find solutions to similar problems, and have independently developed their own programmes. The terminology varies, but the approaches are similar.

Green product design is about reviewing the entire life-cycle of a product or service; from sourcing all the raw materials needed, to ultimate disposal at the end of life. Looking for opportunities to improve performance, reduce costs and reduce environmental impact.

Four innovation strategies have been identified to rethink a product or service.[4] You can:

- Improve the existing product. Making small changes and improvements to a product already in the market to tackle a perceived problem such as an ingredient with known or suspected toxic effects. The product, manufacturing process and functionality remain basically the same.
- Redesign the product. The product purpose, concept and functionality stay the same, but the way product is put together changes completely.
- Provide new functionality. Rather than improving the existing product concept, find new ways of meeting the customer needs. For example, by converting the product into a service.
- Redesign the business system. Create a new business model in which product, production system, delivery system, supply chain and customer care may all be changed.

Table 3.2 Potential for reduction in environmental impact for different innovation strategies.[4]

Strategy	Reduction in impact
Improve existing product	×2
Redesign the product	×5
New function	×10
New business system	×20

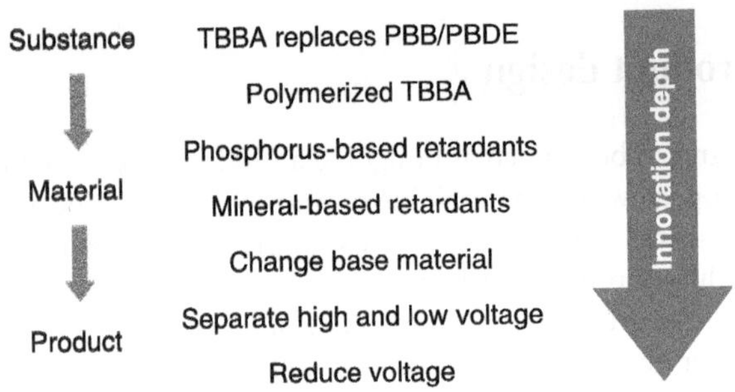

Figure 3.2 Innovation depth in electronic flame retardants.[6]

It is argued that the potential to reduce the environmental impact varies dramatically across these strategies as shown in Table 3.2. Simply improving the existing products is unlikely to yield more than a factor 2 reduction in impact. As more fundamental changes are made to the product and business system, the potential reduction in impact rises; however, the timescale and risk in the project also rise. There is only anecdotal evidence connecting these levels of reduction in impact with the specific strategies, but many have argued that in order to deliver the factor 10 and factor 20 improvements required for a sustainable economy it is essential to find new types of products and services and new business models to meet our current and future needs.[5]

Another way to think about this is in terms of shallow and deep innovation. Shallow innovation focuses on incremental changes to existing products and services. Shallow innovation can be quick to deliver and relatively low risk. The benefits may not be great, but they are easy to obtain. Deep innovation takes a hard look at the real user needs. It uses a thorough understanding of the way in which a product or service is used, and the benefits it delivers, to produce new concepts that can transform a market.

A good example of different innovation strategies and depth of innovation can be found in the challenges of designing fire retardant printed circuit boards.[6] Some of the different strategies adopted by industry are shown in Figure 3.2.

The challenge was to avoid the use of hazardous brominated fire retardants such as polybrominated biphenyl (PBB) and pentabromodiphenyl ether (PBDE). The first strategy was to replace PBB and PBDE with tetrabromo-bisphenol-A (TBBA). This has a much lower toxicity, but is still a brominated fire retardant with safety concerns. An improvement on the strategy is to use polymerised TBBA. This is much less mobile, and therefore less likely to escape into the environment.

Phosphorous-based fire retardants carbonised the circuit boards surface, preventing fresh material becoming available for burning. Mineral fire retardants, such as aluminium hydroxide, dilute the flammable organic compounds in the bulk material, cool the material and release water on heating.

Although epoxy resins are the standard base materials for printed circuit boards, other materials with reduced flammability such as polysiloxane can be used.

Since the main risk factor for fire in electronic devices is high voltages, less fire retardant is needed if the high voltage and low voltage systems can be separated, for example by using an external power supply for a laptop computer. Ultimately, it may be possible to redesign a product so that only low voltage systems are used.

There is a continuum of ideas from substituting specific substances, through changing the materials used in the product, to redesigning the product itself. These levels represent an increasing depth of innovation and the potential for increased customer benefit and market differentiation.

Shallow innovation can appear to be an attractive strategy. Particularly where a specific ingredient is a cause for concern, finding a substitute looks like the quickest and most cost-effective solution. Organisations like the Environmental Protection Agency in the US have collated tables of known substitutes for materials where the environmental or health and safety profile have caused concern. This approach is particularly used for solvents, and has been quite effective.[7]

However, the cost of simple substitution and shallow innovation can be substantial. The new EU regulations on Registration, Evaluation and Authorisation of Chemicals (REACH) are expected to result in a large number of chemicals that are currently freely available to vanish from the market. Formulated commercial products such as inks, adhesives and paints can contain up to 60 individual chemicals in one formulation. If one of these is withdrawn as a result of REACH, the potential costs of reformulation can be very high.

The British Coatings Federation has estimated that if it was necessary to reformulate the resin that binds an aircraft topcoat to the primer because of the withdrawal of one ingredient under REACH, the effort required for substitution would be:

- 4 man years work on reformulation and testing for the aircraft coatings manufacturer;
- 1.2 man years on assessment and production for the aircraft component manufacturer;
- 2.5 man years on assessment and monitoring for the aircraft constructor;
- 1.2 man years on retraining for airlines in maintenance and repainting.

It would take approximately 9 man years of work to replace a single application of a withdrawn substance, and little of that cost would be recoverable from the market.[8] That is a substantial investment in an R&D programme whose outcome leaves you exactly where you were before you started. It may be necessary to do this to preserve market position, but it is not an attractive option.

Deep innovation is a key part of green product design thinking. It offers the biggest gains in environmental, technical, economic and social performance.

3.3 Strategies for green product design

This section will describe approaches to identifying and implementing opportunities for green product design, with practical industrial examples.

3.3.1 Developing the mindset

Successful green product design is as much about mindset as it is about technical expertise. There are common assumptions which need to be challenged, and common areas of opportunity. Starting with assumptions that block thinking on green product design:

- Environmental issues are *not* just overhead

Many companies treat environmental issues as a threat, not an opportunity. A cost to be minimised, not a chance to create competitive advantage. When an external pressure is applied, the tendency is to 'circle the wagons' and attempt to fight off the perceived threat. The company becomes stuck in a defensive posture where they cannot consider alternatives. Smart companies see environmental issues as an opportunity. If the existing product/solution is under pressure, there is a business opportunity in solving the problem.

- Don't just *fix* the problem

There is a temptation to fix problems when they occur rather than to eliminate the source of the problem. End-of-pipe or end-of-life solutions are superficially attractive because you don't have to go back and repeat work which has already been done. However, it is often both environmentally and financially inefficient. Often technical fix is piled upon technical fix until the entire system collapses. Deep innovation thinking is more demanding but offers the chance to leapfrog competition.

- There is *no* waste

Once you have developed a waste management strategy for your product or process, you have designed waste into the system. It is better to start with the assumption that there is no waste. Any material left over from your products and processes is a feedstock for another technical process or a biological process. The goal should be that every material is as close as possible to a closed loop. This approach has been championed by William McDonagh and Michael Braungart as 'cradle-to-cradle' instead of the more usual 'cradle-to-grave'.[9]

- You can achieve reduced impact *and* higher value

A common assumption is that reduced environmental impact must be paid for by higher prices or lower profit margins. This is a dangerous assumption because it reinforces the idea that environmental issues are a problem and cost. With good design you can achieve reduced impact and higher value at the same time.

The key areas for finding new opportunities are:

- Think service *not* product

Chemical companies tend to think about products not the service that the customer is buying. Forcing yourself to think about it from the customer perspective, focusing on the service or functionality required to meet their needs, opens up new innovation space. If you concentrate on the product, you will improve the product, upgrade the product, or redesign the product. If you think about the service, completely new business opportunities will open up.

- Think lifecycle

The chemical industry has a strong tendency to think from factory gate to factory gate. Their historic focus on environmental issues has been in minimising waste and emissions from their own operations. Many chemical companies are weak at looking at what is going on upstream of their factory or downstream. Thinking about the entire life cycle of a product allows many new opportunities for improved performance to be identified.

- Look for benefits downstream

Chemical companies are also weak at looking for opportunities downstream of their own operations. Whilst most would recognise the need to understand the business of their customers, few look further downstream and try and understand their customer's customer. However, if your materials can reduce energy consumption or waste production in a final product, you reduce the costs further downstream and potentially change the value distribution across the entire supply chain. If you understand how your products are used downstream, you can share that value with your customers.

3.3.2 Describing impact

To successfully carry out green product design, it is important to be able to properly describe the baseline case, the environmental impact of current methods of achieving a particular goal, and to be able to compare alternative options for achieving that goal.

All useful methods attempt to capture and describe all of the impacts during the entire life of the product or service; from 'cradle to grave'. This is the basis of the cluster of measurement tools known collectively as Life Cycle Assessment (LCA).

LCA also uses the key concept of the unit of service. The unit of service is a convenient measure of what is actually being delivered to the user. For an automobile the unit of service could be a passenger kilometre. For an adhesive it could be 1 cm^2 of bonded materials at a defined strength, or for a photocopier one A4 copy of 5% coverage. By focusing on the benefit that the user receives, the unit of service allows us to compare products and services to deliver that benefit in very different ways. For example, delivering an insecticide inside a plastic trap requires the use of more material in total than spreading it loose in the target area. However, the trap may deliver the insecticide precisely where it is needed, prevent spread throughout the environment, reduce risks to other animals, and reduce the amount of active ingredient required. Overall, if we focus on the required benefit, the destruction of pests, a product that is safer and targets more effectively will probably have a lower impact.

The whole topic of Life Cycle Assessment is dealt with in detail in Chapter 7. However, there are some important points about LCA as it relates to green product design. It is important to be able to compare different solutions, it is also important to be aware of the limitations of LCA. These include:

- Sensitivity to system boundaries – you can change the result of the LCA completely depending on what is included within the analysis and what is excluded.
- Availability of data – a lot of life-cycle data are available in both public and proprietary databases. However, the data that are available will usually be industry aggregate data

and may well be out of date. It may also not be relevant to the specific situation where you are trying to innovate.

- Cost – the most reliable LCA tools rely heavily on human expertise. This means that the cost of an assessment can be very high. It is usually necessary to trade off confidence in the assessment method against cost and time to deliver.
- Comparing apples and pears – LCA methods generate a list of impacts in a number of different categories. How are these categories to be compared? Which is worse, a certain amount of release of a greenhouse gas, or a slight increase in risk to watercourses from a mildly eco-toxic material?

In order of decreasing sophistication and cost, the main methods for assessing the life-cycle impact are:

- Life Cycle Assessment under ISO 14000

This is the gold standard of environmental impact assessments. It has been developed over more than 20 years, and is governed by international standards ISO 14040 and ISO 14044. The process is very detailed, and each LCA is a project in its own right. It is far too expensive and time-consuming for screening innovation ideas. It is most effectively used for large-scale policy questions, and for justifying specific environmental claims when a new product or service is launched.

- Software-based LCA

A range of software tools have been developed using streamlined LCA methods. These generally use data from existing databases of material impacts and process step impacts. They can be used to identify the major environmental impacts of the product throughout its life-cycle, and to determine the environmental priorities that should be addressed in the design process. They are widely used in the 'assembly' manufacturing industries, such as electronics, but often lack the data required for the chemical and chemicals using industries.

- Matrix-based LCA

An even more streamlined approach scores each stage of the life cycle for impact on a number of environmental indicators. Typical indicators include resource depletion, global warming potential, smog production, acidification, eutrophication, toxic waste production and biodiversity impact. Impact is estimated using a simple numerical scale. The completed matrix is used to focus attention on areas for improvement.

- Proxy measures

Streamlined LCA methods are still too complicated for many applications. So proxy measures have been developed that use a single value to represent the environmental impact of a product or material. Examples include:

  - Embodied energy – this is a commonly available indicator and works well for systems that are dominated by energy use, such as packaging and construction.
  - Material input per unit of service (MIPS) – overcomes some of the issues of embodied energy by accounting for all material movements, but does not really discriminate between materials with different environmental risks.
  - Ecological footprint – measures the total land area that is required to support the production of the service, product or lifestyle. Most commonly used to communicate

the impact of different lifestyles on the total amount of land required to support each society, it is not sufficiently detailed for green product design purposes.

- Eco-indicators – an attempt to model a wide range of impacts that are then weighted against each other and summed into a single value. Because of the value judgements built into weighting the different impact categories, these are weak measures of actual impact. However, they are very popular, and the eco-indicator 99 model has been widely used by designers and has software to support it (EcoIT).

Given the known problems, and a wide range of methods available, what is the best way of comparing environmental impacts in green product design? The answer will depend on what information is readily available, and what stage the design project is at. If you are screening ideas, you need a very fast, low cost, method. If you are finalising a product which will be launched soon, you may require more detailed and high-quality information that will become part of your marketing story.

For most people attempting green product design, the main problem is at the idea screening stage. How can we make sure we are making a realistic assessment of the potential impact? The first step is to think very carefully about the unit of service. A poorly chosen unit of service can block innovative ideas by focusing attention in the wrong place. This frequently happens when the unit of service is defined in terms of the current solution, instead of the benefit that the user is seeking. The second step is to make the impact assessment method as simple as possible. Unless there is a specific safety or toxicity issue, the lifecycle impact is usually dominated by energy. So a simple proxy measure can be effective during screening.

The main purpose of life-cycle data in green product design is to help select attractive opportunities for further development. LCA is not an end in itself, only a guide to decision making. So the advice attributed to the economist John Maynard Keynes is relevant: 'it is better to be vaguely right than precisely wrong'.

3.3.3 Tools for finding innovation opportunities

There are many ways you can prime the search for innovation opportunities in green product design. Two with a proven track record are described in detail below; the life-cycle profile and the eco-innovation compass.

3.3.3.1 Life-cycle profile

The life-cycle records the relative environmental impact of a product or service in each of five stages of the life-cycle:

1 Raw materials extraction and production
2 Manufacturing
3 Distribution
4 Use
5 End of life disposal.

The percentage of the overall impact allocated to each stage of the life-cycle can then be displayed as a simple bar chart. This allows attention to be focused on the stage with maximum environmental impact, where green product design will have most effect.

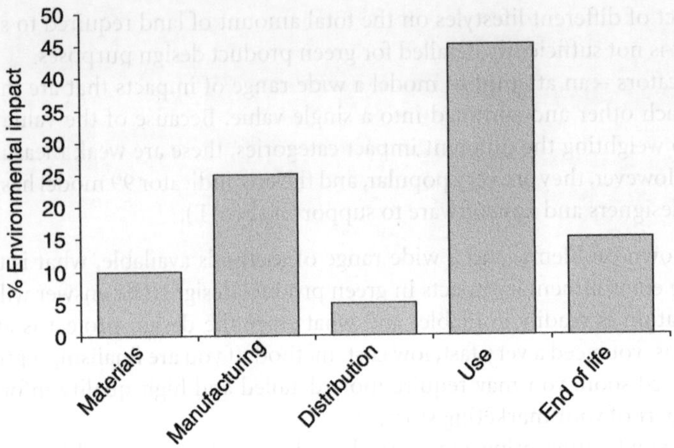

Figure 3.3 Life-cycle impact profile.

Life-cycle profiles are better suited to thinking about improving and re-designing existing products. Thinking about the areas of maximum impact tends to push thinking towards 'tweaking' the existing model. This tool is less useful for deeper innovation. Detailed information on the use of life cycle profiles in green product design can be found in the excellent book by Wimmer and Züst and the extensive website that supports it.[10] Many of the strategies described below are taken from this source.

Figure 3.3 shows an illustrative case where the major impacts are during the manufacturing and the use phase. This would be typical of a consumer durable such as a refrigerator or dishwasher.

The stage of the life cycle which has the most environmental impact depends on the product type:

- Raw material intensive product

Most of the impact is created by providing the materials contained in the product. This includes production of waste and consumption of energy in producing the raw materials. Typical high impact materials would include rare metals, natural extracts such as perfume ingredients, and energy intensive materials such as bricks and concrete. Electronic and electrical equipment are typical of products in this category.

- Manufacturing intensive product

Processing the raw materials during manufacture causes the greatest impact. Typical manufacturing intensive products use materials which undergo extensive processing during manufacture, or which produce large volumes of waste. Examples include many consumer products and chemicals.

- Distribution intensive product

The overall environmental impact is dominated by the system of distribution. This category includes products which are transported over long distances, are heavy, and use of a lot of packaging. Examples include fresh out of season vegetables.

- Use intensive product

Environmental impacts in the use phase dominate the profile. Products with high durability and which go through many cycles of use are found in this category. Examples include automobiles, dishwashers and laser printers.

- Disposal intensive product

Products where the main environmental impact comes at end of life. Products including hazardous materials are often expensive and difficult to dispose of safely. Batteries are an example of this class.

The next sections describe strategies for innovation to reduce overall environmental impact for each of the five different categories described above. These can be used to assist the brainstorming process.

Design strategies for raw material intensive products. Ideas for reducing environmental impact when the current solution is raw material intensive:

- Reduce the hazard of the materials used – avoid the use of materials that are persistent, bio-accumulative, or are named on any list of high-risk materials published by a safety regulator.
- Use renewable raw materials – for example carbohydrates, lipids or biopolymers.
- Use recycled materials for the product.
- Use raw materials obtained from a waste stream of another process.
- Reduce the number of different raw materials used.
- Minimise transportation of raw materials – can locally produced raw materials be used, or can manufacturing be shifted to where the raw materials are available?
- Avoid raw materials whose production generates a lot of waste.
- Avoid raw materials whose production is particularly energy intensive.
- Avoid raw materials produced in an environmentally damaging way.

As an example of good choice of starting materials, NEC were interested in making wide use of biopolymers instead of petrochemically derived plastics and composites in their consumer electronics. There is a desire to use renewable materials to reduce reliance on non-renewables and to avoid the predicted rising cost of petrochemically derived materials. A range of biopolymers already exist, but they lack the performance required for products such as mobile phones, portable entertainment devices and mobile computers.

NEC were particularly interested in using polylactic acid (PLA) for mobile phone casings, but the commercially available polymer lacked strength, heat resistance and processability. Working with the polymer company Unitka, NEC developed a composite using kenaf fibre. Kenaf is a fast-growing relative of the hibiscus grown across Africa and Asia. It produces high-quality fibres mostly used for papermaking. The composite has better heat resistance and rigidity than conventional petrochemical ABS polymers, and is much more mouldable than pure PLA.

In 2006 the material was used for the first commercial launch of mobile phones in Japan with a largely bioplastic case. 75% of the surface of the phone is bioplastic with only the screen and keypads made from conventional materials. The product has a good surface feel making it attractive to consumers, and meets all Japanese regulations for green purchasing and green electronic devices.

A larger scale example is the involvement of the cement industry in the growing field of industrial ecology. This takes the idea of an ecosystem with its many intersecting loops of materials and energy and applies it to industrial systems. Why use virgin raw materials and take them through a linear system to produce waste, when an unused material from one industry can be a feedstock for another?

The cement industry is ideally placed for industrial ecology; it uses large volumes of primary extracted raw materials that cannot be sourced sustainably, but utilises chemistry compatible with a wide range of raw materials. With the help of the World Business Council for Sustainable Development the cement industry is actively exploring these possibilities. Fly ash from power stations, furnace slag from steelworks and gypsum from desulphurisation in coal power plants are already used as raw materials, and cement making has been shown to be a potential route for chemically immobilising heavy metals and keeping them out of landfill and the environment. In addition, the cement industry can use a wide range of combustible waste as fuel for the kilns.

Design strategies for manufacturing intensive products. Ideas for reducing environmental impact when the current solution is manufacturing intensive:

- Improve the energy efficiency of the manufacturing process.
- Use manufacturing technologies that minimise the production of waste and emissions.
- What environmental risks are represented by a failure or accident in the production process? What can be done to mitigate these risks?
- Use renewable energy in product manufacture.
- Use renewable materials in the process.
- Avoid the use of hazardous process and auxiliary materials.
- Use closed-loop manufacturing processes. Recycle auxiliary and process materials back into the manufacturing process.
- Choose disposal methods for process waste which minimise environmental impact.
- Use any waste as the feedstock for another process.

An unusual example of a process that produces a lot of waste is the intensive rearing of pigs. They need additional phosphorus in their feed for healthy growth, and this is usually added as inorganic phosphorus in the form of monocalcium phosphate (calcium dihydroxy-oxido-oxo-phosphorane). Unabsorbed phosphorus passes through into the manure, and if spread onto fields as a fertiliser can lead to excess phosphorus run-off into rivers and lakes leading to eutrophication.

Phytic acid (inisitol hexakisphosphate) is the main storage form of phosphorus in plants. The phosphorus is not bioavailable to non-ruminants as they lack the enzymes to break it down. Novozyme has developed a commercial enzyme, phytase, that can be added to animal feed to release the phosphorus. No inorganic phosphorus needs to be added. This shift in the source of phosphorous has a large impact on the environmental footprint of pig farming.

The annual impact of shifting all the pig herd in Denmark to a phytase supplemented diet instead of MCP supplemented would be a reduction in greenhouse gases by 27 kT CO_2 equivalent, reduction in acidification by 450 T SO_2 equivalent, reduction in nutrient enrichment by 1200 T phosphate equivalent, and avoiding mining of 20 kT phosphate bearing rock.

Another material that risks becoming a problem waste is glycerol. Glycerol is a co-product from the manufacture of soaps, fatty acids and biodiesel from oils and fats. For every tonne of oil processed about 100 kg of glycerol is created. As the demand for biodiesel grows, existing markets for glycerol have become saturated and the price has fallen. Alternative markets are urgently needed.

Two options are being developed at the moment. The first is to produce 1,2-propanediol (propylene glycol) from glycerol. 1,2-Propanediol has a number of industrial uses, including as a less toxic alternative to ethylene glycol in anti-freeze. Conventionally, 1,2-propanediol is made from a petrochemical feedstock, propylene oxide. The new process uses a combination of a copper-chromite catalyst and reactive distillation. The catalyst operates at a lower temperature and pressure than alternative systems; 220°C compared to 260°C and 10 bar compared to 150 bar. The process also produces fewer by-products, and should be cheaper than petrochemical routes at current prices for natural glycerol. The first commercial plant is under construction and the process is being actively licensed to other companies.[11]

A second option is the production of epichlorohydrin. Solvay have developed a process for the production of epichlorohydrin from bio-diesel glycerol by reacting the glycerol with hydrochloric acid. Epichlorohydrin is used in epoxy resins and coatings and demand is growing rapidly. The existing process uses petrochemical-derived propylene as a feedstock, and the new process will be cost competitive with the petrochemical route. Solvay are currently building a 10 kT p.a. plant in France; expected to come on-line in 2007. In a reversal of history, synthetic glycerol used to be manufactured from epichlorohydrin when natural glycerol was in short supply.

Design strategies for distribution intensive products. Ideas for reducing environmental impact when the current solution is distribution intensive:

- Minimise the packaging required per unit of service.
- Use returnable packaging.
- Make the packaging part of the product so that there is no packaging waste.
- Avoid the use of 'problem' materials in packaging – for example PVC.
- Make the packaging material from renewable feedstocks.
- Use recyclable packaging.
- Manufacture the product at the point of use.
- Minimise number of transport miles per unit of service.
- Use a lower impact form of transportation.
- Reduce risk associated with loss of containment during transportation.
- Make the packaging strong enough to minimise losses during transportation.

The opportunities in the chemical industry to directly change distribution impacts are less obvious. However, a great deal is going on that suggests possible routes.

In a joint project between the Waste and Resources Action Programme (WRAP) and Coca Cola Enterprises in the UK it proved possible by redesigning bottle production to reduce the weight of 500 ml PET bottles by 8%. These lighter bottles meet all of the required performance standards, and when production reaches the full scale of 700 million p.a. will save 1400 tonnes of PET. This may seem to be a small improvement, but lightweighting of drinks bottles has been a continuous process for many years, and yet progress is still possible.

The limit on lightweighting of packaging in general is the strength and performance of the materials being used. Novel, cost effective materials with improved protective performance would find an immediate market in the packaging world.

A different concept is the idea of manufacturing at point of use rather than transporting materials. A simple and familiar example is the office water cooler. The majority that you see around use large bottles of water. These need to be transported and stored and the empty bottles recycled. More than 600 kT of bottled water were imported into the UK in 1999, creating a significant transport impact. The alternative is to chill and filter mains water on-site. The water needs to be filtered and purified to meet consumer perceptions and expectations. These systems use a mixture of purification techniques including advanced filters, UV-disinfection and silver-treated surfaces to give users the confidence they require.

There is a growing interest in manufacturing a wide range of chemicals at point of use. In some cases, such as phosgene, chlorine and hydrogen cyanide there is a clear risk in transporting and handling these chemicals. A great deal of research has been carried out on the use of micro-reactors of various types to produce toxic and hazardous chemicals on demand, avoiding transport and keeping the inventory to a minimum.

Design strategies for use intensive products. Ideas for reducing environmental impact where the current solution is use intensive:

- Maximise energy efficiency per unit of service.
- Can the product be made safer for the user and the environment?
- Are there negative environmental impacts from improper use? Can these be designed out?
- Can the amount of product used be automated to prevent user error?
- Design out waste and emissions in use.
- Make any emissions or waste environmentally benign.
- Minimise the use of auxiliary and process materials per unit of service. Make these materials renewable, safe and recyclable.
- Collect and recycle any waste.
- Find an environmentally acceptable route for disposal of waste from the use stage.
- Package the product so that waste from unused and unusable product is minimised.

An application where the dominant impact is in the use phase is automobiles. These have a long life and relatively high impacts per unit of service. Conventional tyres use large quantities of carbon black and silica to improve durability and performance. However, they also add to the weight and increase rolling resistance and fuel consumption. In 2001 Goodyear introduced their BioTred range of tyres using corn starch as the filler. This reduced rolling resistance and weight, and so improved fuel efficiency, but not enough to be really attractive to users. Goodyear are now working with Novamont to find an improved filler. Using Novamont's MaterBi blend of starch and polycaprolactone they achieved a further 5% reduction in fuel consumption, and in 2006 Goodyear, Novamont and BMW received EU funding for a project targeting tyres with even lower rolling resistance, shorter braking distances and better durability.

An application where toxicity is the issue is pesticides. Commercial agriculture needs effective pesticides, particularly fungicides to deal with a range of plant pathogens. Most currently available actives are highly toxic and a risk to both operator and the environment.

There must be a waiting period between application and harvesting, and few of these products can be used in organic farming.

To meet the market needs AgraQuest Inc developed a bio-fungicide, Serenade®, from a naturally occurring strain of *Bacillis subtilis.* This organism produces three different groups of lipopeptides, iturins, agrastatins/plipastatins and surfactins, that act synergistically to disrupt spore germination, germ tube formation and mycelial growth. They also inhibit attachment of the pathogen to the plant leaf. Because of the multiple modes of action, it is hard for the pathogen to develop resistance.

Serenade® has low human and environmental toxicity outside the target organisms, is broad spectrum and suitable for organic farming, there is no build up in the soil or groundwater, and it can be applied right up to harvesting.

Design strategies for end of life intensive products. Ideas for reducing environmental impact where the current solution is end of life intensive:

- Design so that the manufacturer can recover and reuse the product at end of life.
- Design for ease of disassembly and separation of components.
- Can individual components be recycled at end of life?
- Design for reuse, upgrading or refurbishment at end of life.
- Avoid harmful substances that could be released at end of life.
- Ensure any harmful substance can be easily extracted from the product.
- Design the product so that only benign materials are left at the end of life – biodegradable and compostable materials with low eco-toxicity.

Paper recycling is plagued by contaminating adhesives on the used paper. These 'stickies' cause major problems. They degrade the recycled paper leaving spots and holes, and block machinery that must be regularly stopped and cleaned using VOC generating solvents. Some grades of waste paper are so contaminated that they cannot be economically recycled, adding to waste disposal costs.

The main adhesive causing the problem is poly(vinyl acetate) (PVA). Buckman Laboratories have developed an enzyme treatment called Optimyze® using an esterase to hydrolyse the polymer to poly(vinyl alcohol). This is water soluble, not sticky and easy to remove.

In one paper mill the new treatment was responsible for:

- Reducing solvent use by about 750 litres per day
- Reducing chemical use by about 270 tonnes per annum
- Increasing production by about 6% through reduced downtime
- Generating about $1 million financial benefits in the first year of use.

Sometimes it is impossible to effectively design out waste, and safe means have to be found to dispose of it. Many industrial processes produce relatively high levels of waste in a finely divided or dispersed form, such as ash, contaminated soil, treatment sludges, and so forth. This contaminated material is difficult to handle and process. A common approach to tackling this problem is to 'stabilise and solidify' the waste using a binder that immobilises contaminants within a hard matrix. This does not destroy the contaminants, but keeps them from moving into the surrounding environment.

Carbon8 Systems Ltd have developed a process called accelerated carbonation to treat waste. The waste is mixed with a cementitious binder and reacted in a carbon dioxide atmosphere to produce a solidified material. Silicate bearing wastes may not need a binder as they can react directly with the carbon dioxide. The process can use waste CO_2 from industrial processes helping with carbon sequestration.

The process can handle a wide range of materials, including heavy metal contaminated wastes. It is very fast compared to conventional processes and produces an easily handled product that can be used in construction. Waste CO_2 can be captured and converted into carbonate to yield carbon credits. Target applications include pre-treatment of waste prior to disposal, recycling and reuse of industrial waste, and remediation of contaminated brownfield sites.

3.3.3.2 Eco-innovation compass

The eco-innovation compass was developed by Dow and the World Business Council for Sustainable Development (WBCSD), see Figure 3.4.[12,13]

The eco-innovation compass identifies seven directions in which you can innovate to improve the environmental impact of a product or service:

- Convert the product into a *service* so that more value comes from the intangible part of the product.
- Improve the *durability* of the product so that the financial and environmental costs of replacement and disposal are reduced.
- Design the product for *reuse*, remanufacture and recycling to reduce the amount of virgin raw material used, and the impact on disposal at end of life.
- Reduce the *mass* intensity per unit of service of the product.
- Reduce the *energy* intensity per unit of service of the product.
- Improve the human and environmental *safety* of the product by reducing toxic materials and avoiding emissions.
- Improve *resource use* by conserving depleting resources and using renewables or recycled materials where possible.

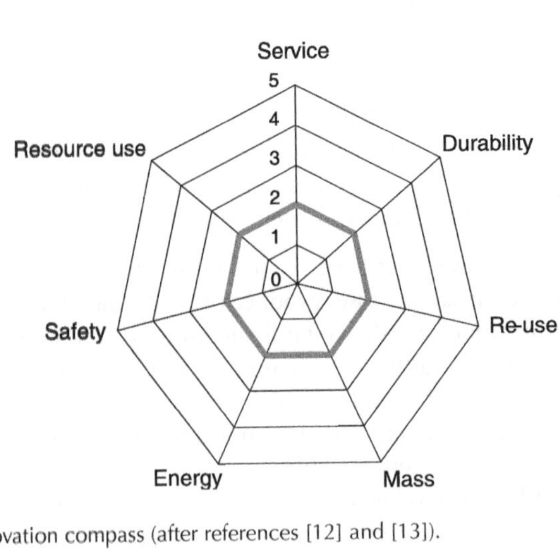

Figure 3.4 Eco-innovation compass (after references [12] and [13]).

For comparison purposes, the performance of the baseline solution against which new ideas are to be compared is set to 2 on all dimensions. Alternative concepts are evaluated in comparison with the baseline, and can be better or worse. Moving outward from the baseline towards the edge of the compass means improving the performance; moving in towards the centre from the baseline means a worse performance.
The eco-innovation compass has a scale running from 0 to 5. The scale is:

0 = Factor 0.5 – a reduction by 50% or more
1 = A reduction of uncertain size, or less than 50%
2 = No change compared to the baseline solution
3 = An improvement of uncertain size, or less than 100%
4 = Factor 2 – doubling of performance
5 = Factor 4 – fourfold increase in performance.

When identifying possible areas for innovation the seven dimensions provide a useful checklist. Is it possible to deliver the user needs through improving durability? Have we paid enough attention to the potential in reducing the mass intensity?

New product concepts are tested against the baseline to check whether there is a genuine improvement or not and can be plotted in the form of a spider-diagram (Figure 3.5). This allows alternative solutions to be easily compared, and specifically highlights the trade-offs that have to be made. It may not be possible to ensure that a new idea has a lower or equal impact on every dimension to the existing solution. There may be a price to pay on one dimension to secure a range of benefits on other dimensions.

The following sections contain ideas for innovation along each of the seven dimensions, and examples of successful green product design from a range of industries.

Service. The basic idea of service extension is to capture more value in the intangible part of the product. This means that increasing revenue does not necessarily mean increasing the physical resources used, with their related costs and environmental impacts. This 'dematerialisation' is a strong trend across industry, even within manufacturing. 'Solution

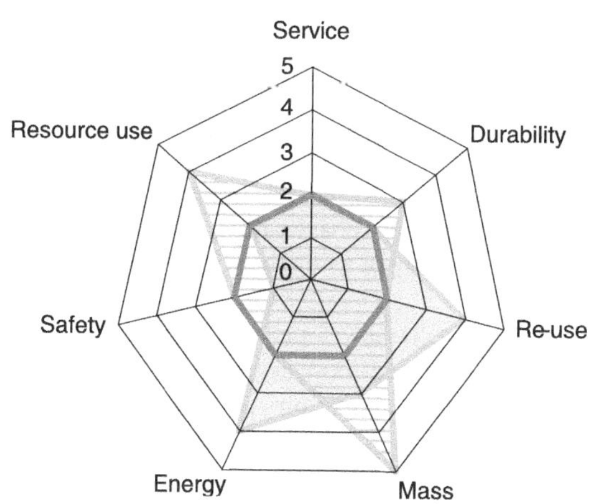

Figure 3.5 Comparing options with the eco-innovation compass.

selling', 'value added services' and 'product plus' are common marketing concepts that imply dematerialisation. The economist Brian Arthur said 'In the old world we bought and sold congealed resources. . . . In today's economy we buy and sell congealed knowledge', and Dr Claus-Peter Herold of Henkel commented 'today you do not sell chemicals any more, but suggestions about solving problems'.

From the environmental impact point of view service extension gives the supplier a greater involvement in the product life-cycle, with the potential to help manage the product through the life-cycle.

There are several ways of converting a product into a service:

- Take over some of the customer's operations and integrate them into your product offering. Many large manufacturers use suppliers to provide chemical management systems and look after their entire chemical inventory. Some companies have turned commodities ingredients into services by installing and maintaining metered tanks on the customer site. Telemetry enables monitoring of usage, and the tanks can be refilled at the convenience of the supplier. The product becomes a utility like electricity from the wall socket, and the customer never needs to make an order. They are just charged for what they use each month. Once set up this is a powerful incentive not to change suppliers.
- Provide application guidance and support to the customer that teaches them how to use the product efficiently. This reduces their costs, solves a knowledge problem for them, and helps lock them into the supplier.
- Think about the product in terms of the leasing model. This is already widely done in the catalyst industry for precious metal catalysts, and is being promoted by the UN Industrial Development Organisation (UNIDO) as part of its Cleaner Production Initiative.
- Provide design services with the product.
- Provide formulation guides to smaller customers to help them with their own product development. Since the various formulations are based around the supplier product, a good formulation guide is likely to enhance sales.
- Provide design and use tools that are locked to the supplier product. Some catalyst companies provide sophisticated computing models to their customers that enable their customers to optimise their processes and achieve maximum efficiency. These computer models are provided with extensive data sets that only relate to the supplier's own brand of catalyst.

Chlorinated solvents are extremely efficient cleaning agents. Unfortunately, significant toxicological and environmental hazards have caused many companies to stop using them altogether. Could suppliers find a way of using them safely and responsibly for applications where they are the best choice?

SafeChem, a subsidiary of Dow, has developed a handling system for chlorinated solvents that allows them to be used in closed-loop degreasing systems. The Safe-Tainer system uses two dedicated double wall containers; one to hold fresh solvent and the other used solvent. The containers are connected to the cleaning equipment with zero dead volume, leak-free connections that prevent spills, leaks or vapour emissions during use. Used solvent is collected for recycling and professional disposal of any residues. The system minimises solvent use and release to the environment. A study carried out by Dow during a trial in

Austria showed a 60% reduction in solvent used by customers. Virtually all of that solvent would have been released into the environment or would have become a hazardous waste stream to be managed.

However, this is not just a safe container. It actually switches the business model from sale of a chemical to management of a chemical process throughout the life cycle. SafeChem is fundamentally a service company.

Managing the life cycle of chemicals for a customer, instead of just supplying their needs, is proving to be a very successful business model. There are major cost savings to be made by the user as it is estimated that for every $1 spent on chemicals, between $1 and $10 are spent on managing the acquisition, inventory and disposal. A specialist company is in a much better position to carry out these tasks cost-effectively than the end-user. This approach is widely used in the automotive sector where a surprising number of chemicals are used. About three-quarters of the main car and truck manufacturers use chemical management services.

A good example is the relationship between GM at their truck and bus plant in Janesville Wisconsin, and one of their Tier 1 suppliers BetzDearborn. The agreement covers water treatment chemicals, paint, lubricants, commodity chemicals and solvents. For a fixed fee per vehicle manufactured, plus some additional service fees, BetzDearborn provides the following services:

- Acquisition and inventory control of chemicals.
- Monitoring and coordinating chemical usage.
- Improving chemical performance.
- Maintains a laboratory for chemical and regulatory testing.
- Participates in product and process engineering development.
- Provides environmental health and safety training, and ensures compliance.
- Minimises waste through continuous improvement.

The benefits from this approach have included:

- Over $1 million in savings.
- An 8% decrease in chemical costs per vehicle during the first 3 years.
- Improved inventory control and reduced inventory costs.
- A 78% reduction in paint inventory and a 50% reduction in paint use.
- Reduced VOC emissions.
- Improved paint quality.

Chemical management services are a good example of how by increasing the service content of the product, both supplier and customer can benefit.

Durability. Durability can be a useful way of improving eco-efficiency. A product that has a long useful life minimises the impact per unit of service in the overall life cycle. However, there needs to be a balance, as a durable product with a poor environmental performance may block the introduction of a more eco-efficient technology. In many cases, governments have been driven to use fiscal incentives to encourage the adoption of a more environmentally friendly product because of the installed base of older but durable technologies.

Durability is not normally a high priority for the chemical industry, but our materials go into many products where it is a very significant issue.

Ideas for designing for durability include:

- Design the product to have an operational life that matches the user need.
- Provide for easy repair, refurbishment or upgrading of the product. Click on casings allow users to change the style of their mobile phone without changing the core product.
- Ensure that the product cosmetics are maintained through its life. High-performance coatings or body colour can ensure that the product does not become scuffed, chipped or visually impaired before the end of life.
- Design for long life in the target environment. What temperature, humidity, radiation conditions and so forth does it have to withstand? For example the use of photo inhibitors can dramatically increase the life of plastic conservatory glazing.
- Make sure that components of the product with radically different lifetimes are easily separable so that repair, refilling or upgrading are feasible.

The chemical industry has most impact on durability through design of materials. Urban Hygiene is a small company that produces the easy-on™ anti-graffiti coating. The product is a mix of silicone resins and epoxy resins, and produces a permanent, easy clean surface. It can be used on a wide range of surfaces including metals, brick, concrete, tile, mortar, timber and plaster. Alternatively, it can be applied as a protection over existing coatings.

The coating will provide protection for up to 22 years, and significantly increases the lifetime of a surface before refurbishment. Because it allows the use of mild cleaning agents, it significantly reduces the amount of industrial chemicals that might be released into the environment through aggressive graffiti removal processes.

Powder coating is another process that improves product durability. Powder coating is a dry process in which pigment and resin particles are electrostatically charged and then sprayed on to a grounded surface. The substrate, with its adhering resin and pigment, is then put through an oven to fuse the coating. The resulting coating is tough, long-lived, attractive, and can protect the substrate from scratches, corrosion or other damage. The process uses no solvents, and waste coating powder is easily recovered for reuse. The whole process has a good environmental profile.

A specific example of its use is in high-tech consumer electronics products that are increasingly using magnesium structural components to make products lighter and more durable. Unfortunately, magnesium alloys need protection otherwise they quickly become damaged. AT&T's service teams use a specially road-hardened laptop manufactured by Itronix. This needs to survive being casually thrown into a car, dropped, rained on, and other insults. The powder coating is an important part of the product protection and durability.

Re-use, remanufacture and recycle. Focusing on re-use, remanufacture and recycling comes from a number of trends and drivers in society. Because of problems with waste disposal, many companies are trying to work towards a 'zero to landfill' policy. As waste disposal costs rise, this trend will accelerate. At the same time, where raw materials are difficult or expensive to obtain, there is a real commercial incentive to recycle and reuse. Regulators also continue to push for the minimisation or elimination of the release of hazardous materials into the environment, and for recycling wherever feasible. Finally, as some resources become scarce, pushing up prices, the idea that 'there is no such thing as waste' gains ground.

Ideas for reuse, remanufacture and recycling include:

- Design the product for a secondary use so that it has a market value at end of life. Composting of municipal green waste to produce garden compost is an attempt to balance waste disposal costs by producing a valuable product.
- Create a closed-loop system so that the industry recycles waste. Many copier and laser printer manufacturers have now done this, including the recovery and recycling of used toner cartridges as part of the overall service.
- Design the product for easy disassembly at end of life.
- Use single materials rather than composite materials to aid separation for recycling.
- Identify how a product could become a feedstock for another industry at end of life. Producing thermal fleeces from PET bottles and resilient playground surfaces from used tyres are examples.
- Avoid contaminating the materials with toxic residues at the end of life.
- Make sure the product is accurately labelled with the materials it contains to aid recycling.

National Starch has successfully developed and launched a range of adhesives for high-speed labelling of PET bottles. PET has a low surface energy, and traditionally required solvent-based adhesives to successfully label on a high-speed filling line. This exposed workers to VOCs. A range of starch-based water-borne adhesives was developed that are compatible with existing filling lines. As well as avoiding solvents, it turned out that the labels can be removed much more easily saving energy and time. This increased the value of PET scrap. Cleaning of the packaging and labelling equipment was also faster and cheaper.

Traditionally carpet tile backings have been made from bitumen, PVC or polyurethane. There are problems with the environmental impact of these materials and they are difficult to recycle. Shaw industries developed their EcoWorx® carpet tile backing using a polyolefin resin. This has superior environmental profile to PVC and is easier to recycle than bitumen and the cross-linked polyurethane backings. EcoWorx® has lower toxicity, better adhesion and good dimensional stability. It is easy to recycle and can be separated from any fibre type used in the carpet tile. It works particularly well with nylon-6 fibres. EcoWorx® tiles contain 40% recycled content and Shaw have established a collection and return scheme to close the loop.

Mass intensity. The idea behind focusing on the mass intensity of the product or service is dematerialisation: using less physical material to produce the required service.

The quantity of materials required to sustain the lifestyle of the developed world is large and increasing. In the 1990s, Germans discovered that they consume about 76 tonnes of solid materials and 60 tonnes of domestic water per capita per annum to support their lifestyle. Only a small fraction of this is retained for any length of time. Most of it results in sewage, solid waste and atmospheric pollution.

Producing, processing, transporting and disposing of these materials creates massive pollution. Reducing material intensity gives an immediate environmental benefit, and the potential for a significant economic benefit.

Ideas for reducing mass intensity include:

- Using stronger materials or engineered composites to produce lighter products.
- Improving the atom efficiency of chemical reactions.
- Designing waste out of the product or service life cycle.
- Designing delivery systems that ensure that the active materials are delivered to the right place, at the right time and in the right quantity.
- Questioning every component of a product to see if it is adding value. Detergent powders were revolutionised when it was realised that some of the ingredients in the formulation were there to make processing easier, and changing the manufacturing process made them redundant. By the time you include the packaging, this redesign reduced the amount of materials used per wash by 30%.
- Designing multifunctional materials. In metalworking, hydraulic oils that leak from the machinery into the cutting lubricants destroy the performance of these lubricants so they need to be replaced. By designing hydraulic oil that is miscible with, and functioned as, cutting oil you can avoid having to replace the cutting oil so frequently.
- Designing the product for ease of reuse, remanufacturing and recycling.

A famous example of reducing the mass intensity per unit of service is the Dow Sentricon™ Termite Colony Eliminator.[11] Termites cause over $1.5 billion damage to buildings in the US alone each year. Once they have become established, termites are notoriously difficult to eradicate. Traditionally the only effective defence were chemical barriers containing powerful pesticides. However, any gaps in a barrier around the building would allow the termites through. This led to the ground around buildings being saturated with highly toxic pesticides that were potentially dangerous to humans, animals and the environment in general.

The solution that Dow developed was a growth regulator, hexaflumaron. If this is incorporated into a bait, termites that take the bait cannot moult properly. However, finding an active is only part of the problem. It is also necessary to deliver the pesticide to the termites. The Sentricon™ system achieves this by placing the bait inside hollow plastic housing like a giant nail which is pushed into the ground near the colony. The packaging prevents access by humans or animals and has a long life in all soil and weather conditions.

The system is more effective than traditional barriers, and uses 1/10 000 of the active material, and 1/10 of the total materials required for barrier treatments.

Another area where delivery mechanisms are important is in the use of pesticide sprays in agriculture. The process of actually getting to the point where pesticide can have a biological effect is complicated. First of all it has to be atomised to produce a spray, the spray has to be transported to the target plant, it must impact on the plant and stick to form a deposit, and normally must be translocated to the site of action. Throughout this process there are opportunities for loss of efficiency. The spray can drift away from the target, the droplets may bounce off the plants and fall onto the ground, and the active ingredient may fail to penetrate the plant. All of these processes not only waste at the expense of active ingredient, but they significantly increase the risk of adverse environmental impact.

To solve this problem, modern pesticide formulations use a variety of additives (adjuvants) to improve the mass efficiency. Surfactants and polymeric rheology modifiers are used to reduce spray drift, surfactants are used to modify surface tension and reduce droplet 'bounce', and other ingredients are added to promote transport of the active into the plant. In modern pesticide systems, the physical chemistry of the spray system is nearly as important as the biological chemistry of the active ingredient. Using these techniques, the rate of application of toxic material to achieve the desired level of plant protection can be dramatically reduced; saving on expensive actives and reducing the health and environmental risk.

Energy intensity. Energy intensity is an obvious target for green product design. This is the easiest dimension to identify commercial benefits. Lower energy intensity can clearly either save manufacturing costs, or total cost of ownership for the user.

Energy is consumed throughout the life cycle of products and services. Enormous quantities of pollution and waste are generated in its production and consumption.

However, it is important to look at the energy intensity across the lifecycle. Simply focusing on one particular stage can give a false impression. For example, hybrid cars like the Toyota Prius use significantly less energy per passenger mile than conventional vehicles. However, at this stage of development of the engineering and materials, the energy costs of manufacturing these more complicated machines, and the anticipated lower lifetime of components like the batteries, means that the full life cycle energy cost may not be as attractive as the advertising would suggest.

Ideas to reduce energy intensity include:

- Developing processes that operate at less extreme conditions; for example, using enzymes to catalyse reactions.
- Modelling products and processes on biological systems – biomimetics.
- Using raw materials that have lower embodied energy; for example, using recycled aluminium instead of virgin aluminium.
- Using raw materials that have a high biological content. Much of the embodied energy comes from renewable solar radiation rather than environmentally damaging fossil fuels. Using natural plant fibres for reinforced composites instead of glass fibre will significantly reduce the embodied energy.
- Redesigning the products to have higher energy efficiency in use. For example, adhesives that can be cured at low temperatures or using an energy source that can be produced efficiently such as UV radiation.
- Efficient insulation for products that must maintain a temperature difference across a boundary.
- Self-lubricating surfaces for moving parts.
- High efficiency lubricants.
- Lightweighting moving parts to reduce energy consumption.
- Designing products so that disposal of end of life is easy and energy efficient.
- Designing products so that the embodied energy can be recovered at end of life. For example, producing plastics that can be burned for energy recovery without producing toxic emissions.

The use of enzymes is a common approach to reducing energy consumption during manufacture, and they are often found in unlikely products. Providing jeans with soft feel and attractive colouration of the 'stonewashed' look is an expensive business.

New jeans are washed with crushed pumice and acid to scuff, soften and fade them. This requires open-cast mining of pumice, a lot of energy input to the washing and disposal of difficult solid waste. The alternative is to use cellulases to degrade the cotton fibres of the fabric. Genencor discovered a cellulase derived from a new species of strictly alkaliphilic *Streptomyces.* The bacterium was isolated from the soda mud flats on the shores of the highly alkaline Lake Nakuru in Kenya. For economic production, the endocellulase gene was cloned and expressed in *Streptomyces lividans*, and commercialised as IndiAge Neutra in 1999. Novozyme has similar products – also obtained from extremophiles. The benefits are reduced use of non-renewable pumice, lower energy and water consumption, reduced fabric damage and a softer fabric, and significantly reduced costs.

An example of reducing energy consumption in use is found in 4-stroke automobile lubricants.[14] Only about 20–35% of the energy released by the combustion of fuel is turned into useful mechanical work. The rest is lost as heat and friction. Lubricants play an important role in the overall efficiency of the vehicle.

Recently, the so called 'synthetic' oils based on fatty acid esters have been used to produce lubricants with a low viscosity when cold, a flat viscosity profile so that they do not become too thin as the engine reaches operating temperature, and a very good thermal stability. This helps to tackle a particular problem in Northern developed countries, where many vehicles used in commuting never reach their designed operating temperature. This means that with the older lubricants there was not only increased fuel consumption, but an increase in a variety of tailpipe emissions. These new lubricants cut both emissions and fuel consumption. Ford has estimated that this may result in a small improvement in fuel efficiency of 0.6%, but points out that this is equivalent to eliminating 190 kT of CO_2 emissions p.a. in the US, and saving nearly 80 million litres of road fuel.

Similarly, the use of ultra-low viscosity lubricants in compressors for refrigerators and freezers can cut energy consumption by 10%. With refrigerators, freezers and air-conditioning units representing somewhere between 5 and 10% of the UK's total electricity consumption, this is a very interesting commercial opportunity.[13]

Human and environmental safety. Human and environmental safety is already the major focus of most company environmental programmes. The key here is to substitute less harmful for more harmful materials, and to avoid the release of hazardous materials into the environment at all stages of the life cycle.

Ideas for improving human and environmental safety include:

- Avoid the use of persistent and bio-accumulating materials.
- Avoid the use of materials listed as risky by any national regulator.
- Actively try to include materials that are known to be environmentally benign.
- Estimate the risk of any proposed new material using computer-based models for eco-toxicity and human toxicity.
- Avoid the release of organic vapours or gases at any stage in the lifecycle.
- Design products to only release benign materials into the environment. This is not easy. For example, both brake pad dust and material from car tires is highly eco-toxic.
- Design the product for containment or destruction before end of life of any hazardous materials.

- Design a product for safe disassembly and recovery of any hazardous materials.
- Design manufacturing processes to avoid the release of any hazardous material; including ozone depleting or greenhouse gases.

Sometimes dramatic improvements can be made in overall safety during manufacture. Taxol is an important anti-cancer drug. The active ingredient is paclitaxel, originally extracted from the bark of the Pacific yew *Taxus brevifolia*. Unfortunately, paclitaxel is present in small concentrations and the supply would not be sustainable. A new route was urgently needed. Direct chemical synthesis involved about 40 steps and an overall yield of 2% and was not considered commercially viable. Bristol-Myers Squibb developed a route based on plant cell fermentation. Calluses of a specific cell line from yew are grown in an aqueous media using nutrients from renewable sources. The paclitaxel is directly harvested from the cells and then purified by chromatography and crystallisation. Compared to the best semi-synthetic route, that starts with a closely related compound that can be sustainably obtained from European Yew, the plant cell culture process eliminates production of 6 intermediates, 10 solvents and 6 drying steps. In its first 5 years it will avoid the production of 32 tonnes of difficult waste. It reduces solvent and other chemical use, reduces waste and is much cheaper.

In other cases the risk to people is more direct. Flexographic printing is a widely used technique. Photopolymerisable material on a flexographic printing plate cross-links when exposed to light and captures an image. After exposure, flexographic printing plates are immersed in a 'washout' solvent, typically chlorinated hydrocarbons or Xylene, to remove the unpolymerized material. These washout solvents are hazardous VOCs subject to stringent reporting requirements and create problems with recycling and disposal. North America alone uses 7.5 million litres of washout solvents each year with a market value of $20 million. Alternative solvents and processing systems were required to replace the toxic VOCs with less hazardous materials at similar cost.

NuPro Technologies Inc. developed a series of safer washout solvents based on renewable methyl esters, terpene derivatives and highly substituted cyclic hydrocarbons. All the solvents are designed to be recycled in NuPro's Cold Reclaim System™, a combination of filtration and centrifugation. These solvents have higher flash points reducing explosion potential, worker exposure and VOC emissions. Methyl esters and terpene derivatives are sustainable, biodegradable and low toxicity. Solvent recovery using the Cold Reclaim System™ is low energy in comparison to vacuum distillation, and solid waste recovered after solvent recovery is a non-hazardous polymeric material.

Conserve scarce or non-renewable resources. The mass and energy dimensions of the eco-innovation compass look at the quantity used per unit of service. However, they do not focus on the source of the energy or materials. Many of the resources that we have been using in the industrial world are under pressure in one way or another. The resource may be severely limited, and being rapidly depleted. The resource may incur very high environmental costs to win it. There may be competition between different uses, for example water for agriculture, industry or human consumption. There may be high environmental costs associated with the use of the resource, for example fossil hydrocarbons and global warming.

These issues of scarcity, high environmental costs and sustainable use will ultimately be reflected in the market price of resources. It is therefore economically sensible to plan for significantly higher costs for some of these materials in the future. Trying to avoid the use of scarce, damaging or depleting resources therefore brings both environmental and financial benefits.

Ideas for conserving scarce or non-renewable resources include:

- Using renewable energy sources at all stages of the life cycle.
- Minimising the use of water, particularly potable water, in the manufacture and use of the product.
- Using renewable and sustainable materials instead of non-renewables. For example, shifting from hydrocarbon to carbohydrate feedstocks.
- Using recycled materials where possible.
- Using waste products from your own or another industry as a feedstock.
- Avoiding the use of materials produced with significant damage to ecosystems.
- Switching product design away from rare and hard to win metals and towards more common metals.
- Although biologically derived materials may be renewable, they may not be sustainable at an economic rate of exploitation. Alternative synthetic or bioprocessing routes to materials may be necessary alongside bio-prospecting.

The switch to renewable feedstocks is being found in many areas of industry. The Ford Motor Company uses between 20 and 30 million litres of soluble metal cutting fluid globally every year; primarily in the manufacture of engines and transmissions. Until 1999 over 99% of this fluid was based on mineral oil technology, creating significant waste disposal problems. There were over 25 suppliers of fluids to Ford Powertrain Operations, providing over 200 different products worldwide. There were few common products or management systems, representing a major chemicals and waste management problem. Ford wanted to dramatically reduce the complexity of the operations whilst meeting all the complex technical requirements for metalworking fluids; including tool-life, surface finish and quality, environmental impact, stability, compatibility with the work environment and cost.

Working with an established Tier 1 supplier Houghton International, Ford selected a two-phase metalworking lubricant/coolant based on vegetable oil. In trials at the Dagenham Engine Plant Ford found many benefits:

- Lower operating concentrations leading to lower costs and lower exposures.
- Safer compared to single phase semi-synthetics.
- Superior technical performance.
- 20% overall cost reduction even though the product had a higher cost per litre.
- Use of a renewable feedstock.
- The waste can be burned as a renewable energy source.
- Compatible with waste water treatment plants.
- 2 million litre reduction in emulsion use.
- Cleaner swarf and a cleaner shop-floor.
- Longer system life.

Extending this to other plants Ford predicts savings of up to 40% at the Bridgend plant and a potential saving of $250 000 p.a. at the Bordeaux transmission plant.

Use of renewable feedstocks is most likely where they can compete economically with petrochemically derived materials. This already happens in many areas, and it is sometimes forgotten that even in a world that seems to be dominated by chemicals and materials from fossil carbon and other non-renewable sources, industry already uses annually: 19.8 MT of vegetable oils, 22.5 MT starch, 28.4 MT of plant fibres and 42.5 MT of wood pulp. These all compete on price and performance with synthetic alternatives.

Poly(trimethylene terephthalate) (PTT) is a polymer with very useful properties. As a textile fibre it has excellent softness, stretch and recovery. As a resin it has excellent barrier properties. Developed over 60 years ago, PTT has not been very widely used compared to poly(ethylene terephthalate) (PET) as one of the key monomers 1,3-propanediol (PDO) has been expensive.

DuPont has developed a new bio-based source of PDO for its Sorona® range of PTT. Working with Genencor, they have engineered metabolic pathways from wild organisms into a commercial host to produce PDO from glucose. Genes from a yeast that can convert glucose into glycerol, and genes from a bacterium that converts glycerol into PDO were used. The pathways of the host were also altered to maximise production of PDO and minimise production of biomass.

A joint venture plant is under construction by DuPont and Tate & Lyle that is expected to produce PDO at a significantly lower cost than the petrochemical route. DuPont's 45 ktpa plant uses 40% less energy than petrochemical production of PDO.

References

1. Facts and Figures: The European chemical industry in a worldwide perspective. CEFIC, 2006. Online report: www.cefic.org/factsandfigures/
2. Responsible Care: Focusing on the future – Annual report Europe 2005/6. CEFIC, 2006.
3. Sheldon, R.A. (2000) Atom Efficiency and Catalysis in Organic Synthesis. *Pure Applied Chemistry*, **72**(7), 1233–1246.
4. Brezet, J.C. (1997) Dynamics in Eco-Design Practice. *UNEP Industry and Environment*, **20**(1), 22.
5. Hawken, P., Lovins, A., Lovins, L.H. (1999) *Natural Capitalism – The Next Industrial Revolution*. Earthscan Publications, London. ISBN 1-85383-461-0.
6. Lohse, J., Wirts, M., Ahrens, A. *et al.* (2003) *Substitution of Hazardous Chemicals in Products and Processes*. Ökopol GmbH and Kooperationsstelle, Hamburg.
7. Envirosense Solvent Substitution Data Systems. http://es.epa.gov/ssds/ssds.html.
8. UK Government Parliamentary Office of Science and Technology, Postnote, No. 229, September 2004.
9. McDonough, W. & Braungart, M. (2002) *Cradle to Cradle – Remaking the Way We Make Things*. North Point Press, New York. ISBN 0-86547-587-3.
10. Wimmer, W. & Züst, R. (2001) *ECODESIGN Pilot*. Kluwer, Dordrecht. ISBN 1-4020-1090-7. And also www.ecodesign.at.
11. Dasari, M.A., Kiatsimkul, P., Sutterlin, W.R., Suppes, G.J. (2005) Low-Pressure Hydrogenolysis of Glycerol to Propylene Glycol. *Applied Catalysis A: General*, **281**, 225–231.
12. Fussler, C. with James, P. (1996) *Driving Eco-Innovation – A Breakthrough Discipline for Innovation and Sustainability*. Pitman Publishing, London. ISBN 0-273-62207-2.

13. Brown, D., Green, J., Hall, F., Rocchi, S., Rutter, P., Dearing, A. (2000) *Building a Better Future – Innovation, Technology & Sustainable Development – A Progress Report.* World Business Council for Sustainable Development.
14. Boyde, S.R. (2002) Green Lubricants. Environmental Benefits and Impacts of Lubrication. *Green Chemistry,* **4**, 293–307.

Chapter 4
Application of Green Metrics Analysis to Chemical Reactions and Synthesis Plans

John Andraos

4.1 Introduction

Over the last decade there has been a vigorous attempt to answer quantitatively the simple question 'is a given reaction or synthesis plan "green"?' using the so-called 'green metrics'.[1–12] It is now accepted that reaction 'greenness' encompasses many factors including material efficiency, energy efficiency, environmental impact, toxicity data for all materials used, safety and hazard procedures, life cycle analysis of all products including target, side and byproducts, and cost efficiency in all its forms. It is also clear from the question posed that 'greenness' is a relative term rather than an absolute one. This means that no stand-alone reaction or plan to a target molecule can be called 'green'; it must be compared to and ranked against at least one other route to the same target according to some set of criteria. In effect any endeavour to devise a 'green' reaction or synthesis plan is actually a standard optimization problem with well-defined boundary conditions. Such problems are encountered routinely in the day-to-day workings of any sector of the chemical industry. Basically, a set of parameters are identified and these are optimized in such a way that some are minimized and others are maximized as appropriate given a set of constraints. When applying these ideas to the 'greenness' optimization problem the direction of optimization with respect to the relevant variables is often self-evident. Hence, it is obvious that a reaction, sequence of reactions, or process is chosen so as to maximize material and energy efficiencies toward the target product and to minimize overall cost, solvent usage, waste production, and toxicities of materials used in the entire operation. The real problems encountered in the selection of an optimum synthetic route are that not all variables to be optimized are known quantitatively to begin with, the scale of the operation brings about unforeseen and unexpected challenges in the reaction and post-reaction phases that are not predictable from experimental results obtained at small scales, and that the final chosen route often does not have all relevant parameters optimizing synergistically in the same direction. In industry the strongest controller of optimization is cost. Hence, it may be the case that the overall cheapest plan or reaction may also be the one producing the most waste or utilizing the most toxic reagents or utilizing the most solvents. Costs can be driven down significantly regardless of the material or energy efficiency of a plan if starting materials are cheap and readily available, if labour costs are low, and if the price of electricity is kept as low as possible. This means that it is often difficult to coincide the merits of a chosen plan with the philosophical ideals of 'green' chemistry according to the Anastas–Warner protocol.[13–15]

Coupled with the above optimization problem faced by the chemical industry there is also a parallel academic endeavour to understand and quantify what it means to achieve the 'ideal synthesis' for a given target molecule.[16–23] This is often couched in the language of synthetic elegance or art. Work in this area precedes that of 'green chemistry' and is also based on philosophically pleasing ideals such as minimization of the number of steps in a plan and maximization of individual and overall yields for all chemical reactions. In fact, many of the goals of 'green chemistry' and designing 'ideal syntheses' are consistent with one another and are obvious. However, the difficulty in both cases lies in translating philosophical ideals into rigorous quantification of well-defined parameters so that intelligent and directed optimization can be achieved while at the same time dipping into the ever-growing database of known transformations to find the best sequence of steps to get to a target molecule.

The purpose of this chapter is to introduce and unify key reaction metrics referring to material efficiencies of chemical reactions and entire synthesis plans into an integrated whole that can be applied to any chemical reaction or synthesis plan of any degree of complexity. The goal is to make rigorous the characterization of a reaction or synthesis plan as green and to reconcile the aims of 'green chemistry' and 'ideal synthesis design' in a quantitative manner so that intelligent choices can be made in the selection of an optimum plan over other candidates. In the first section we begin by giving a complete definition of reaction mass efficiency based on the conservation of mass law applied to balanced chemical equations. As an aside, it is interesting to point out that balancing chemical equations has fallen out of fashion among organic chemists since they are more focused on synthesizing their target molecules of interest and determining their correct structures. However, the venerable collected volumes of *Organic Syntheses* and the older German chemical literature provided properly balanced chemical equations for all transformations. Moreover, the latter used the equal sign notation for chemical equations which is more in tune with the algebraic notation of mathematical equations instead of the modern arrow symbolism. The balancing of chemical equations is the crux of all the analysis presented in this work and forms the basis of assessing properly the 'greenness' of chemical transformations. A true green chemist who wishes to minimize waste production must be able to account for the fate of all atoms in a chemical reaction as did Antoine Lavoisier in 1775 when he formulated the conservation of mass law.[24] We then introduce a pictorial representation of the master reaction mass efficiency relationship using a radial pentagon to assess the green material footprint of any given chemical reaction. Next, we introduce a powerful concept based on probabilities which allows the likelihood of achieving a given threshold reaction mass efficiency to be determined. These ideas are then applied to the database of organic reactions in order to elucidate patterns among the various reaction classifications. In the second part we introduce synthesis trees as a new representation of traditional synthesis plans to facilitate the quantitative determination of material metrics for linear, convergent, divergent and mixed routes. This allows unbiased ranking of synthesis plans to a common target according to various criteria based on material efficiency and synthetic elegance. About 130 examples of synthesis plans to 25 classic and recent important target molecules are analysed in an effort to gain a deeper understanding of the key strategies that are employed in achieving 'ideal synthesis design'.

Metrics are categorized under three classes: (a) materials efficiency, (b) energy efficiency, (c) synthesis strategy and elegance. Under materials efficiency we have reaction yield with respect to limiting reagent, atom economy, stoichiometric factor, materials recovery parameter, reaction mass efficiency and environmental impact factor. Under energy efficiency we have the absolute magnitude of input energy and the fraction of total energy input directed to producing a given target product. Under synthesis strategy we have degree of convergence, degree of asymmetry or skewness, rate of convergence, relative rate of convergence, molecular weight first moment or building up parameter and hypsicity (oxidation level) index. The reader is referred to key background references on which these ideas were first reported.[25–31]

4.2 Green metrics applied to individual reactions

4.2.1 *Complete expression for reaction mass efficiency (RME)*

The general expression for reaction mass efficiency (RME),[1–3] derived in Appendix B, for any individual chemical reaction is given by

$$\begin{aligned} RME &= (\varepsilon)(AE)\left(\frac{1}{SF}\right)(MRP) \\ &= (\varepsilon)(AE)\left(\frac{1}{SF}\right)\left(\frac{1}{1+(\varepsilon(AE)[c+s+\omega])/((SF)(m_P))}\right) \end{aligned} \tag{4.1}$$

where ε is the reaction yield with respect to limiting reagent ($0 < \varepsilon < 1$); AE is atom economy ($0 < AE < 1$); SF is the stoichiometric factor that takes into account the use of excess reagents ($SF = 1$ for stoichiometric reactions carried out with no excess reagents; $SF > 1$, otherwise); MRP is the material recovery parameter that takes into account other materials used in the reaction and post-reaction phases (work-up and purification) such as solvents and washings for extractions, solvents for chromatography, and so on ($0 < MRP < 1$); c, s and ω are the masses of reaction catalyst, reaction solvent and all other post-reaction materials respectively; and m_P is the mass of the collected target product. Atom economy is given by the well known definition[32–36]

$$AE = \frac{MW_{product}}{\sum MW_{reagents}} \tag{4.2}$$

and the stoichiometric factor, by definition, is given by[25]

$$SF = 1 + \frac{\sum mass_{excess\ reagents}}{\sum mass_{stoichiometric\ reagents}} = 1 + \frac{(AE)\sum mass_{excess\ reagents}}{theoretical\ mass_{product}} \tag{4.3}$$

If a balanced chemical reaction involves more than one reagent, one of them will be the reference limiting reagent which will define the scale of the entire reaction. The numerator in the stoichiometric factor term takes into account the sum of the masses of all excess reagents used as appropriate. For any balanced chemical reaction in which all byproducts are identified, equation (4.1) may be used to determine RME under a variety of scenarios.

Table 4.1 Expressions for RME for any chemical reaction under various scenarios.

RME Expression	Conditions
$(\varepsilon)(AE)\left(\frac{1}{SF}\right)\left(\frac{1}{1+((\varepsilon(AE)[c+s+\omega])/((SF)(mp)))}\right)$	Excess reagents used Reaction catalyst destined for waste Reaction solvent destined for waste Work-up and purification materials destined for waste
$(\varepsilon)(AE)\left(\frac{1}{1+((\varepsilon(AE)[c+s+\omega])/(mp))}\right)$	No excess reagents used Reaction catalyst destined for waste Reaction solvent destined for waste Work-up and purification materials destined for waste
$(\varepsilon)(AE)\left(\frac{1}{1+((\varepsilon(AE)[c+\omega])/(mp))}\right)$	No excess reagents used Reaction catalyst destined for waste Reaction solvent recovered Work-up and purification materials destined for waste
$(\varepsilon)(AE)\left(\frac{1}{1+((\varepsilon(AE)[c])/(mp))}\right)$	No excess reagents used Reaction catalyst destined for waste Reaction solvent recovered Work-up and purification materials recovered
$(\varepsilon)(AE)\left(\frac{1}{SF}\right)$	Excess reagents used Reaction catalyst recovered Reaction solvent recovered Work-up and purification materials recovered
$(\varepsilon)(AE)$	No excess reagents used Reaction catalyst recovered Reaction solvent recovered Work-up and purification materials recovered

Table 4.1 summarizes reduced expressions for RME depending on whether or not reaction solvents or other post-reaction materials are recovered. When such a material is recovered it does not appear in the MRP factor. Note that the entry in the last row corresponding to the simplest scenario has appeared in all elementary treatments of green metrics for raw material usage. Since this entry corresponds to the basic RME value and is characteristic of the intrinsic performance of a chemical reaction, it is referred to as a kernel metric. Its value is therefore the maximum achievable reaction mass efficiency for a given reaction. The most important points to recognize about equation (4.1) are that (a) RME can be factored into four independent factors each ranging in value between 0 and 1; (b) each of these factors acts to attenuate RME; and (c) the MRP factor is the strongest attenuator of RME since solvents normally account for the bulk mass of reaction materials used in a chemical reaction. It is clear that to have a decent chance of achieving a high enough value of RME one must have reactions that perform with high reaction yields *and* that produce no byproducts, or else low molecular weight byproducts with respect to that of the target product. These two factors contributing to the kernel *RME*, ε and *AE*, are therefore crucial.

The quantities *RME* and E_m (Sheldon environmental impact factor or *E*-factor based on mass)[37–40] are related by a simple expression given by equation (4.4) which allows easy calculation of either parameter once one of them is known. It is often simpler to determine E_m first and then to use equation (4.4) to calculate *RME*. The derivation of equation (4.4) is linked to that of equation (4.1) and is also given in Appendix B.

$$RME = \frac{1}{1 + E_m} \tag{4.4}$$

A similar expression for *AE* in terms of E_{mw}, environmental impact factor based on molecular weight, is given by

$$AE = \frac{1}{1 + E_{mw}} \tag{4.5}$$

One can see that the forms of equations (4.4) and (4.5) are identical. It is clear that RME (*AE*) and E_m (E_{mw}) describe material efficiency from different points of view, the former with respect to the target product and the latter with respect to the waste products. Figure 4.1 shows the interconnections between the key material green metrics presented above.

4.2.2 Radial pentagon representation of RME

A Microsoft Excel (Version 5.0 or higher) spreadsheet template form has been developed which allows the calculation of the complete reaction mass efficiency (RME) according to equation (4.1) and raw material cost (RMC) for any chemical transformation.[30] Lines are numbered and line instructions are embedded in the same manner as a personal income tax form. Green metrics are evaluated to determine the 'greenness' of the experiment in a rigorous quantitative way and to determine the 'bottom line' cost of carrying out the experiment. Formula entries are inserted in appropriate cells to facilitate computation. Any

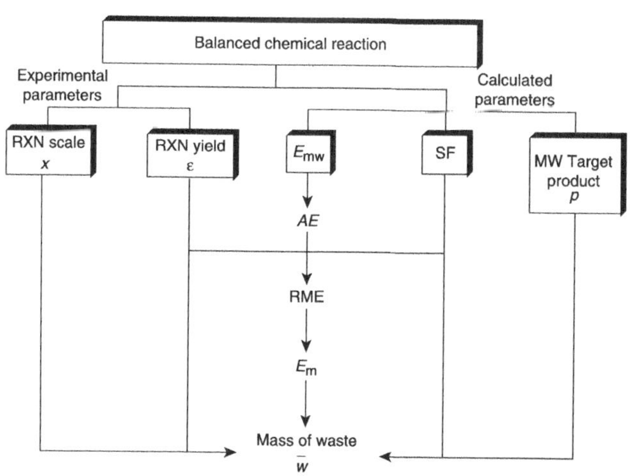

Figure 4.1 Flowchart showing interconnections between reaction metrics for a single general chemical reaction.

changes made in any input data cell for materials used or cost per gram result in automatic updates of formula entry cells and in the final graphical output. This powerful feature of Excel facilitates analysis of effects of changing any input parameters.

One begins by writing out fully balanced chemical equations for the transformation that will be performed in the laboratory. Quantities of all chemicals used in the reaction and their associated costs are entered directly in the appropriate cells. Volume measures in millilitres are entered for liquids and mass quantities in grams are entered for any solid materials used. The template form is divided into three parts: raw materials usage (Part 1), green metrics analysis (Part 2), and assessment criteria (Part 3). The raw materials usage section in Part 1 is in turn divided into three subsections: (A) reaction stage, including reagents, catalysts and reaction solvents; (B) work-up stage; and (C) purification stage. The total mass and total cost of input materials used are determined. From the mass of product collected the cost of product in \$/g (total input materials cost/mass of product collected) is determined. In Part 2, the kernel green metrics atom economy (AE) and environmental impact factor based on molecular weight (E_{mw}) are first evaluated. Mass of waste, RME, environmental impact factor based on mass (E_m), and wasted input costs are evaluated under three scenarios: (a) reclaiming reaction solvents, catalysts, byproducts and all post-reaction materials; (b) committing all reaction solvents, catalysts, byproducts and all post-reaction materials to waste; and (c) partial reclaiming of materials as appropriate. In Part 3, an overall assessment is made of the experimental procedure according to the following criteria in tabular format: (1) potential for side reactions, (2) potential for separation of multiple products, (3) potential for recyclability of byproducts back to reagents, (4) potential or actual use of a 'green' technology, (5) toxicity concerns for reagents, (6) toxicity concerns for byproducts, (7) toxicity concerns for solvents, (8) hazard concerns for reagents, (9) hazard concerns for byproducts, (10) hazard concerns for solvents, and (11) energy demands outside of standard temperature pressure conditions of 25°C and 1 atm.

The following undergraduate laboratory procedure for the synthesis of diphenylmethanol using the Grignard methodology (Box 4.1) according to Scheme 4.1 is given as an illustrative example where the above analysis is employed.

Figure 4.2 shows the results of Parts 1 and 2 for the reaction metrics form in Excel format for this procedure where equation (4.1) is used to check the overall RME for production of diphenylmethanol according to all materials used. Under conditions of using excess reagents (benzaldehyde is the limiting reagent), the kernel RME for production of diphenylmethanol is 28.2% with a reaction yield of 80%. The cost of producing 2.18 g of this material amounts to \$2.45 or about \$1.12/g. It is clear that the raw material cost on a per gram basis is inversely related to the reaction yield. If one does not recover reaction solvents and all other post-reaction washings and solvents used in the procedure, the analysis reveals that the overall RME for production of the target product decreases dramatically to 1.1%. In terms of input costs of all materials, this translates into \$2.42 of the \$2.45 spent, or 99%, as destined for waste. If diethyl ether and petroleum ether are recovered at appropriate stages in the synthesis without mixing them with other waste materials and with no added input materials, then \$0.93 of the \$2.45 spent, or 38%, is destined for waste though the overall RME nearly doubles to 1.8%.

The five parameters in equation (4.1) (RME, AE, ε, 1/SF and MRP) can be displayed graphically in the form of a radial pentagon as shown in Figure 4.3 depicting a 'materials usage footprint' so that it can be recognized at once which of the four factors on the

Box 4.1 Experimental protocol for synthesizing diphenylmethanol via a Grignard reaction.

To a 25-mL round-bottomed flask charged with 0.4 g dry magnesium turnings is added dropwise a solution of 1.8 mL bromobenzene in 9 mL dry ether over 20 min. The reaction solution is gently refluxed for a further 20 min. A second solution of 1.5 mL benzaldehyde in 4 mL dry ether is added dropwise over a period of 20 min. After addition is complete the mixture is refluxed for 15 min then cooled. The reaction mixture is then poured over 10 g crushed ice followed by addition of 3 mL of 5% aqueous HCl solution. The ether layer is separated and washed successively with water (30 mL), saturated sodium bisulphite ($NaHSO_3$) solution (30 mL), and again with water (30 mL). After drying with 5 g $MgSO_4$, filtration, and evaporation of the solvent, the crude product is recrystallized from petroleum ether (100 mL) to afford 2.18 g of pure diphenylmethanol.

Notes:

i Densities (g/mL): diethyl ether (0.708), petroleum ether (0.64), bromobenzene (1.495), benzaldehyde (1.046), water (1), 5% HCl solution (1.02), saturated $NaHSO_3$ solution (1.345).
ii Costs from 2004 to 2005 Aldrich Chemical Catalogue ($CAD): Mg turnings ($170.90/2.5 kg), bromobenzene ($543.45/18 L), benzaldehyde ($269/18 kg), 37% HCl ($125.60/10 L), diethyl ether ($325/16 L), sodium bisulphite ($416.50/12 kg), petroleum ether ($247/12 L), magnesium sulphate ($618.80/10 kg).
iii No cost is assigned to water.

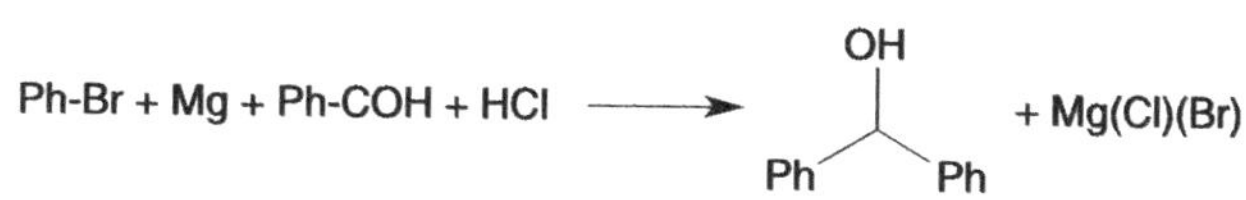

Scheme 4.1 Grignard reaction.

right-hand side of equation (4.1) are contributing to an attenuation of RME. Each axis corresponding to one of the five parameters emanates from the centre and ranges in value between zero and one. The values of these parameters are depicted as dots and these are connected to form a pentagonal figure. The ideal 'green' situation is depicted by a regular pentagon of unit radius where each parameter is equal to one. The less 'green' a reaction is, the more the resultant pentagon is distorted toward the centre. These diagrams may therefore be used to compare the RME performances of different classes of reactions so that chemists can ascertain which reaction classes are inherently 'green' and which are not by visual inspection. The degree of distortion of the radial pentagon from its regular ideal shape may be directly linked to which parameters are responsible for that distortion. These diagrams therefore inform us as to what to do if we wish to 'green up' our experimental procedure, whether it is to reduce reaction solvent usage, cut down on unnecessary washes and extractions, avoid using excess reagents unless there is a chemical reason for doing so as in driving equilibria toward product, select lower mass reagents to effect improved atom economical performance, or optimize reaction yield by tweaking such parameters as reaction time, reaction temperature, reaction pressure, or the use of catalysts.

Figure 4.4 shows the resultant radial pentagon for the Grignard reaction corresponding to the numerical results given in Figure 4.2. From this visual representation it is clear that the low overall RME is due to a modest AE of 60%, the use of 61% excess reagents, and a

REACTION METRICS FORM	
DATE:	
NAME OF TARGET PRODUCT:	*trans*–Benzalacetophenone
REACTION CLASSIFICATION	Carbon–carbon bond forming

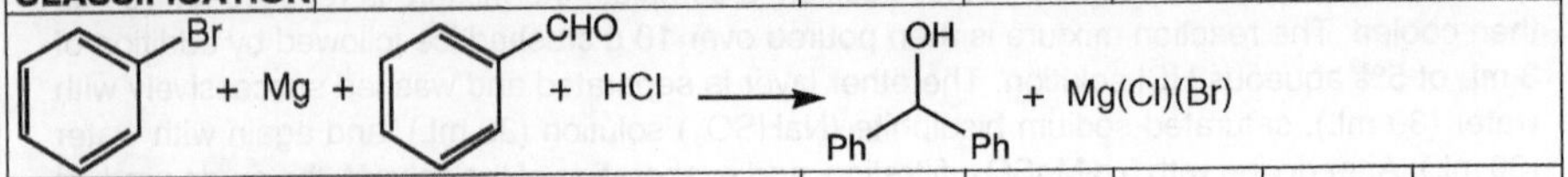

BALANCED CHEMICAL EQUATIONS:

PART 1: RAW MATERIALS USAGE

(A) REACTION STAGE:

(i) REAGENTS	MW (g/mol)	Density (g/mL)	Volume (mL)	Moles	Mass (g)	Cost ($/g)	Cost ($)	
Mg	24.3			0.0165	0.4	0.0684	0.027	12
Ph-Br	156.9	1.495	1.8	0.0172	2.691	0.0202	0.054	13
Ph-CHO	106	1.046	1.5	0.0148	1.569	0.0149	0.023	14
5% HCl (aq)	36.45	1.02	3	0.0840	3.06	0.0014	0.004	15
TOTAL REAGENTS	323.65			Add lines 12 to 15	7.72		0.110	**16**

(ii) CATALYSTS	MW (g/mol)	Density (g/mL)	Volume (mL)	Moles	Mass(g)	Cost ($/g)	Cost ($)	
None	0	0	0	0	0	0	0.000	19
							0.000	20
TOTAL CATALYSTS				Add lines 19 to 20	0		0.000	**21**

(iii) SOLVENTS	Density (g/mL)	Volume (mL)		Mass(g)	Cost ($/g)	Cost ($)	
Et_2O	0.708	13		9.204	0.0287	0.264	24
H_2O	1			10	0	0.000	25
TOTAL SOLVENTS			Add lines 24 to 25	19.204		0.264	**26**
Reaction Materials Subtotals			Add lines 16, 21, 26	26.924		0.374	**29**

(B) WORK-UPSTAGE:

MATERIAL	Density (g/mL)	Volume (mL)		Mass(g)	Cost ($/g)	Cost ($)	
H_2O	1	60		60	0	0.000	33
sat. $NaHSO_3$ (aq)	1.345	30		40.35	0.0347	0.532	34
						0.000	35
						0.000	36
TOTAL WORK-UPMATERIALS			Add lines 33 to 36	100.35		0.532	**37**

(C) PURIFICATION STAGE:

MATERIAL	Density (g/mL)	Volume (mL)		Mass(g)	Cost ($/g)	Cost ($)	
Petroleum Et_2O	0.64	100		64	0.0193	1.235	41
$MgSO_4$				5	0.0619	0.309	42

Figure 4.2 Microsoft Excel spreadsheet output of green metrics analysis for the synthesis of diphenylmethanol using the Grignard methodology.

						0.000	43	
						0.000	44	
TOTAL PURIFICATION MATERIALS			Add lines 41 to 44	69		1.544	**45**	
Post-reaction Materials Subtotals			Add lines 37, 45	169.35		2.077	**47**	
				Mass (g)	**Cost ($)**			
TOTAL INPUT MATERIALS			Add lines 29,47	196.274	2.450	**50**		
	MW (g/mol)	**Moles**	**Yield**	**Mass (g)**	**Cost ($/g)**			
OUTPUT TARGET PRODUCT	184	0.0118	0.8004	2.18	1.124	**53**		
PART 2: GREEN METRICS ANALYSIS								
Limiting reagent:	Ph-CHO							
PARAMETER	VALUE							
Reaction scale	0.0148	moles		**58**				
E(mw)	0.759	MW byproducts/MW product		**59**				
AE	0.569	MW product/ΣMW reagents		**60**				
(i) Under reclaiming reaction solvents, catalysts, and byproducts, and all post-reaction materials								
Mass of waste (line 16–53)	5.54	g		**63**				
E(m)	2.541	g waste/g product		**64**				
RME	0.282	g product/Σ g reagents		**65**				
SF	1.611			**66**				
Wasted input costs ($)	0.079							
(ii) Under committing all reaction solvents, catalysts, and byproducts, and post-reaction materials to waste								
Mass of waste (line 50–53)	194.094	g		**69**				
E(m)	89.034	g waste/g product		**70**				
RME	0.011	g product/Σ g reagents		**71**				
Wasted input costs ($)	2.423							
Check formula	0.011							
(iii) Under reclaiming ether from reaction and petroleum ether from purification								
Mass of waste	120.890	g		**76**				
E(m)	55.454	g waste/g product		**77**				
RME	0.018	g product/Σ g reagents		**78**				
Wasted input costs ($)	0.934							

Figure 4.2 Continued.

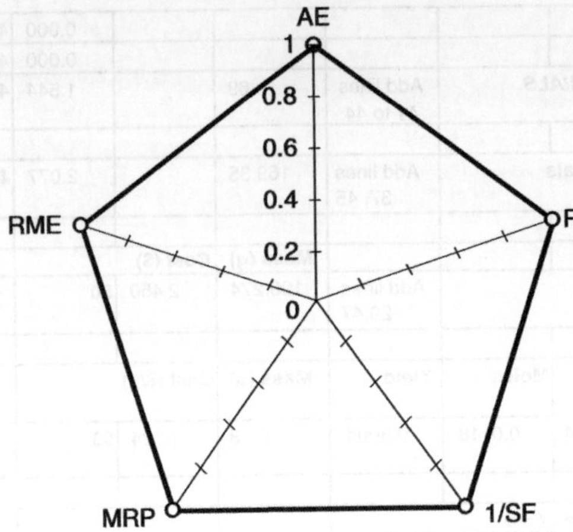

Figure 4.3 Radial pentagon representation of reaction mass efficiency (RME) showing dependence on four independent parameters given in equation (4.1).

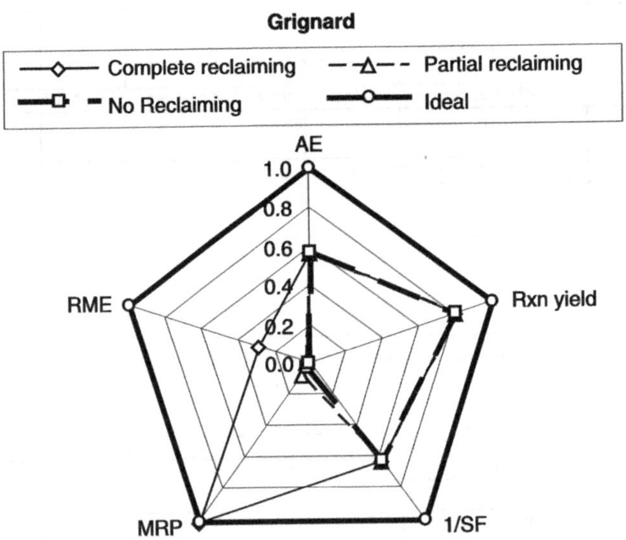

Figure 4.4 Radial pentagon representing RME values for the synthesis of diphenylmethanol using the Grignard methodology under various scenarios. See Scheme 4.1 for balanced chemical reaction.

minimum MRP of about 4%. The best RME possible for synthesizing diphenylmethanol by this procedure is 28.2%.

Scheme 4.2 shows various organic reaction types and Figure 4.5 shows their corresponding radial pentagons based on actual experimental data. When these diagrams are compared side by side it is possible to visualize at a glance which types of reactions are better atom economical and reaction mass efficient performers than others. In particular, it is possible to pick out which of the four factors in equation (4.1) is responsible for a low or high RME

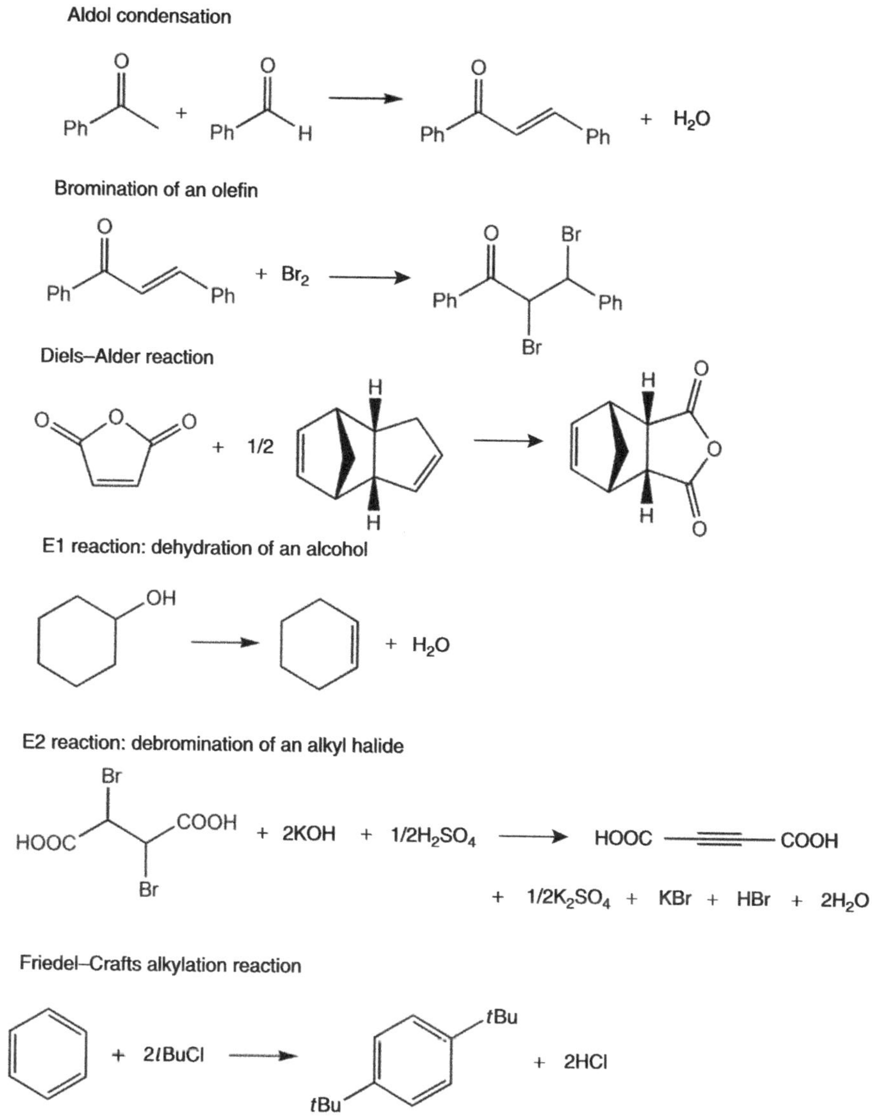

Scheme 4.2 Set of reactions used in a typical undergraduate laboratory.

outcome, and hence point the way to 'greening up' a procedure to a given target. One can see for example that the E2 reaction to produce acetylenedicarboxylic acid and the oxidation of cyclohexanol to cyclohexanone by the Jones procedure are by far the worst performers. In both cases this is because of low atom economies, low MRP values due to the use of excessive volumes of solvents and washings for extractions, and the use of excess reagents, though the former reaction exhibits a good reaction yield. The Beckmann rearrangement material performance is poor mainly because of excessive washings used in the work-up and purification steps though it has a 100% atom economy. The aldol condensation, olefin

Michael 1,4-addition reaction

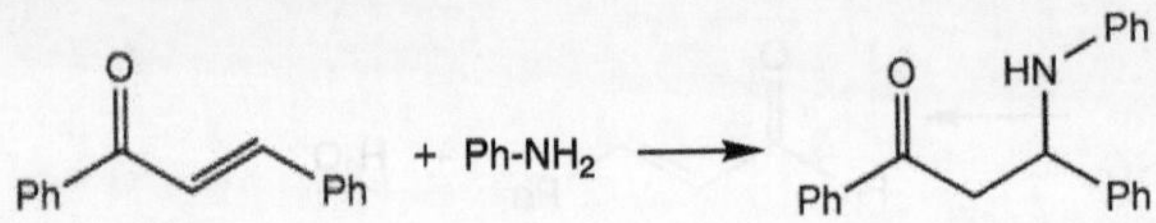

Aromatic nitration

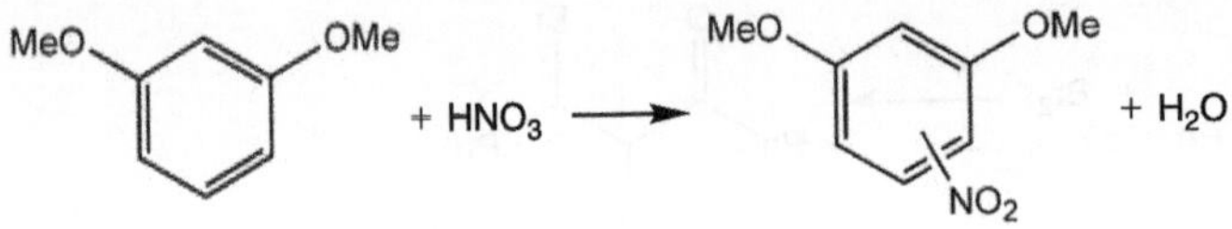

Oxidation of secondary alcohol

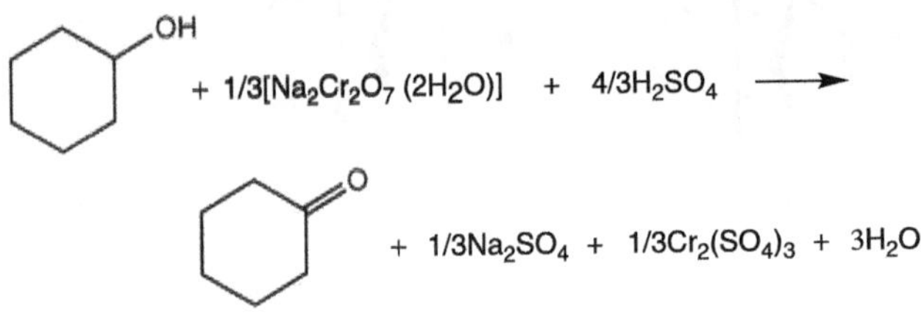

Oximation

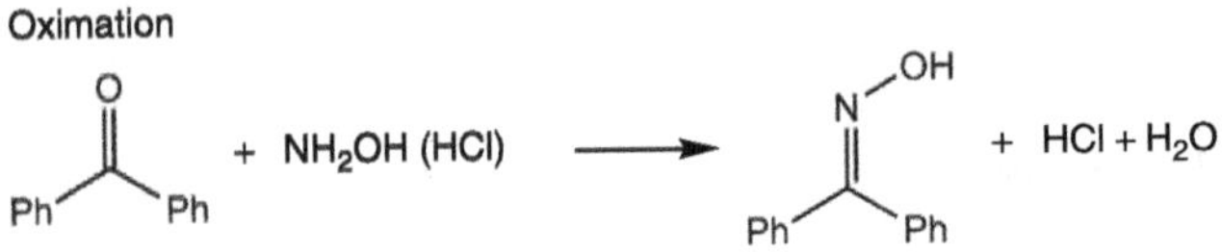

Beckmann rearrangement

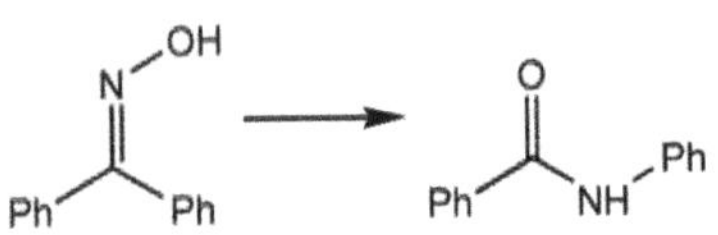

SN2 reaction: bromination of a primary alcohol

Scheme 4.2 Continued.

bromination, Diels–Alder, Michael 1,4-addition, and aromatic nitration reactions are all good atom economical transformations. The low values for MRP in all reactions severely reduce the overall RME for each reaction.

4.2.3 Special reactions

We now examine some special reaction types which yield more than one useful product to illustrate how the general expression for RME may be used.

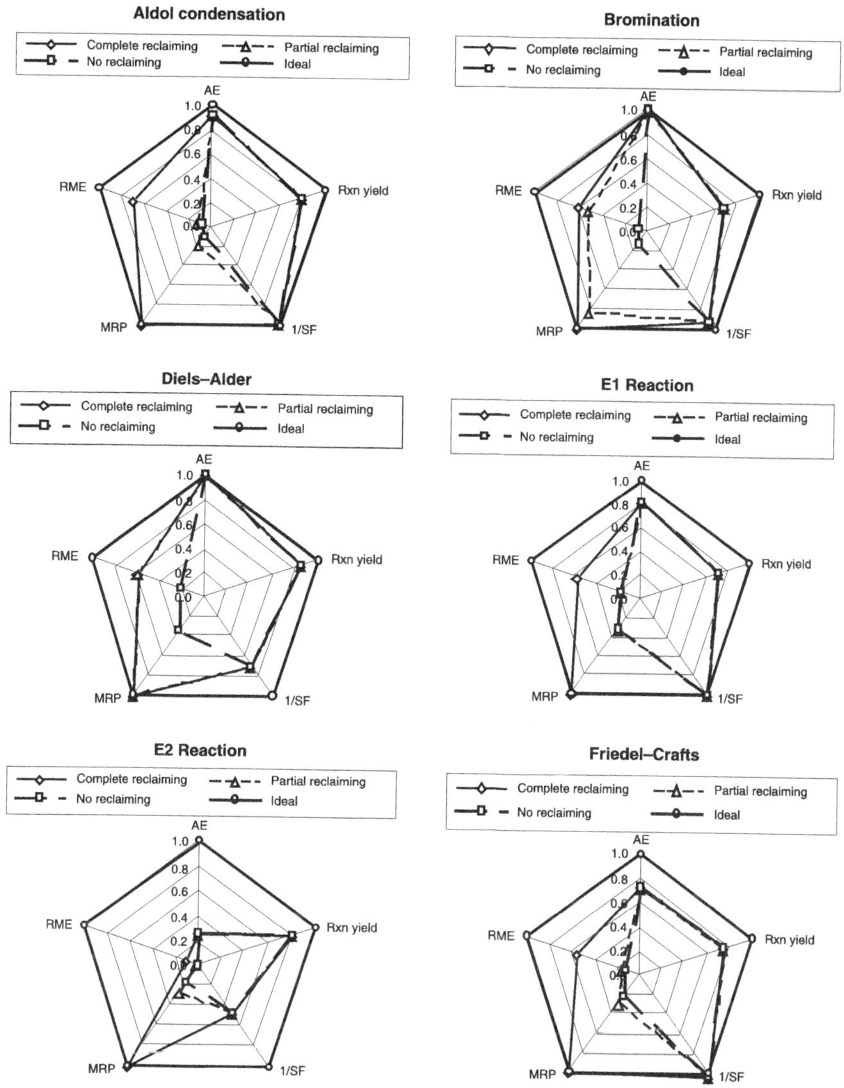

Figure 4.5 Radial pentagons representing RME values under various scenarios for a variety of reaction types. See Scheme 4.2 for balanced chemical reactions.

4.2.3.1 Reactions leading to isomeric products

Scheme 4.3 shows example reactions that produce more than one geometric or stereoisomeric product. If we consider all isomers as the global target product then the reaction yield with respect to limiting reagent is designated as $\varepsilon_{overall}$ and the expression for RME as given in equation (4.1) is used with this modification. If we wish to evaluate the RME for any one specific isomer then we multiply the overall reaction yield for all isomeric products by the fraction of the total isomeric product that represents the specific isomer and substitute that value for ε in equation (4.1). This fraction may be easily obtained from the product ratio.

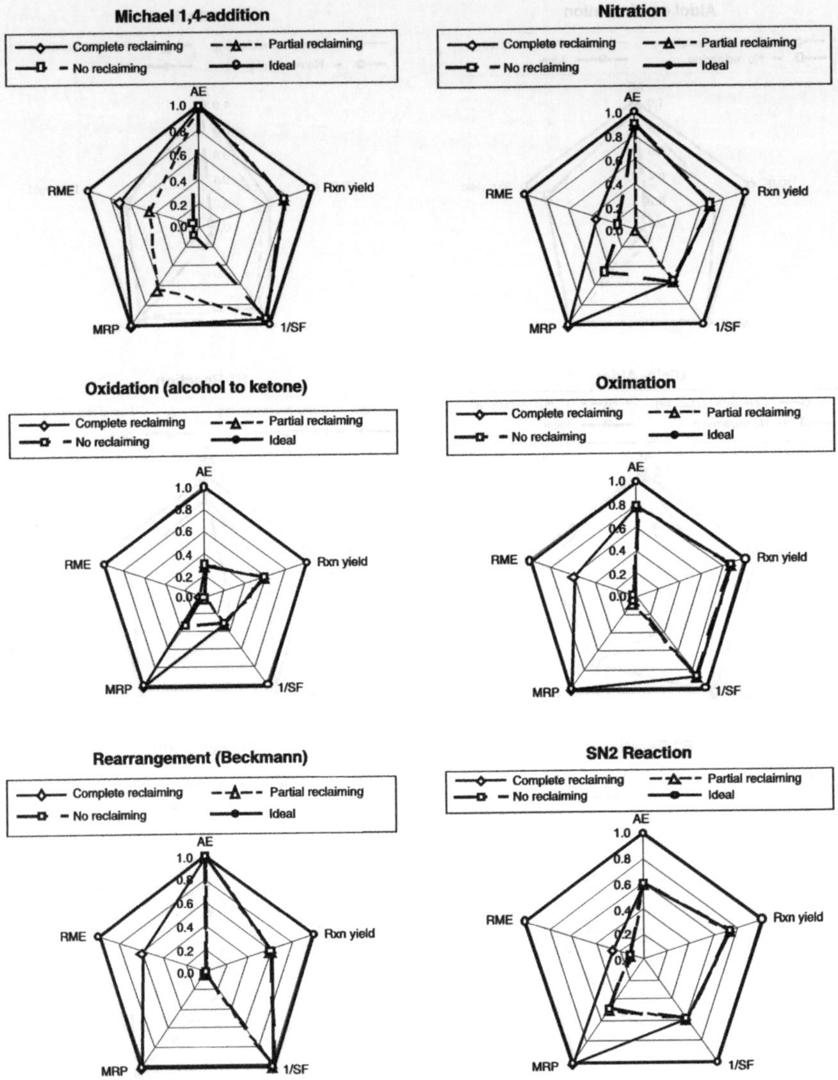

Figure 4.5 Continued.

For example, in the case of aromatic substitution reactions the RME for production of para product is given by

$$(RME)_{para} = \varepsilon_{para}(AE)\left(\frac{1}{SF}\right)\left(\frac{1}{1 + (\varepsilon_{para}(AE)[c + s + \omega])/((SF)(m_{para}))}\right) \quad (4.6a)$$

where

$$\varepsilon_{para} = f_{para}\varepsilon_{overall} = \left(\frac{m_{para}}{m_{para} + m_{meta} + m_{ortho}}\right)\varepsilon_{overall} \quad (4.6b)$$

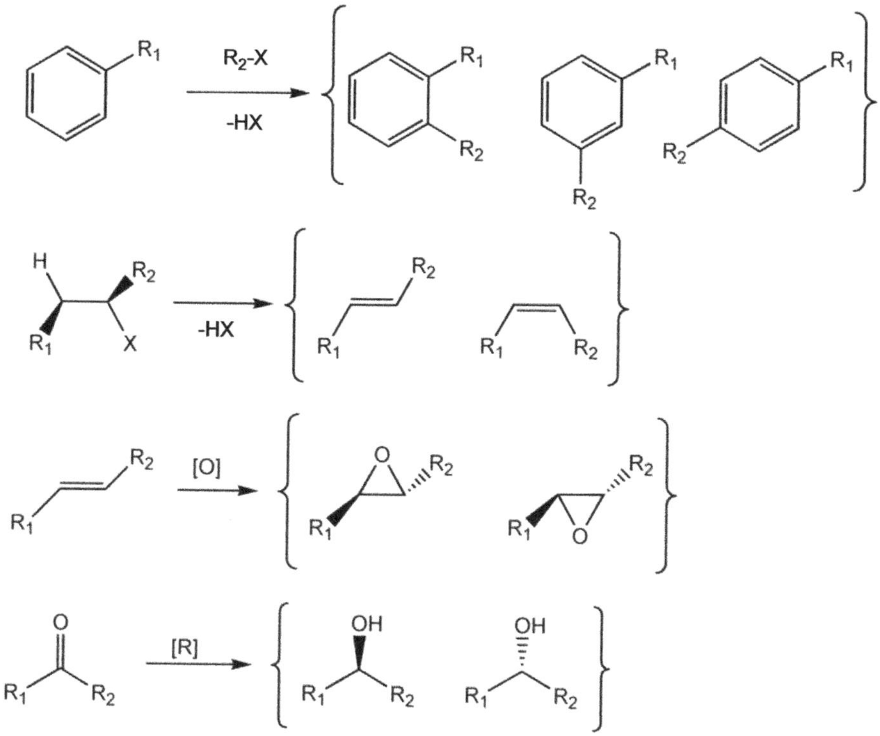

Scheme 4.3 Example reactions leading to isomeric products.

It is understood that the ortho and meta products form part of the waste produced. In determining AE, the balanced chemical equation is written with a generalized structure for the product indicating all possible isomers and since the molecular weights of all isomers are identical equation (4.2) is used without change. In the above example, the atom economy for the production of para, meta or ortho products is the same.

When dealing with reactions leading to stereoisomeric products we have the additional complication that descriptors such as enantiomeric (diastereomeric) excess and enantiomeric (diastereomeric) ratio are used to describe product purities. The evaluation of RME for a specific stereoisomer, say the R enantiomer, is exactly as above using the connecting relationships for the fraction of each product shown below.

$$f_R = \frac{R}{R+S} = \frac{er_R}{er_R+1} = \frac{1+ee_R}{2} \tag{4.7a}$$

$$f_S = \frac{S}{R+S} = \frac{1}{er_R+1} = \frac{1-ee_R}{2} \tag{4.7b}$$

where $er_R = R/S$ is the enantiomeric ratio and $ee_R = (R-S)/(R+S)$ is the enantiomeric excess with respect to the R enantiomer, respectively.

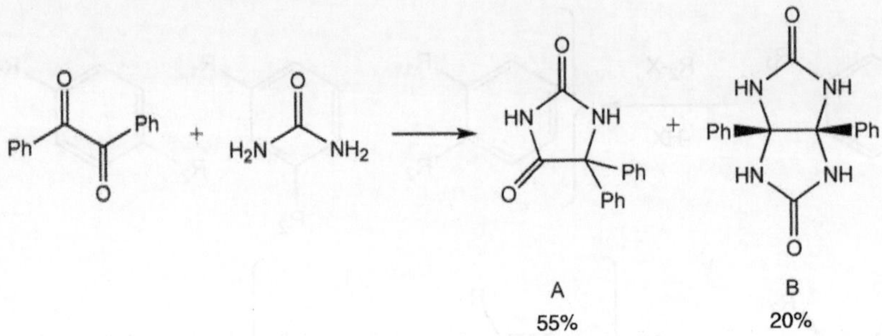

Scheme 4.4 Synthesis of phenytoin (product A) along with its side product (product B).

4.2.3.2 Reactions leading to more than one target non-isomeric product

Scheme 4.4 shows an example reaction that leads to two different products where the product yields are determined with respect to benzoin since this is the reference limiting reagent.[41] Here we must make explicit distinctions between what is designated as the target product of a reaction, any side products, and associated byproducts. A target product is the intended product sought after in a given reaction. As a consequence of producing this target product, byproducts and side products may also arise. A byproduct is a product that arises as a consequence of a reaction mechanism that leads to the target product. A side product is another end product arising in a reaction from a different mechanism than the intended target product. The production of a side product may also give rise to its own byproducts. In the present example, if we treat product A, phenytoin, as the target product then the balanced chemical equation is as shown below. One equivalent of water is the byproduct of this reaction and therefore the dimeric product B is the side product of the reaction since it arises from a different mechanism than phenytoin.

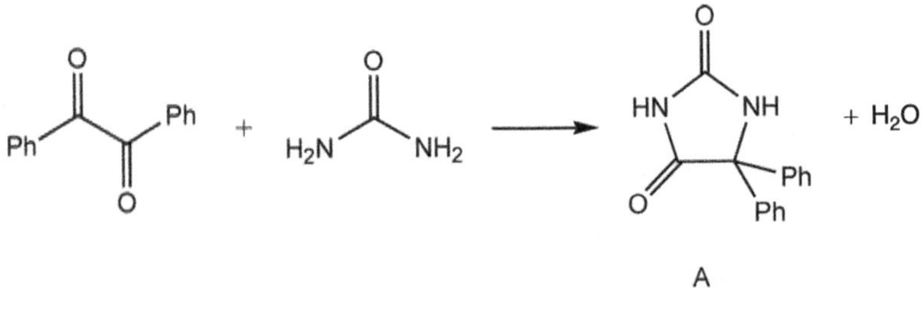

From this we determine that $AE = 0.933$ and RME with respect to product A is given by

$$(RME)_A = \varepsilon_A(AE)_A\left(\frac{1}{SF}\right)\left(\frac{1}{1+((\varepsilon_A(AE)_A[c+s+\omega])/((SF)m_A))}\right)$$
$$= (0.55)(0.933)\left(\frac{1}{SF}\right)\left(\frac{1}{1+((0.55)(0.933)[c+s+\omega])/((SF)m_A)}\right) \tag{4.8a}$$

where $SF = 1 + ((z-x)MW_{urea})/(x(MW_{benzoin} + MW_{urea}))$, x is the number of moles of benzoin, z is the number of moles of urea, and $z \geq x$. Alternatively, we may write $(RME)_A$

in terms of masses as $(RME)_A = m_A/(m_{benzoin} + m_{urea} + c + s + \omega)$. If we treat product B as the target product then the balanced chemical equation is as shown below. In this case the equivalents of water are byproducts of the intended reaction and therefore product A, phenytoin, now is the side product.

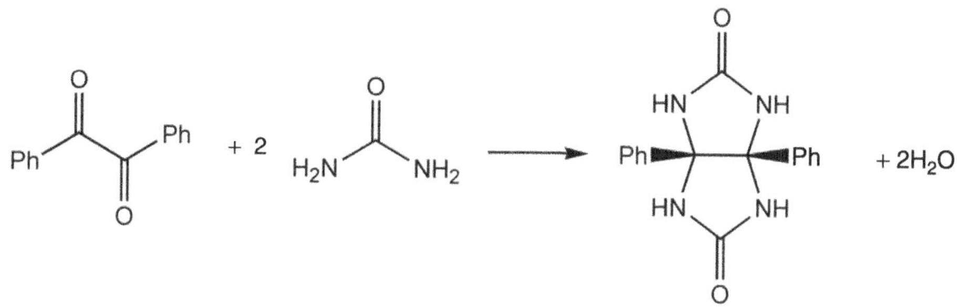

B

From this we determine that $AE = 0.891$ and RME with respect to product B is given by

$$(RME)_B = \varepsilon_B(AE)_B\left(\frac{1}{SF}\right)\left(\frac{1}{1+(\varepsilon_B(AE)_B[c+s+\omega])/((SF)m_B)}\right)$$
$$= (0.20)(0.891)\left(\frac{1}{SF}\right)\left(\frac{1}{1+((0.20)(0.891)[c+s+\omega])/((SF)m_B)}\right) \quad (4.8b)$$

where $SF = 1 + ((z - 2x)MW_{urea})/(x(MW_{benzoin} + 2MW_{urea}))$, x is the number of moles of benzoin, z is the number of moles of urea, and $z \geq 2x$. Again, we may also write $(RME)_B$ in terms of masses as $(RME)_B = m_B/(m_{benzoin} + m_{urea} + c + s + \omega)$.

4.2.4 Dealing with byproducts: Retrieval versus recycling

In dealing with reactions that necessarily give byproducts in addition to the target product one is faced with evaluating the merits of recycling versus retrieving versus disposal of the byproducts. The above formalism can be used to determine conditions when any one of these options may be advantageous. For the general sequence shown in Scheme 4.5. where P is the target product, Q is the byproduct and C is the reagent used to recycle Q back to reagent A we can derive expressions for the total mass of waste produced and E_m for each of these scenarios. These are summarized in Table 4.2. Simplified kernel expressions may be

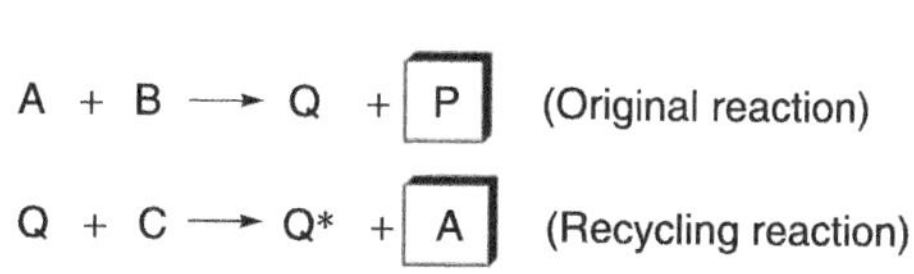

Scheme 4.5 General sequence showing an original reaction between reagents A and B and an accompanying recycling reaction that recycles byproduct Q back to reagent A.

Table 4.2 Summary of expressions for mass of waste produced, $\bar{w}$, and environmental impact factor based on mass, E_m, for the three options dealing with byproduct Q.

Condition	Total mass of waste, $\bar{w}$ (g)	E_m
Disposing Q[a]	$x\left(\frac{p}{AE}\right)[(SF)+\varepsilon(AE)(f_1-1)]$	$\frac{SF}{\varepsilon(AE)}+f_1-1$
Retrieving Q[b]	$x\left(\frac{p}{AE}\right)[(SF)+\varepsilon((AE)f_1-1)]$	$\frac{SF}{\varepsilon}+f_1(AE)-1$
Recycling Q back to reagent A[c]	$x\left[\frac{p(SF)}{AE}+\varepsilon\left(C+p(f_2-1)+a\left(\frac{(SF)^*-1}{(AE)^*}-\varepsilon^*\right)\right)\right]$	$\frac{(p(SF)/AE)+\varepsilon(c+pf_2+a(((SF)^*-1)/(AE)^*))}{\varepsilon(p+a\varepsilon^*)}-1$

[a] Conservation of mass law: $\bar{w}=ax+bz_1-py_1(1-f_1)$.
[b] Conservation of mass law: $\bar{w}=ax+bz_1-py_1(1-f_1)-qy_1$.
[c] Conservation of mass law: $\bar{w}=ax+bz_1+Cz_2-py_1(1-f_2)-ay_2$.

Symbol definitions: ε is reaction yield with respect to limiting reagent A for original reaction equal to y_1/x; ε^* is reaction yield with respect to limiting reagent Q for recycling reaction equal to y_2/y_1; y_2 is number of moles of Q* and A produced in the recycling reaction; z_1 is moles of reagent B; z_2 is moles of reagent C; a, b, C, p, and q are the molecular weights of reagents A, B, C, P, and Q; $f_1=(c_1+s_1+\omega_1)/py_1$; $f_2=f_1+[(c_2+s_2+\omega_2)/py_1]$; c_1, s_1, and ω_1 are the masses of catalyst, reaction solvent, and post-reaction solvents used in the original reaction; c_2, s_2, and ω_2 are the masses of catalyst, reaction solvent, and post-reaction solvents used in the recycling reaction; AE is atom economy of original reaction; $(AE)^*$ is atom economy of recycling reaction; SF is stoichiometric factor for original reaction; $(SF)^*$ is stoichiometric factor for recycling reaction.

Table 4.3 Summary of advantageous conditions for three options dealing with byproduct Q[a].

Condition	Mass of waste criterion[a]	E_m criterion[b]
Retrieving Q over disposing Q	$qy_1 > 0$, always true	$qy_1(ax + bz_1 + \Delta_1) > 0$, always true
Recycling Q back to reagent A over disposing Q	$ay_2 > cz_2 + \Delta_2$, or $\varepsilon^* > \left(\frac{C}{a}\right)\left(1 - \frac{\Delta_2 + c\phi_2}{ay_2}\right)^{-1}$ $\varepsilon^* > \frac{C}{a}$ (kernel condition)	$\frac{ax + bz_1 + \Delta_1}{py_1} > \frac{\Delta_2 + cz_2}{ay_2}$, or $\varepsilon^* > \left(\frac{C}{a}\right)\left(\frac{\Delta_1}{py_1} + \frac{SF}{\varepsilon(AE)} - \frac{\Delta_2 + c\phi_2}{ay_2}\right)^{-1}$ $\varepsilon^* > \left(\frac{C}{a}\right)\varepsilon(AE)$ (kernel condition)
Recycling Q back to reagent A over retrieving Q	$ay_2 > cz_2 + \Delta_2 + qy_1$, or $\varepsilon^* > \left(\frac{C+q}{a}\right)\left(1 - \frac{\Delta_2 + c\phi_2}{ay_2}\right)^{-1}$ $\varepsilon^* > \frac{C+q}{a}$ (kernel condition)	$\frac{ax + bz_1 + \Delta_1}{(p+q)y_1} > \frac{\Delta_2 + cz_2}{ay_2 - qy_1}$, or $\varepsilon^* > \frac{1}{a}\left[\frac{p\varepsilon(\Delta_2 + C(\phi_2 + y_1))}{\varepsilon\Delta_1(AE) + py_1(SF)} + q\right]$ $\varepsilon^* > \frac{C\varepsilon + q}{a}$ (kernel condition)

[a] Stronger criterion.
[b] Weaker criterion.

Symbol definitions: $\Delta_1 = c_1 + s_1 + \omega_1$; $\Delta_2 = c_2 + s_2 + \omega_2$; ϕ_1 is excess moles of reagent B used in original reaction equal to $z_1 - x$; ϕ_2 is excess moles of reagent C used in recycling reaction equal to $z_2 - y_1$.

obtained for these if both reactions are run under stoichiometric conditions with no excess reagents and reaction solvents, catalysts, and post-reaction materials are recovered by setting $SF = 1, (SF)^* = 1, f_1 = 0$, and $f_2 = 0$. Table 4.3 summarizes three advantageous conditions according to lower total mass of waste produced (stronger condition) and lower E_m values (weaker condition). It is obvious that retrieving byproduct Q over disposing it is always going to reduce the amount of waste produced. Recycling byproduct Q back to reagent A is advantageous over disposing Q if the mass of recovered A in the recycling reaction exceeds that of the sum of the masses of reagent C and the masses of solvents, catalysts, and post-reaction materials used in the recycling reaction. On the other hand, recycling byproduct Q back to reagent A is advantageous over simply retrieving Q if the mass of recovered A in the recycling reaction exceeds the sum of the masses of reagent C, byproduct Q, and the masses of solvents, catalysts, and post-reaction materials used in the recycling reaction. All of these conditions of course depend on the existence and feasibility of recycling reactions that interconnect chemical structures Q and A. Additionally, an important point to note is that inclusion of energy consumption in recycling and retrieving or recovery procedures needs to be considered. This can change dramatically the choice of possible outcomes. It is often the case that when energy demands and their associated costs are factored in existing recycling options may be abandoned in favour of disposal. The viability of recycling, therefore, is strongly linked to the invention of energy efficient recycling reactions and processes. The value of the above analysis is that various options may be evaluated quantitatively at least within the constraint of raw materials consumption and mass of waste generated. In this context an example series of calculations is given for the recycling of dicyclohexylurea

(DCU) back to dicyclohexylcarbodiimide (DCC) via various dehydration reactions. DCC is a preferred reagent used to carry out esterification reactions and peptide syntheses under very mild conditions. The drawback is that large quantities of insoluble DCU are produced. Scheme 4.6 shows a reaction[42] producing DCU as a byproduct along with five recycling options.[43–47] Under kernel best-case scenario conditions, the mass of waste criterion in Table 4.3 for recycling DCU back to DCC over disposal of DCU is applied by calculating

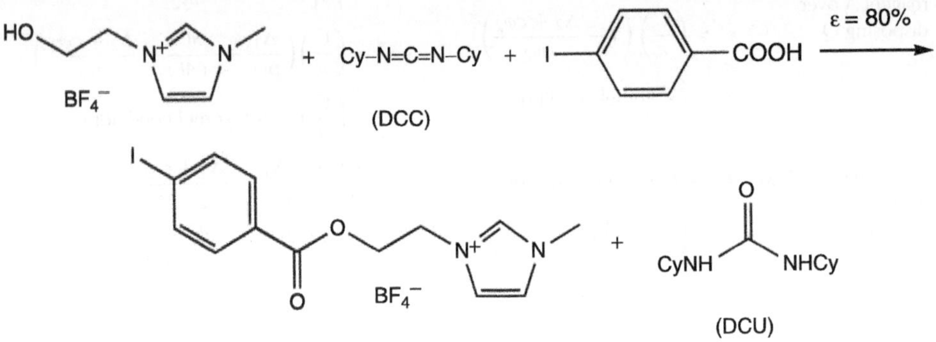

Option 1

$$\mathrm{DCU} + \text{p-TsCl} + 2\mathrm{C_5H_5N} \xrightarrow{\varepsilon^* = 82\%} \mathrm{DCC} + \mathrm{C_5H_5NH^+\,Cl^-} + \mathrm{C_5H_5NH^+TsO^-}$$

190.45 2(79)

Option 2

$$\mathrm{DCU} + \mathrm{P_2O_5} + 2\mathrm{C_5H_5N} \xrightarrow{\varepsilon^* = 76\%} \mathrm{DCC} + 2\mathrm{C_5H_5NH^+PO_3^-}$$

142 2(79)

Option 3

$$\mathrm{DCU} + 1/3\mathrm{O{=}PCl_3} + 2\mathrm{NaOH} \xrightarrow{\varepsilon^* = 89\%} \mathrm{DCC} + 1/3\mathrm{Na_3PO_4} + 2\mathrm{H_2O} + \mathrm{NaCl}$$

1/3(153.35) 2(40)

Option 4

$$\mathrm{DCU} + 1/2\mathrm{O{=}CCl_2} + 2\mathrm{NH_3} \xrightarrow{\varepsilon^* = 98\%} \mathrm{DCC} + 1/2\mathrm{(NH_4)_2CO_3} + \mathrm{NH_4Cl}$$

1/2(98.9) 2(17)

Option 5

$$\mathrm{DCU} + \mathrm{(pTs)_2O} + 2\mathrm{C_5H_5N} \xrightarrow{\varepsilon^* = 80\%} \mathrm{DCC} + 2\mathrm{C_5H_5NH^+TsO^-}$$

326 2(79)

Scheme 4.6 Various reactions that recycle dicyclohexylurea (DCU) back to dicyclohexylcarbodiimide (DCC).

the difference $\varepsilon^* - (C/a)$, where C is taken as the sum of the molecular weights of reagents used in the recycling reaction and $a = 206$ for the molecular weight of DCC. The best option based only on the kernel mass of waste produced due to byproducts would be the one with the largest positive value for this quantity. The values for options 1 through 5 are -0.872, -0.696, 0.253, 0.575, and -1.550 respectively. Hence, the best choice is the fourth recycling reaction.

4.2.5 Optimization of green metrics

In this section we generalize chemical reactions so that worst-case scenarios for atom economy and waste production may be determined. For example, it is possible to ask what is the lowest possible atom economy that might be expected for a given chemical reaction, or the highest possible environmental impact factor. We begin by generalizing reactions by writing out balanced chemical equations accounting for all byproducts using Markush structures for substrates, reagents, and products so that variable chemical groups or atoms are distinguished from those that need to be specified. In any kind of chemical transformation some fragment of the substrate chemical structure is operated upon whereas other parts remain untouched. The relevant atoms involved in the transformation are thus specified while the spectator atoms are designated with the customary R group notation. From this Markush-type chemical equation, general expressions for AE and E_{mw} as functions of R group size may be obtained under the assumption of 100% reaction yield, stoichiometric operating conditions, and complete recovery or elimination of solvents. It is then possible to estimate the minimum atom economy, $(AE)_{min}$, or the maximum unit waste expected per unit target product, E_{max}, according to the dependence of these metrics on R group size. These estimates of course represent 'best-case' scenarios since we already know that once these constraints are lifted the maximum E_{mw}, E_{max}, and the minimum AE, $(AE)_{min}$, estimates will increase and decrease respectively. In fact, the maximum E_{mw} becomes the maximum E and the minimum AE becomes the minimum RME. It is argued that in order to have the best chance to achieve 'greenness' the kernel reaction metrics E_{mw} and AE derived from the intrinsic nature of a given chemical reaction need to be optimized from the outset. This in turn is directly linked to the synthetic design or plan of a given target product. Also evident from the inverse relationships in equations (4.4) and (4.5) is that maximization of $E_{mw}(E)$ automatically implies minimization of AE (RME), and vice versa. Hence, estimation of an upper bound for $E_{mw}(E)$ coincides with estimation of a lower bound for AE (RME).

In order to better quantify what 'greenness' is we introduce the concept of a threshold criterion. If a reaction is found to have an environmental impact factor below a certain threshold that defines 'greenness', then it is possible to rank the 'greenness' of the reaction using this criterion against others that are of the same or different chemical class. Recognizing that $0 \leq AE \leq 1$ we may set a threshold value of α between 0 and 1 such that a reaction is considered 'green' if $AE \geq \alpha$. From equation (4.5) we would then have a corresponding environmental impact factor based on molecular weight threshold of $E_{mw} \leq (1/\alpha) - 1$. Similarly following the same argument we obtain from equation (4.4) the thresholds $RME \geq \alpha$ and $E \leq (1/\alpha) - 1$, respectively. It is shown that an appropriate threshold value for 'greenness' is $\alpha = 0.618$ for both the estimates of upper bounds for E_{max} and lower bounds for $(AE)_{min}$. This means that a chemical reaction satisfies the 'green' criterion if $(AE)_{min}$ is at least 61.8% or if E_{max} is at most 0.618 mass units waste per mass unit target product on a 1 mole scale. It is argued that satisfying this criterion for the kernel metrics AE and E_{mw} is key to increase the probability that this same threshold is met when considering the more realistic metrics RME and E_m which include reaction yield, the use of excess reagents, and reaction and post-reaction solvents and catalysts. If the threshold criterion is not met at the kernel metric level then there is no hope that it will be met when these other parameters are taken into account. The probabilities of satisfying this criterion as functions of the threshold α are also determinable under best-case scenarios once lower limits of AE and reaction yield are known.

4.2.5.1 Minimum atom economy thresholds

The concept of minimum AE and maximum E_{mw} is illustrated with the generalized sequence[48] shown in Scheme 4.7 under stoichiometric conditions with complete recovery of reaction solvents, catalysts, and post-reaction materials. Markush structures are used to show both variable R groups and necessarily invariant atoms. This analysis is useful in studying combinatorial libraries where a constant scaffold structure is selected and then is decorated with, in principle, an unlimited number of possible R groups.

If r_1, r_2, r_3, r_4 and r_5 represent the molecular weights of fragments R_1 through R_5, then the general expressions for E_{mw} and AE for the Petasis condensation are given by

$$E_{mw} = \frac{120}{141 + \sum_{i=1}^{5} r_i} \tag{4.9a}$$

$$AE = \frac{141 + \sum_{i=1}^{5} r_i}{262 + \sum_{i=1}^{5} r_i} \tag{4.9b}$$

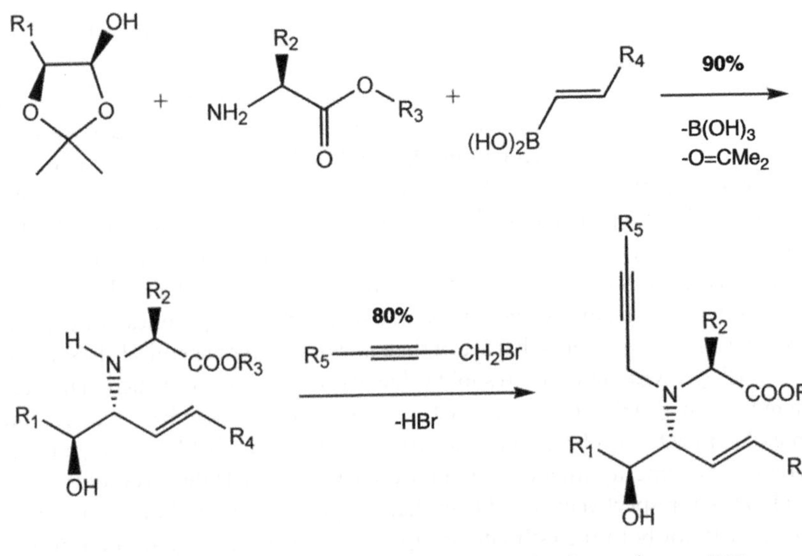

Scheme 4.7 An example diversity-oriented synthesis strategy (taken from reference [48]).

and for the subsequent coupling reaction are given by

$$E_{mw} = \frac{80.9}{178 + \sum_{i=1}^{5} r_i} \tag{4.10a}$$

$$AE = \frac{178 + \sum_{i=1}^{5} r_i}{258.9 + \sum_{i=1}^{5} r_i} \tag{4.10b}$$

Of the two metrics it is easier to determine the limiting values of E_{mw}. Once these have been established the limiting values of AE are at once determined using the connecting relationship given by equation (4.5). It is obvious that the value of E_{mw} decreases as the molecular weights of R_1 through R_5 increase. Thus, as all R groups become large the value of E_{mw} in each case approaches zero and this defines the minimum value, E_{min} for both reactions. If the R groups are as small as possible, that is, if they represent hydrogen atoms, then the largest possible values of E_{mw} are 0.828 and 0.442 for the first and second reactions respectively when $r_i = 1$. From the relationships in equations (4.11a) and (4.11b)

$$(AE)_{min} = \frac{1}{1 + E_{max}} \tag{4.11a}$$

$$(AE)_{max} = \frac{1}{1 + E_{min}} \tag{4.11b}$$

the corresponding ranges of AE are $0.547 < AE < 1$ and $0.693 < AE < 1$, respectively.

A simple graphical analysis of the connecting relationship between AE and E_{mw} given by equation (4.5) indicates how one may define a threshold for 'greenness' for chemical reactions. A plot of AE versus E_{mw} indicates that the graph is divided into two distinct regions, one in which the magnitude of AE exceeds E_{mw} and the other in which the opposite is true. By setting $AE = E_{mw} = x$ in equation (4.5) one obtains the simple quadratic expression $x^2 + x - 1 = 0$ whose positive valued solution is $(\sqrt{5} - 1)/2 = 0.61803\ldots$. This value corresponds to the borderline case when the magnitudes of the atom economy and environmental impact factor based on molecular weight for a given reaction match. This number is recognizable as the 'golden' ratio and has been known since ancient times.[49,50] In the present context this number can be used as a benchmark to identify truly atom economical chemical reactions. This means that a true 'green' reaction is one which produces at least 0.618 mass units of target product per mass unit of all reactants used ($AE > 0.618$), or conversely, one that produces at most 0.618 mass units of waste byproducts per mass unit of target product ($E_{mw} < 0.618$). It should be emphasized that this threshold arises directly from the inverse relationship between AE and E_{mw}, which in turn arises directly from a simple analysis of a balanced chemical equation, and therefore is not an arbitrary designation.

4.2.5.2 Probabilities of achieving RME thresholds

In terms of kernel reaction mass efficiency, the 'golden' threshold for a single step reaction is given by

$$RME = \varepsilon(AE) \geq 0.618 \tag{4.12}$$

when all reaction and post-reaction solvents and any catalysts used are recycled and that the reaction is run with no excess reagents. Figure 4.6a shows a plot of atom economy

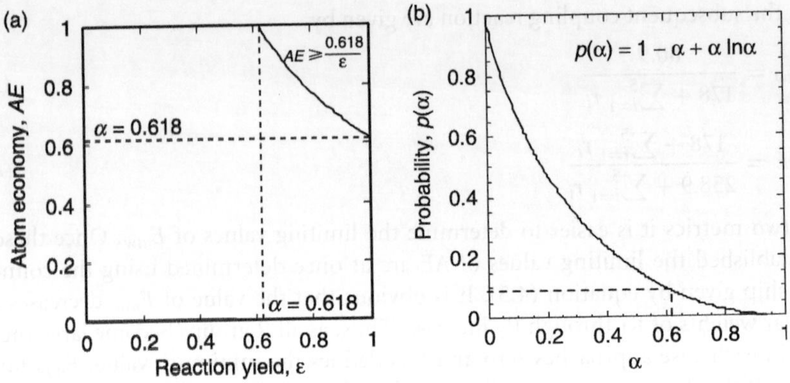

Figure 4.6 (a) Graph showing full domains of atom economy, reaction yield and region where reaction mass efficiency exceeds a threshold value of α for a reaction whose minimum atom economy is zero. (b) Probability that such a reaction can achieve a threshold RME of at least α. If $\alpha = 0.618$, $p(\alpha) = 0.08$ (8%) as indicated by dotted lines.

versus reaction yield illustrating the domains of each parameter as well as the 'green' zone defined by the upper right corner that satisfies inequality (4.12). One can think of the unit square as a probability domain space for all possible reaction outcomes run under stoichiometric conditions and complete solvent recovery and/or elimination with atom economies and reaction yields each falling in between 0 and 1. The question being asked is what is the probability that the multiplicative product of 'co-ordinates' (ε, AE) for a given reaction, which is its reaction mass efficiency, will have a value of at least 0.618. For a general threshold criterion α with a magnitude between 0 and 1, the probability that such a given reaction has an RME above α is given geometrically by the area above the curve $AE = \alpha/\varepsilon$ and bounded by the limits $AE = 1$ and $\varepsilon = 1$. Hence,

$$\begin{aligned} \text{probability} &= \frac{\text{area of 'green' zone}}{\text{area of unit square}} \\ &= 1 - \alpha - \alpha \int_{\alpha}^{1} \frac{d\varepsilon}{\varepsilon} \\ &= 1 - \alpha + \alpha \ln \alpha \\ &= 0.0845 \end{aligned} \tag{4.13}$$

when $\alpha = 0.618$. Hence, the probability that a general chemical reaction run under stoichiometric conditions with complete recovery and/or elimination of reaction and post-reaction solvents and catalysts meets the 'golden' threshold criterion is an abysmally low 8.45%! This value is in fact a minimum probability since we have not specified a particular reaction. Figure 4.6b shows the dependence of the probability function in equation (4.13) on the threshold value α. If we consider a specific reaction and determine its minimum atom economy according to a Markush analysis previously discussed, then the magnitude of this probability is markedly improved. With no restrictions on the reaction yield, there are three cases to consider depending on whether the magnitude of $(AE)_{min}$ is greater than, equal to, or less than the threshold parameter α. These are shown graphically in Figures 4.7–4.9.

Table 4.4 summarizes general expressions for the probability of achieving an RME exceeding a threshold value of α given lower limits for both AE and reaction yield. Setting the minimum reaction yield to be zero reduces these seven cases to the three given graphically in Figures 4.7–4.9.

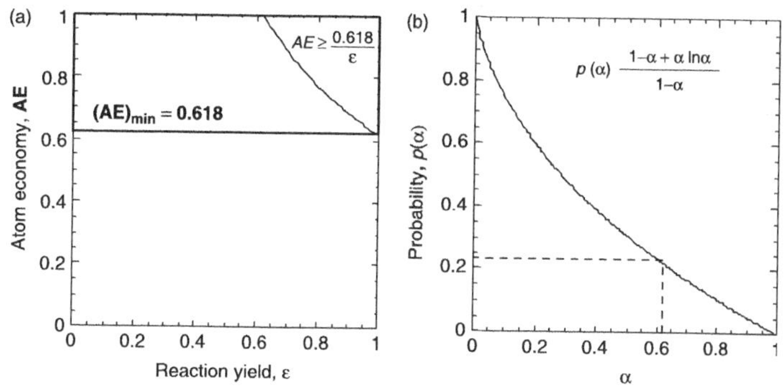

Figure 4.7 (a) Graph showing domains of atom economy, reaction yield and region where reaction mass efficiency exceeds a threshold value of α for a reaction whose minimum atom economy matches α. (b) Probability that such a reaction can achieve a threshold RME of at least α. If $\alpha = 0.618$, $p(\alpha) = 0.22$ (22%) as indicated by dotted lines.

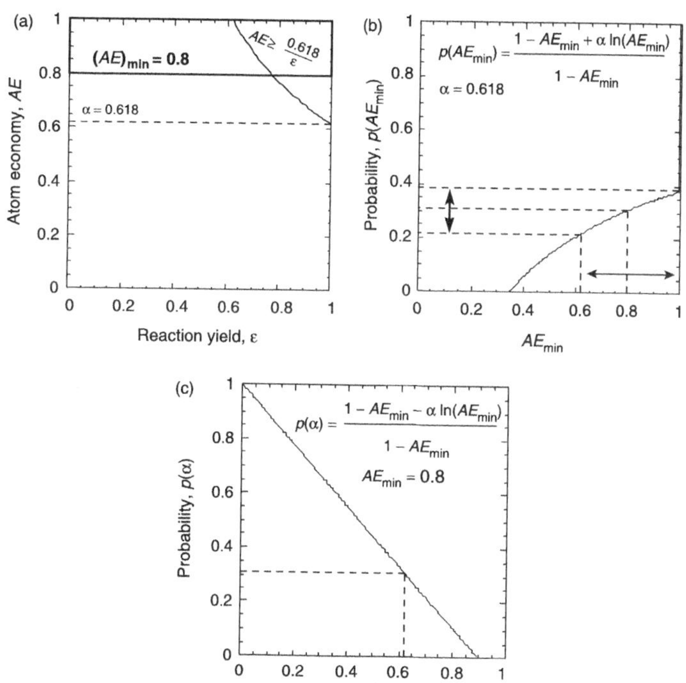

Figure 4.8 (a) Graph showing domains of atom economy, reaction yield and region where reaction mass efficiency exceeds a threshold value of α for a reaction whose minimum atom economy exceeds α. (b) Probability that such a reaction can achieve a threshold RME of at least α, where the probability is given as a function of minimum AE. (c) Probability as in (b) given as a function of α. If $\alpha = 0.618$, $p(\alpha) = 0.31$ (31%) as indicated by dotted lines.

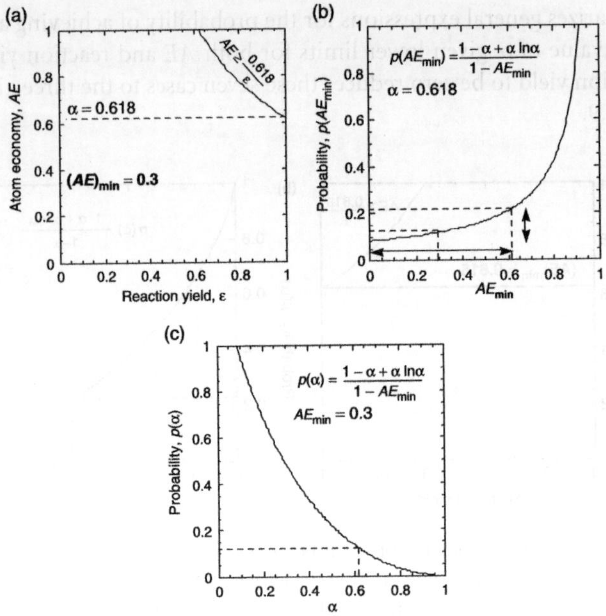

Figure 4.9 (a) Graph showing domains of atom economy, reaction yield and region where reaction mass efficiency exceeds a threshold value of α for a reaction whose minimum atom economy is below α. (b) Probability that such a reaction can achieve a threshold RME of at least α, where the probability is given as a function of minimum AE. (c) Probability as in (b) given as a function of α. If $\alpha = 0.618$, $p(\alpha) = 0.12$ (12%) as indicated by dotted lines.

Table 4.4 Summary of expressions for the probability of achieving a threshold RME value of α given various boundary conditions for minimum atom economy and minimum reaction yield.

Boundary conditions	Probability expression, $p(\alpha)$
Minimum atom economy above threshold RME value	
$(AE)_{min} \geq \alpha$ and $0 \leq \varepsilon_{min} \leq \alpha$	$\dfrac{1-(AE)_{min}+\alpha\ln[(AE)_{min}]}{[1-(AE)_{min}][1-\varepsilon_{min}]}$
$(AE)_{min} \geq \alpha$ and $\alpha \leq \varepsilon_{min} \leq \dfrac{\alpha}{(AE)_{min}}$	$\dfrac{1-(AE)_{min}+\alpha-\varepsilon_{min}-\alpha\ln[\alpha/((AE)_{min}\varepsilon_{min})]}{[1-(AE)_{min}][1-\varepsilon_{min}]}$
$(AE)_{min} \geq \alpha$ and $\dfrac{\alpha}{(AE)_{min}} \leq \varepsilon_{min} \leq 1$	1
Minimum atom economy at threshold RME value	
$(AE)_{min} = \alpha$ and $0 \leq \varepsilon_{min} \leq \alpha$	$\dfrac{1-\alpha+\alpha\ln\alpha}{[1-\alpha][1-\varepsilon_{min}]}$
$(AE)_{min} = \alpha$ and $\alpha \leq \varepsilon_{min} \leq 1$	$\dfrac{1-\varepsilon_{min}+\alpha\ln(\varepsilon_{min})}{[1-\alpha][1-\varepsilon_{min}]}$
Minimum atom economy below threshold RME value	
$(AE)_{min} \leq \alpha$ and $0 \leq \varepsilon_{min} \leq \alpha$	$\dfrac{1-\alpha+\alpha\ln\alpha}{[1-(AE)_{min}][1-\varepsilon_{min}]}$
$(AE)_{min} \leq \alpha$ and $\alpha \leq \varepsilon_{min} \leq 1$	$\dfrac{1-\varepsilon_{min}+\alpha\ln[\varepsilon_{min}]}{[1-(AE)_{min}][1-\varepsilon_{min}]}$

Returning to the example sequence in Scheme 4.7 and using the minimum values of AE determined and setting the reaction yields also as minimum values, it is possible to evaluate the probabilities that each reaction will have an RME of at least 0.618 and also the probability that both reactions will achieve it simultaneously. Figures 4.10 and 4.11 illustrate the relevant regions in the graphs. We can conclude that the probabilities for the Petasis condensation and the coupling reaction are 77% and 94%, respectively. Since the reactions are independent of each other, that is, the individual probabilities are mutually exclusive, the combined probability that both reactions will have RME values of at least

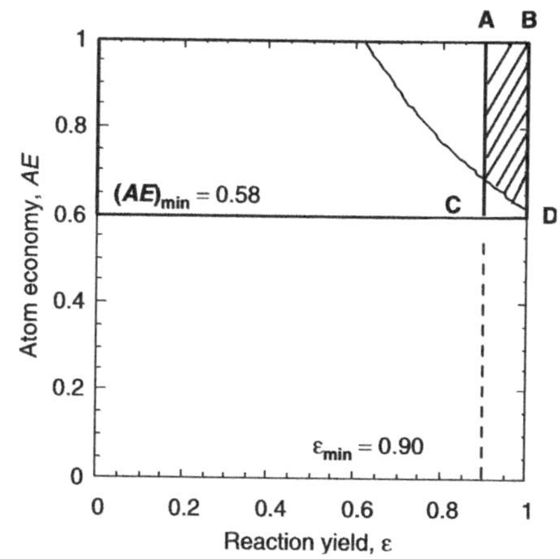

Figure 4.10 Graph showing domains of AE and reaction yield for first reaction in Scheme 4.7. Shaded area versus rectangle ABCD indicates probability that this reaction has an RME of at least 0.618 ($p = 0.77$ (77%)).

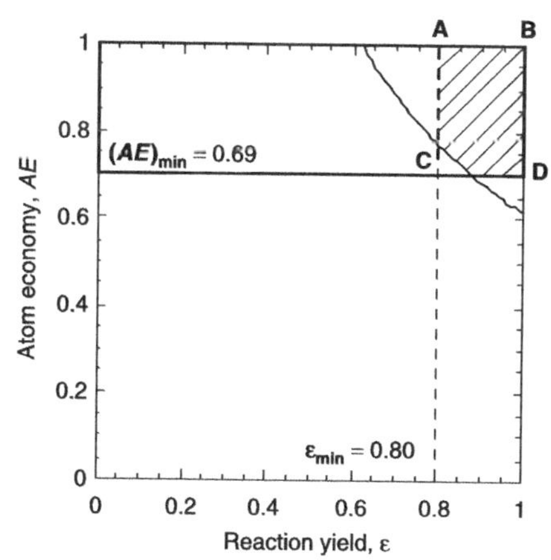

Figure 4.11 Graph showing domains of AE and reaction yield for second reaction in Scheme 4.7. Shaded area versus rectangle ABCD indicates probability that this reaction has an RME of at least 0.618 ($p = 0.94$ (94%)).

0.618 is 0.77 × 0.94 = 0.72 (or 72%). This is a nice example illustrating how the odds of achieving true green reactions increase dramatically if the minimum values of both AE and reaction yield are as high as possible.

The above analyses clearly demonstrate that chemical ingenuity in devising highly atom economical reactions with high reaction yields from the outset is vital for increasing the odds of achieving the 'golden' threshold. *One therefore needs to be ahead of the threshold as early as possible at the kernel metrics level since other waste contributing factors such as excess reagents, solvents, and catalysts will necessarily rapidly diminish RME and amplify the Sheldon E-factor.* For example, if a reaction run under stoichiometric conditions has a borderline AE of 0.618, then in order for it to have a RME of at least 0.618, then its yield must be 100% and all reaction and post-reaction solvents must be reclaimed and/or eliminated from the process. If a reaction has an AE of less than 0.618 from the outset then there is no hope that its RME will be at least 0.618 since the other factors will guarantee attenuation. We may summarize that for a reaction to be called 'green' it must satisfy all of the following conditions simultaneously:

(a) $AE \geq 0.618$ so that $AE \geq E_{mw}$,
(b) $RME \geq 0.618$ so that $RME \geq E_m$, and
(c) reaction solvents, catalysts, and all post-reaction materials used in work-up and purification stages *must* be reclaimed and/or eliminated.

It should be pointed out that claims of 'green' reactions reported in the literature based on the premise that they were run in 'green' solvents such as water or ionic liquids often fail these criteria. Achieving true green reactions is based on redesigning reactions with material and energy efficiency in mind and not on merely substituting a branded nasty solvent for a more benign one while carrying out the old kind of chemistry leading to the same wasteful byproducts.

4.2.6 Survey of organic reactions by class

Concepts presented in the preceding section have established the importance of optimizing the related kernel metrics AE and E_{mw}. The focus is therefore on the intrinsic chemistry and design of chemical reactions that will really determine whether 'green' reactions are possible. It has recently been reported that it is possible to implement these ideas to the set of named organic reactions that form the primary library and database of reactions that synthetic organic chemists draw upon in designing synthetic sequences to important target molecules.[26] Over 400 of named organic reactions[51–56] classified according to the following categories were examined in detail: multicomponent reactions, carbon–carbon bond-forming reactions, non-carbon–carbon bond-forming reactions, condensations, oxidations with respect to substrate, reductions with respect to substrate, rearrangements, substitutions, fragmentations and eliminations, and sequences. For each reaction a general balanced chemical equation showing all byproducts was determined based on appropriate Markush structures. From this equation estimates of AE_{min} and E_{max} were calculated. Figure 4.12 presents the fraction of organic reactions in each category that meet the 'golden' atom economy threshold and Figure 4.13 shows a plot of the cumulative growth of all named organic reactions discovered over time that are above or below it since Wöhler's urea synthesis[57,58]

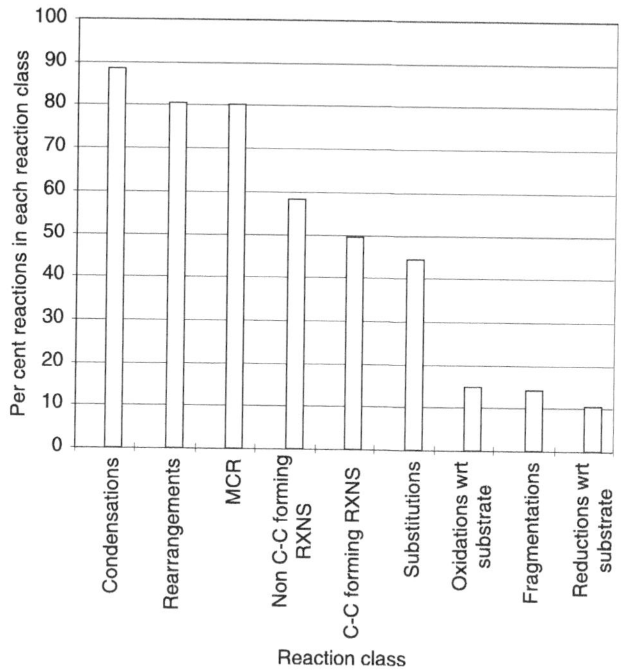

Figure 4.12 Histogram showing the proportion of 'golden atom economical' reactions in each class of organic reaction.

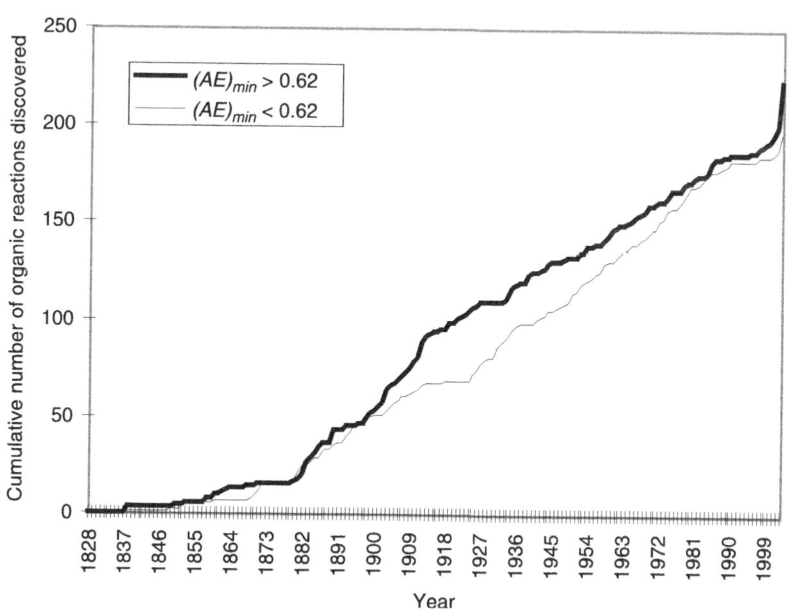

Figure 4.13 Plot showing the cumulative growth of organic reactions discovered that are above and below the 'golden' atom economical threshold with the year of their discovery since Wöhler's synthesis of urea in 1828.

Table 4.5 Summary of atom economical efficiency trends for various reaction classifications.

Reaction classification	General trends
Multi-component reactions	• Most highly atom economical reaction type of all
Carbon–carbon and non-carbon–carbon bond-forming reactions	• Atom economy increases as the molecular weights of the combining fragments increases
Condensations	• Highly atom economical since small molecules of water or alcohol are liberated • Atom economy increases as the molecular weights of the combining fragments increases • For cyclization reactions such as the Dieckmann condensation and the synthesis of cyclic ethers from straight chain diols the atom economy increases with increasing ring size
Oxidations or reductions with respect to substrate	• Worst atom economical performance of all (exceptions are catalytic hydrogenation and oxidation with molecular oxygen or hydrogen peroxide) • Characterized by the production of significant waste byproducts which are the result of oxidation or reduction of reducing and oxidizing reagents, respectively • Recycling of byproducts back to the original oxidizing or reducing reagents necessarily involves at least another redox couple
Rearrangements	• Rearrangements of substrates always have atom economies of 100% • Some rearrangement reactions involve rearrangements of intermediates along their reaction pathways and so their corresponding atom economies are less than 100%
Substitutions	• Atom economy increases if the incoming group is heavier than the leaving group, otherwise it will decrease • The *caveat* is that good leaving groups tend to have higher molecular weights
Fragmentations/eliminations	• Proportion of high atom economical reactions is low since these reactions are the reverse of skeletal building up reactions • Atom economy decreases as the molecular weight of the leaving fragment increases

in 1828 which is widely accepted as the dawn of organic synthesis. This analysis proved to be instructive since it provides organic chemists with an opportunity to gauge the progress of reaction discovery and to identify areas of research in organic reaction development that may need attention according to current interests in improving the 'green' performance of organic transformations. Table 4.5 summarizes key trends in atom economy among the reaction classes. The next sections briefly elaborate further on these.

4.2.6.1 Multicomponent reactions (MCRs)[59–84]

Although this class of reactions has a modern name such reactions have been documented as early as 1850 with the discovery of the Strecker synthesis of α-cyanoamines as noted in Ugi's historical review.[60] Other nineteenth century MCRs include the Hantzsch dihydropyridine synthesis (1882), Radziszewski imidazole synthesis reaction (1882), Riehm quinoline synthesis (1885), Doebner reaction (1887), Pinner triazine synthesis (1890), Hantzsch synthesis of pyrroles (1890), Biginelli reaction (1893), Guareschi–Thorpe condensation (1896), and Thiele reaction (1898). Of these the Biginelli reaction has received considerable attention in recent years.[69] It is interesting to note the two 'bursts' of research activity in 1986 and again since 2003 where this class of reactions is now a rapidly developing field of organic synthesis. However, there are few multistep synthetic plans reported in the literature which utilize these reactions as key strategic elements in synthesis design. Reactions in this class are characterized by having at least three substrates reacting together in one reaction vessel either sequentially in a tandem or domino fashion or all at once. In the cases where two of the components are identical molecules, that is, two equivalents of a substrate are used, such reactions are better designated as pseudo-multicomponent as opposed to true multicomponent reactions with distinct substrates. An important *caveat* in this kind of reaction is that the order of addition of substrates may matter in determining the final product of the reaction. It is obvious that the atom economy for the reaction increases as the molecular weights of the combining fragments increases.

4.2.6.2 Condensations

Condensations are characterized by carbon–carbon couplings via straight chain additions or cyclizations that produce as a byproduct some hydroxylic molecule such as water or an alcohol. For this category as in the case of MCRs the atom economy increases as the molecular weights of the combining fragments increases. For cyclization reactions such as the Dieckmann condensation and the synthesis of cyclic ethers from straight chain diols the atom economy increases with increasing ring size. About 90% of reactions in this class have minimum atom economies above the 'golden' threshold. The exceptions are the Nenitzescu indole synthesis (1929), Mukaiyama aldol condensation (1974), and acyloin condensation (1905) which have $(AE)_{min}$ values of 0.42, 0.27 and 0.21 respectively.

4.2.6.3 Carbon–carbon and non-carbon–carbon coupling reactions

Again, these classes of reactions follow the pattern of other skeletal building reactions (MCRs and condensations) and are characterized by increasing atom economies as the molecular weights of the combining fragments increase. It is interesting to note that the progress of discovery of reactions in this class with $(AE)_{min}$ values above the 'golden' threshold is about the same as that with $(AE)_{min}$ values below it.

4.2.6.4 Reductions and oxidations with respect to substrate

Redox-type reactions show by far the worst performance in meeting the 'golden' atom economical threshold. Three reductions meet this criterion with $(AE)_{min}$ values of 1: hydrogenation of olefins using the Lindlar catalyst (1952), Noyori stereoselective hydrogenation reaction (1985), and Zincke disulphide cleavage reaction (1911); whereas, oxidations meeting this criterion include: Harries ozonolysis of olefins with oxidative workup (1905), Willgerodt reaction (1909), Baeyer–Villiger reaction with hydrogen peroxide (1899), epoxidation of olefins with hydrogen peroxide (2001), Cannizzaro reaction with respect to carboxylic acid as the target product (1853), Uemura oxidation (1998) and Griess diazotization reaction of anilines (1858).

This class of reaction is particularly characterized by the production of significant waste byproducts which are the result of oxidation or reduction of reducing and oxidizing reagents, respectively. Recycling of byproducts back to the original oxidizing or reducing reagents necessarily involves at least another redox couple. It should be noted that the feasibility of recycling oxidizing and reducing reagents is strongly linked to energetics; that is, to appropriately chosen reagents with the right reduction potentials of half reactions so that the recycling process is made energetically favourable. Examples of the use of cascade redox reactions where the ultimate waste product is water have recently been developed.[85–87] The optimum oxidizing agents are molecular oxygen from air and hydrogen peroxide which both produce water as a byproduct. Noyori and others have recently developed 'green' catalytic oxidations using aqueous hydrogen peroxide and other oxidants.[88–90] On the other hand, the optimum reducing agent is hydrogen gas which produces no byproducts. The main objective of current research is to find suitable catalysts to effect redox reactions using any one of the optimum redox reagents. In this regard the Lindlar and Noyori catalytic hydrogenation reactions are good examples where this has been achieved. Progress in developing 'green' redox reactions has been slow but is now gaining attention.

4.2.6.5 Substitution reactions

Substitutions have the unique property that the atom economy increases if the in-coming group is heavier than the leaving group, otherwise it will decrease. The *caveat* is that good leaving groups with the correct electronic requirements tend to have high molecular weights as in the case of tosylate and iodide. It is this class of reactions that is employed in protection–deprotection strategies in organic synthesis and it is not surprising, therefore, that given the behaviour of AE with leaving group size minimization of this kind of strategy in synthetic sequences was named as one of the 12 Principles of Green Chemistry.[13] However, if it is not possible to avoid using this technique in a synthesis plan, then as far as 'greenness' is concerned it may still be possible to minimize wastes if protecting groups are carefully chosen so that recycling of liberated groups generated in deprotection steps to appropriate protecting group reagents is shown to be viable and advantageous by an analysis such as that proposed in Section 4.2.4.

4.2.6.6 Fragmentation and elimination reactions

For fragmentation or elimination reactions it is expected that the proportion of high atom economical reactions is low since these reactions are the reverse of skeletal building up reactions. The atom economy decreases as the molecular weight of the leaving fragment increases. The Tiffeneau–Demjanov reaction is the only one in this category with $(AE)_{min}$ values meeting the 'golden' threshold. The value of $(AE)_{min}$ increases with ring size: 0.69 for ring size 8, 0.66 for ring size 7, 0.64 for ring size 6 and 0.60 for ring size 5.

4.2.6.7 Rearrangement reactions

Rearrangements of substrates always have atom economies of 100%, however about 20% of reactions in this class have atom economies less than unity. This is because they involve rearrangements of intermediates along the reaction pathway yet they are still categorized as rearrangement reactions. An example is the Wagner–Meerwein rearrangement which involves the rearrangement of an intermediate carbenium ion and yet produces a net byproduct in the reaction.

4.3 Green metrics applied to synthesis plans

After presenting the evaluation of the complete material efficiency for any chemical reaction in Section 4.2, we now present the evaluation of the material efficiency for synthesis plans which are in effect sets of reactions that are linked according to some sequential progression from starting materials to final target product. To streamline and facilitate the calculation of green metrics for synthesis plans we first introduce a new representation of traditional synthesis schemes as tree diagrams.[28] It will be shown that these representations give a better 'big picture' view of synthesis plans where paths from starting materials to a final target are easily traceable by a simple connect-the-dots approach. Tree diagrams are able to handle synthesis plans of any degree of complexity including linear, convergent and divergent schemes. This characteristic robustness imparts several advantages that are itemized as follows:

(a) Easy bookkeeping of all reactions in a plan in a compact way is possible.
(b) Tree diagrams are an excellent checking device to ensure correct balancing of all chemical equations in a plan.
(c) Experimental protocol is built into the construction of tree diagrams so that intermediates that are isolated are distinguishable from those that are made *in situ*, and the order of input dots matches the order of addition of reagents used in the synthesis plan.
(d) The number of reaction steps, reaction stages and input materials are precisely determined.
(e) Any points of convergence in a plan may be easily spotted by inspection.
(f) The appearance of horizontal lines connecting sequential reactant and product nodes implies either rearrangement or fragmentation transformations where corresponding equal molecular weights for the nodes indicates the former case and unequal ones the latter.
(g) The required masses and raw material costs of any starting material or intermediate in a plan, including the final target product, may be determined by inspection.
(h) Overall kernel RME is shown to trump 'overall yield' as the key metric to rank material efficiency of a synthesis plan.
(i) Kernel *AE* and *RME* values and mass of waste produced may be determined between any two endpoints in the plan, that is, for a single reaction, sub-sequence of reactions in the plan, or the entire plan.
(j) Histograms tracking the individual and cumulative performances of reactions may be easily constructed according to reaction yield, atom economy, stoichiometric factor,

and reaction mass efficiency so that any material or raw material cost bottlenecks may be easily identified.

(k) Ranking of plans in an unbiased way according to best-case scenario mass, energy and raw material cost throughput efficiencies is possible and this allows for the weeding out of poor performing plans as well as gaining a deeper understanding of what characteristics or strategies make a given synthesis plan good.

(l) Parameters pertaining to 'synthetic elegance' such as degree of convergence and tracking the building up of molecular weight and oxidation number changes (hypsicity) from starting materials to final target product may be determined quantitatively.

4.3.1 Synthesis trees: Construction

Before constructing a tree diagram for a synthesis plan all chemical equations for all steps must be properly balanced so that stoichiometric coefficients for each chemical species are determined as well as all byproducts are identified. Redox reactions are particularly noteworthy in this regard as the formation of byproducts is significant. To maintain consistency stoichiometric coefficients of unity are maintained for all intermediates in a given plan. To get a truer measure of material usage, reactions producing acidic byproducts (e.g., HCl) in a basic medium (e.g., pyridine solvent) or are neutralized by a base in a subsequent workup step should be balanced as the sum of the original and the neutralization reaction. The same is true for reactions producing basic byproducts (e.g., ammonia) in an acidic medium (e.g., sulphuric acid) or are neutralized by an acid in a workup step. Particular attention should be paid to workup conditions that destroy catalysts that were used in the original reaction. Workup reactions should be taken into account and added to the original reaction to get an overall balanced chemical equation. Examples are the Friedel–Crafts acylation or alkylation reaction which uses aluminium trichloride as catalyst which subsequently gets hydrolysed in aqueous workup to aluminium hydroxide, or reactions using boron trifluoride etherate reagent as a Lewis acid catalyst which then is hydrolysed to boric acid and hydrogen fluoride after aqueous workup. Catalysts that remain intact after workup are not counted as inputs in the synthesis tree when kernel RME values are to be determined.

We also introduce the following conventions and assumptions when reading traditional schemes. A chemical species appearing in between square brackets [] is assumed to be made *in situ* and is therefore not isolated. Examples are the production of a Grignard reagent from an organohalide and magnesium, or the production of an organolithiated species after reaction with *n*-butyl lithium. For such cases, the overall chemical equation is the sum of the reaction producing the *in situ* intermediate and the subsequent reaction of this intermediate with whatever other reagents are added in the next step. The reaction yield for the overall transformation is taken with respect to the appropriate limiting reagent. If a reaction produces multiple isomeric products that are not separated and are used as a mixture of starting materials in the next step, these are shown in between curly brackets {}. If a reaction solvent is evaporated to give a crude residue and this is taken up in a different solvent and new reagents added for the next reaction, these are treated as two separate reactions where it is interpreted that the crude product from the first reaction is isolated. If reagents for subsequent reactions are sequentially added to the same reaction vessel without isolation of the initially formed intermediate products in a telescoping manner, then the overall chemical equation is the sum of all reactions in the telescope and it is considered as a one-step reaction with its associated overall reaction yield. Telescoped sequences may be considered as pseudo-multicomponent reactions. If a reaction appears in a conventional scheme with a reaction arrow connecting two structures showing more than one step above it and only an overall yield for the transformation is reported, it is assumed that each intervening intermediate is isolated and the reaction yield for each step is evaluated as the geometric mean of the overall yield.

Once the correct sequence of reactions is properly determined a corresponding tree diagram is drawn on a grid of squares with unit spacing. In constructing a synthesis tree the following conventions are made: (a) the x-axis represents reaction stages with unit spacing; (b) the y-axis represents input structures entered as filled dots beginning at the origin with unit spacing; and (c) isolated intermediate and final products are represented as open and shaded circles, respectively, whose co-ordinates are determined as the centroids of the dots corresponding to their preceding reactant input structures according to equation (4.14) until the final product is reached,

$$centroid = \frac{1}{2^{n-1}}\left[\sum_{j=0}^{n-1} a_{j+1}\binom{n-1}{j}\right] = \frac{1}{2^{n-1}}\left[\sum_{j=0}^{n-1} a_{j+1}\left(\frac{(n-1)!}{(j!)(n-1-j)!}\right)\right] \tag{4.14}$$

where $n \geq 2$, n is the number of points corresponding to the number of reactant input structures, a_{j+1} is the ordinate of the $(j+1)$th input, and $0! = 1$ by definition. Equation (4.14) is well known in classical mechanics where it is used to calculate the centre of mass co-ordinates of equal point masses along a straight line. This protocol of precisely tracking the co-ordinates of each intermediate node to final target product node is important in quantifying parameters such as degree of convergence with respect to the total number of input reactants, rate of convergence and degree of asymmetry for a plan from the actual shape of its tree diagram. In the case of a synthesis plan having multiple linear segments that converge, the synthesis tree is constructed with the longest branch beginning at the origin. If a synthesis plan involves dimerization of a monomer, such as a Diels–Alder step, then the plan is called pseudo-convergent and can be presented in two ways: (a) as a single branch sequential plan with stoichiometric coefficients doubled for all steps preceding the dimerization step, or (b) as a convergent plan with two identical but separate parallel branches that lead to the convergent dimerization step.

A distinction is made here between reaction steps and reaction stages. A reaction step begins with an isolated intermediate and ends with the next isolated intermediate. A reaction stage may be composed of a single reaction step or at least two parallel reaction steps run simultaneously. This designation is important in spotting points of convergence in a synthesis plan or network, in determining the optimum time required to complete a synthesis, and in appropriate resource and time management in the planning of a synthesis. For linear plans, the number of reaction stages is equal to the number of reaction steps. For convergent plans, the number of reaction steps always exceeds the number of reaction stages. Each reaction step has an associated reaction yield with respect to the limiting reagent (ε), which is the multiplicative product of the intrinsic chemical yield, the work-up yield, and the purification yield, an associated reaction time (t) and an associated energy input (Ψ).

Reaction yields for each step are written along the x-axis below the tree diagram in tiers. The top tier of reaction yields refers to the main branch of the tree beginning at the origin and ending at the final target product node tracing a path along the bottom edges of the tree diagram. This is sufficient for a linear plan. For convergent plans which have connecting branches to the main tree, second, third, and so on tiers of reaction yield values are inserted as appropriate below the top tier of reaction yields for the main branch. If a given stage j has more than one reaction in it, the plan is automatically a convergent one and the top tier reaction yield, designated as εj, refers to the main branch, the second tier reaction yield, designated as εj^*, refers to the reaction in the branch immediately above in the same stage, the third tier reaction yield, designated as εj^{**}, refers to the reaction yield in the next higher up branch in the same stage, and so on.

A fundamental assumption underlying the tree construction is that each intermediate product collected is entirely committed as a reactant in the successive step. Essentially the tree, when read from left to right, traces the mass attenuation of input materials, or throughput, until the final product is reached. The reaction sequence is akin to a set of successive sieves each with a given mesh size corresponding to a reaction yield that will dictate how much material passes from one reaction stage to the next. Of course, when the tree is read in the opposite direction from right to left, amplification of scale is implied. It will be shown that reading trees from right to left is computationally much simpler in determining all of the material green metrics for any kind of synthesis plan, particularly convergent ones.

As an illustrated example of these ideas presented thus far we show in Scheme 4.8 and Figure 4.14 the conventional synthesis plan for triclosan and its corresponding synthesis tree diagram.

From Figure 4.14 we note that the plan is convergent with 4 reaction stages and 5 reaction steps. The ordinates of species I1, I1*, I2, I3 and P found using equation (4.14) are 1.5, 4.5, 4.125, 6.53125 and 7.765625, respectively.

4.3.2 Synthesis trees: Material green metrics

In determining the overall kernel RME for a synthesis plan from its synthesis tree one needs to determine the ratio of the output product mass to the sum of all input reactant masses. The calculation is carried out by determining what input masses are required for all reactants so that a target mass of product is obtained given the experimental reaction yields for each reaction step. From the synthesis tree this is easily achieved by defining a target scale for the final product in moles, $x =$ mass of target product/MW target product, and working backwards toward reactant inputs following the lines connecting the dots to determine the required scales at each intermediate and input node. The scale at a given node is given by the quotient of the final product scale, x, and the multiplicative product of the reaction yields corresponding to the reaction steps connecting that node to the target product node as traced by the joining lines in the synthesis tree. At each node the corresponding mass in grams is obtained by multiplying the scale in moles at that node by the molecular weight in grams per mole of the corresponding chemical structure.

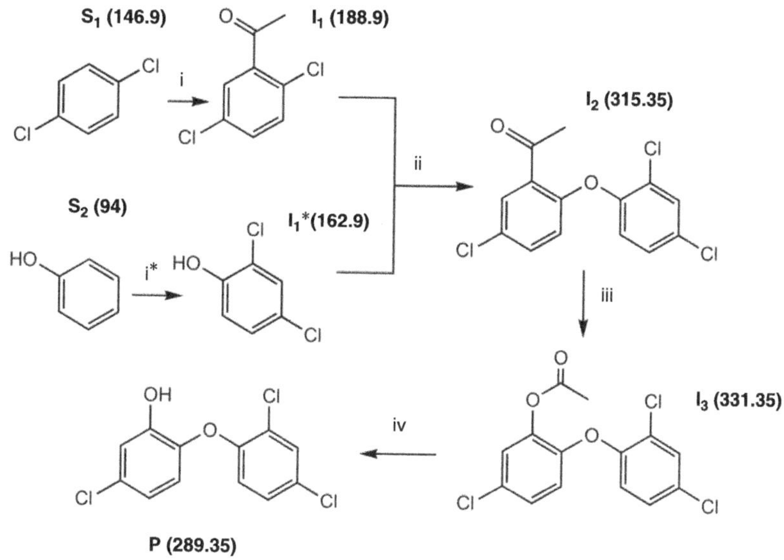

Scheme 4.8 Convergent synthesis plan for triclosan. Reaction conditions: (i) acetyl chloride, $AlCl_3$ catalyst ($\varepsilon_1 = 94.3\%$); (i)* $2Cl_2$ ($\varepsilon_1^* = 81\%$); (ii) $1/2K_2CO_3$, CuCl catalyst, xylenes ($\varepsilon_2 = 48.3\%$); (iii) 62.5% H_2O_2, 1/2 maleic anhydride, CH_2Cl_2 ($\varepsilon_3 = 91.3\%$); (iv) MeOH, 35% HCl catalyst ($\varepsilon_4 = 94.5\%$). Molecular weights in g/mol are given in parentheses.

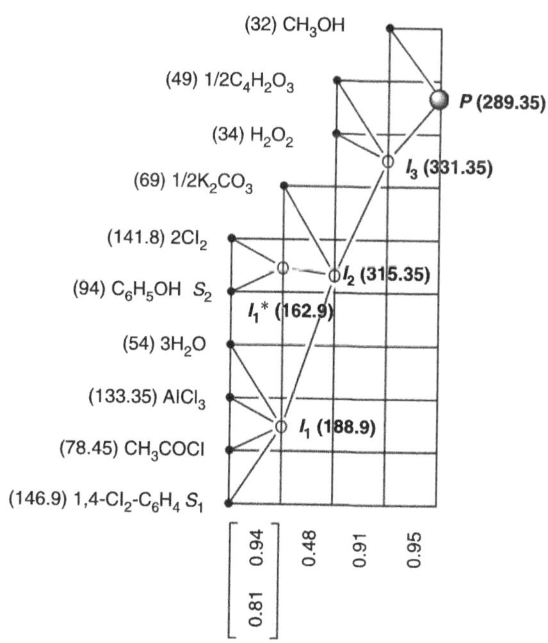

Figure 4.14 Synthesis tree for triclosan synthesis plan shown in Scheme 4.8. Molecular weights for each species are shown within parentheses.

Under the assumption of no excess reagents used and recovery or elimination of all reaction solvent and post-reaction materials, from Figure 4.14 we can deduce at once that the masses of 1,4-dichlorobenzene, acetyl chloride, aluminium chloride, water, phenol, chlorine, potassium carbonate, hydrogen peroxide, maleic anhydride and methanol required to produce 1 mole of triclosan (289.35 g) are $(146.9/\varepsilon_1\varepsilon_2\varepsilon_3\varepsilon_4) = 376.60$, $(78.45/\varepsilon_1\varepsilon_2\varepsilon_3\varepsilon_4) = 201.12$, $(133.35/\varepsilon_1\varepsilon_2\varepsilon_3\varepsilon_4) = 341.87$, $(54/\varepsilon_1\varepsilon_2\varepsilon_3\varepsilon_4) = 138.44$, $(94/\varepsilon_1^*\varepsilon_2\varepsilon_3\varepsilon_4) = 279.66$, $(141.8/\varepsilon_1^*\varepsilon_2\varepsilon_3\varepsilon_4) = 421.88$, $(69/\varepsilon_2\varepsilon_3\varepsilon_4) = 166.28$, $(34/\varepsilon_3\varepsilon_4) = 39.33$, $(49/\varepsilon_3\varepsilon_4) = 56.68$ and $(32/\varepsilon_4) = 33.68$ g, respectively. Therefore, the total mass of input materials is 2055.54 g, the kernel overall RME for the plan is (289.35/2055.54) = 0.141 (14.1%), and E_m is 6.10 g waste/g triclosan; whereas, the overall AE is (289.35/832.5) = 0.348 (34.8%) and E_{mw} is 1.88 g waste/g triclosan. The masses of intermediates I_1, I_1^*, I_2 and I_3 produced along the way are $(188.9/\varepsilon_2\varepsilon_3\varepsilon_4) = 455.22$, $(162.9/\varepsilon_2\varepsilon_3\varepsilon_4) = 392.57$, $(315.35/\varepsilon_3\varepsilon_4) = 364.78$ and $(331.35/\varepsilon_4) = 348.79$ g, respectively. The minimum mass of waste produced in reaction 1 excluding reaction solvents and post-reaction materials is

$$\bar{w}_1 = \frac{146.9 + 78.45 + 133.35 + 54}{\varepsilon_1\varepsilon_2\varepsilon_3\varepsilon_4} - \frac{188.9}{\varepsilon_2\varepsilon_3\varepsilon_4}$$
$$= 1058.04 - 455.22 = 602.82 \text{ g}$$

For reaction 1* it is

$$\bar{w}_1^* = \frac{94 + 141.8}{\varepsilon_1^*\varepsilon_2\varepsilon_3\varepsilon_4} - \frac{162.9}{\varepsilon_2\varepsilon_3\varepsilon_4} = 701.54 - 392.57 = 308.97 \text{ g}$$

Therefore for stage 1 the minimum total mass of waste produced is 602.82 + 308.97 = 911.79 g. For the convergent reaction 2, the minimum mass of waste produced is

$$\bar{w}_2 = \frac{188.9 + 162.9 + 69}{\varepsilon_2\varepsilon_3\varepsilon_4} - \frac{315.35}{\varepsilon_3\varepsilon_4} = 1014.07 - 364.78 = 649.29 \text{ g}$$

The minimum total mass of waste produced for the entire plan is 2055.54 − 289.35 = 1766.19 g. Note that 1766.19/289.35 = 6.10 checks with the value for E_m determined previously.

The raw materials cost function for a plan, representing the minimum materials cost, is determinable directly from the overall kernel RME dependence on all reaction yields and molecular weights of all input materials. This is given by

$$(RMC)_{min} = x\left[\sum_{inputs,j} \frac{(MW)_{input,j}(\$)_{input,j}}{\prod \varepsilon_k}\right] \tag{4.15}$$

where x is the number of moles of target product, $(\$)_{input,\,j}$ is the cost per gram of input j, and $\prod \varepsilon_k$ is the multiplicative chain of reaction yields connecting the target product node to input node j as defined by the synthesis tree diagram for the plan.

Since the synthesis plan has a point of convergence it is not possible to define an 'overall yield' for the entire synthesis by simply multiplying the respective reaction yields as would be correct for a truly linear synthesis. This can be deduced by observing that there is no common yield factor that clears all fractions when it is multiplied by the sum of all terms representing the mass of input materials. In the case of a linear plan this would be possible and thus the resulting numerator in the general expression for overall kernel RME would be the multiplicative product of the molecular weight of the target product, the scale of the entire synthesis, and a reaction yield factor given by $\varepsilon_1\varepsilon_2\ldots\varepsilon_M$ commonly referred to as the 'overall yield'. Previously, 'overall yields' for complex syntheses were commonly determined either by erroneously multiplying all reaction yields in a plan, or more correctly by multiplying reaction yields in the longest branch of a plan usually corresponding to the main branch of its synthesis tree. For the triclosan plan there are two possible values for 'overall yield' that could be used, 0.390 (39.0%) for the lower branch and 0.336 (33.6%) for the upper branch. In general for any synthesis plan of any degree of complexity the present work shows that the overall kernel RME which incorporates reaction yields and atom economies according to the connectivity of reaction inputs, intermediates and final target product is indeed the best measure of its material efficiency. One may define a 'pseudo-overall yield' as

$$\varepsilon_{pseudo-overall} = \frac{RME_{overall\ kernel}}{AE_{overall}} \tag{4.16}$$

For a linear plan the pseudo-overall yield as defined above is numerically close to the multiplicative product of the reaction yields. The difference between the two values diminishes as the reaction yields approach 1 as would be expected. However, for complex plans with several converging branches this alternative definition becomes less useful. For the triclosan branch equation (4.16) yields 0.405 (40.5%) for the pseudo-overall yield.

A numerical algorithm using the right to left amplification convention in reading synthesis tree diagrams is given below for deducing the individual and overall material performances for each reaction and for the entire plan, respectively. In this treatment, all intermediate nodes and the starting material node at the origin are always considered limiting reagents, excess reagents are accounted for using appropriate stoichiometric factors, and reaction solvents and post-reaction materials are recovered or eliminated. This algorithm works for both linear and convergent synthesis plans.

For material green metrics calculation for each reaction beginning with the last step ($j = N$) and working toward the first step ($j = 1$):

1. Balance all chemical equations in the plan taking into workup reactions as appropriate and arrange these in the appropriate reaction stages.
2. Determine $(E_{mw})_j$ using

$$(E_{mw})_j = \frac{\left(\sum MW_{byproducts}\right)}{p_j}$$

3. Determine $(AE)_j$ using

$$(AE)_j = \frac{1}{1 + (E_{mw})_j}$$

4. Determine $(SF)_j$ using

$$(SF)_j = 1 + \frac{\left(\sum mass_{excess\ reagents}\right)_j}{\left(\sum MW_{reagents}\right)_j (x/\prod \varepsilon_k)}$$

5. Determine $(RME)_j$ and $(E_m)_j$ using

$$(RME)_j = \frac{(AE)_j \varepsilon_j}{(SF)_j} \quad \text{and} \quad (E_m)_j = \frac{1}{(RME)_j} - 1$$

6. Determine $\bar{w}_j$ using

$$\bar{w}_j = p_j (E_m)_j \left(\frac{x}{\prod \varepsilon_i} \right), \quad \text{for } j = 1 \text{ to } N-1 \quad \text{and}$$

$$\bar{w}_n = p_n x (E_m)_n \quad \text{for } j = N, \quad \text{or}$$

$$\bar{w}_j = \frac{p_j}{(AE)_j} \left(\frac{x}{\prod \varepsilon_k} \right) [(SF)_j - \varepsilon_j (AE)_j], \quad \text{for } j = 1 \text{ to } N$$

For overall performance of synthesis plan:

7. Determine $(E_{mw})_{overall}$ by summing molecular weights of all byproducts in the plan and dividing by the molecular weight of the final target product.
8. Determine $(AE)_{overall}$ using

$$(AE)_{overall} = \frac{1}{1 + (E_{mw})_{overall}}$$

9. Determine $\sum \bar{w}_j$ by summing all terms found in step 6.
10. Determine $(E_m)_{overall}$ using

$$(E_m)_{overall} = \frac{\sum \bar{w}_j}{x p_n}$$

11. Determine $(RME)_{overall}$ using

$$(RME)_{overall} = \frac{1}{1 + (E_m)_{overall}}$$

Symbol definitions:

x is moles of target product in synthesis plan.
$\prod \varepsilon_k$ is multiplicative chain of reaction yields connecting target product node to reactant nodes for step j as per synthesis tree diagram.
$\prod \varepsilon_i$ is multiplicative chain of reaction yields connecting target product node to product node for step j as per synthesis tree diagram.
p_j is molecular weight of product of step j.
εj is reaction yield with respect to limiting reagent for step j.

Table 4.6 summarizes all metrics referring to material efficiency for each reaction and for the entire synthesis plan for triclosan using this algorithm. The ease of computation using the right-to-left convention is appreciated over the left-to-right convention when evaluating the material efficiencies of convergent plans since the matching of scales problem encountered for intermediates in convergent steps where multiple branches coincide is entirely avoided in the backwards right-to-left convention. For linear synthesis plans both conventions work equally well and there is no advantage in either computation. The only differences in the above algorithm for the left-to-right convention are as follows:

(a) since the analysis begins with the scale of the first starting material at the origin of the tree diagram, x is taken as the moles of starting material;

Table 4.6 Summary of reaction and overall material efficiency performances for the production of 1 mole of triclosan according to the tree diagram shown in Figure 4.14.

Reaction	p_j (g/mol)	ε_j	$(SF)_j$	$(E_{mw})_j$	$(AE)_j$	$(RME)_j$	$(E_m)_j$	$\bar{w}_j$ (g)
1	188.9	0.94	1	1.185	0.458	0.430	1.324	602.82
1*	162.9	0.81	1	0.448	0.691	0.560	0.787	308.97
2	315.35	0.48	1	0.334	0.749	0.360	1.780	649.29
3	331.35	0.91	1	0.202	0.832	0.757	0.321	112.00
4	289.35	0.95	1	0.256	0.796	0.757	0.322	93.12
Overall		0.390[a] 0.336[b]		1.88	0.348	0.141	6.10	1776.2

[a] Lower branch.
[b] Upper branch.

(b) the stoichiometric factors are evaluated using the formula

$$(SF)_j = 1 + \frac{\left(\sum mass_{excess\ reagents}\right)_j}{\left(\sum MW_{reagents}\right)_j x(\prod \varepsilon_i)}$$

(c) the masses of waste are evaluated using the formulas

$$\bar{w}_1 = \frac{p_1}{(AE)_1} x[(SF)_1 - \varepsilon_1 (AE)_1]$$

$$\bar{w}_j = \frac{p_j}{(AE)_j} x \left(\prod \varepsilon_k\right) [(SF)_j - \varepsilon_j (AE)_j] \quad \text{for } j = 2 \text{ to } N$$

or

$$\bar{w}_j = p_j (E_m)_j x \left(\prod \varepsilon_i\right) \quad \text{for } j = 1 \text{ to } N$$

where $\prod \varepsilon_k$ is the multiplicative chain of reaction yields connecting zeroth stage starting material nodes to product node for step $j-1$ as per synthesis tree diagram and $\prod \varepsilon_i$ is the multiplicative chain of reaction yields connecting zeroth stage starting material nodes to product node for step j as per synthesis tree diagram; and

(d) the overall E_m is evaluated using

$$(E_m)_{overall} = \frac{\sum \bar{w}_j}{x p_n (\prod_{j=1}^{N} \varepsilon_i)}$$

It is important to realize that though the formulas for RME explicitly do not depend on reaction scale, x, since this variable cancels out in the computation, RME does in fact implicitly depend on reaction scale because reaction yields are scale dependent and RME in turn depends on reaction yield. Reaction yields are parameters whose magnitude cannot be predicted theoretically but must be verified experimentally. It does not always follow that a reported yield for a given reaction at a given scale will be the same at another scale. This requires experimental verification. Moreover, the direction of change as the scale is changed is also not predictable. All of this means that when the same synthesis plan is run at a different scale, different reaction yields will be determined and hence different material efficiency performance values of *RME*, E_m and mass of waste will be obtained. However, the form of the general expression for RME showing its dependence on reaction yields as given by the synthesis tree diagram will remain the same. The only constants are the magnitudes of AE and E_{mw} since the set of balanced chemical reactions remains the same.

4.3.3 Synthesis trees: Tree parameters

In characterizing a synthesis plan, the geometric shape of its corresponding synthesis tree offers possible measures such as degree and rate of convergence toward making the final target product and degree of asymmetry. The first qualitative description of convergence in synthesis plans was given by Velluz and co-workers[91] and described more quantitatively by others.[20,92,93] These parameters naturally will depend on the number of reactions, M, the number of reaction stages, N, the number of reactant input structures required to build the product chemical structure, I, and the number of intermediate structures along the synthesis path. The convention described earlier in constructing synthesis trees by determining the co-ordinates of intermediate products and ultimately the final target product using the method of centroids now becomes apparent.

We first note the following fundamental relationships for linear and convergent synthesis plans with the following designations: I = number of reactant inputs, M = number of reactions, N = number of reaction stages, L = number of parallel reactions, G = number of stages with parallel reactions.

For linear synthesis plans,

Height of tree $= I - 1$
Width of tree $= N = M$
Number of intermediate products $= N - 1 = M - 1$
Number of convergent reactions $= 0$
$G = 0$
$L = 0$

and for convergent synthesis plans:

Height of tree $= I - 1$
Width of tree $= N < M$
Number of intermediate products $\geq N$
Number of intermediate products $= M - 1$
$G \leq$ Number of intermediate products $- N + 1$ (equality means that no more than two parallel reactions occur in each stage having parallel reactions, otherwise more than two parallel reactions occur)
Number of points of convergence = Number of branches attached to main reduced tree consisting of vertices representing all intermediate structures and final product structure along root of tree.

For plans with parallel reactions with (i) at least one stage having more than 2 parallel reactions, then $L > 2G$ and $M > N+G$, and (ii) all reaction stages having parallel reactions with no more than 2 parallel reactions, then $L = 2G$ and $M = N + G$.

Figure 4.15 shows the same synthesis tree for triclosan as given before in Figure 4.14 with some modifications made and added geometric parameters superimposed. The

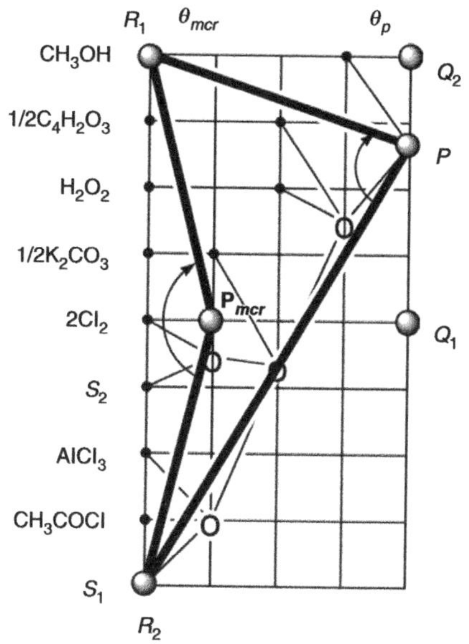

Figure 4.15 Synthesis tree for the synthesis of triclosan according to reactions given in Scheme 4.8. Co-ordinates of P_{mcr} denote position of final product if it were formed in a hypothetical single step 9-component reaction corresponding to a reaction with maximal convergence using the given input reagents. Angles subtended at points P_{mcr} ($\theta_{mcr} = \angle R_1 P_{mcr} R_2$) and P ($\theta_p = \angle R_1 P R_2$) and line segment lengths as shown are used to calculate the degree of convergence, the relative rate of convergence and the asymmetry parameter.

input reactants are aligned vertically along the y-axis in unit intervals to clearly show the height dimension of the tree. The point designated as P_{mcr} describes the co-ordinates of the target product if it had been made in a hypothetical single multi-component reaction (MCR) by using all input reactants. This describes the most convergent synthesis possible given the input structures used to build up the target structure. The more reaction steps and reaction stages are concatenated and telescoped the more this limit is achievable practically. The point designated as P represents the co-ordinates of the target product for a given synthesis plan. The abscissa is equal to the number of reaction stages and the ordinate by the method of centroids is related to the number of reactant inputs and the shape and connectivity of the synthesis tree. The triangles $R_1R_2P_{mcr}$ and R_1R_2P share a common base which represents the height of the tree. An appropriate measure of degree of convergence may be made with respect to a hypothetical I-component MCR (I is the total number of input reactants). This serves as a reference. Hence, we may take the ratio of the angles θ_p and θ_{mcr} subtended at P and P_{mcr}, respectively, to represent this. In terms of the synthesis plan parameters the degree of convergence, δ, is then

$$\delta = \frac{\angle R_1 P R_2}{\angle R_1 P_{mcr} R_2} = \frac{\arctan((I - 1 - \langle y \rangle)/N) + \arctan(\langle y \rangle/N)}{2\arctan((I-1)/2)} \tag{4.17}$$

where I is the number of input reactant structures, N is the number of reaction stages, $\langle y \rangle$ is the ordinate of the target point P, and $0 < \delta < 1$. A completely convergent plan has $\delta = 1$

where all input materials are used to reach the target in a single reaction stage as in a multi-component reaction. A plan with $\delta = 0$ corresponds to one that is the most linear possible consisting of sequential transformations over one or more reaction stages with no input reagents incorporated into the starting substrate. Such a plan would have a corresponding tree diagram appearing as a sequence of horizontal lines connecting the starting material node with intermediate nodes until the final target node is reached. Examples of such a plan include a sequence of rearrangements, intramolecular cyclizations, electrocyclic reactions or some combination of these. It is important to note that with this definition sequential plans that are traditionally classified as 'linear' but which introduce contributing reagents in each step have an actual degree of convergence that can be measured. In the case of the triclosan plan the relevant parameters are $\langle y \rangle = 7.765625$, $N = 4$ and $I = 9$ which result in a degree of convergence of 0.435.

The relative rate of convergence, ρ_{rel}, for a given synthesis is given by the ratio of slopes of the lines R_2P and R_2P_{mcr}

$$\rho_{rel} = \frac{slope(R_2P)}{slope(R_2P_{mcr})} = \frac{\rho_{actual}}{\rho_{I-MCR}} = \frac{(\langle y \rangle / N)}{((I-1)/2)} = \frac{2\langle y \rangle}{N(I-1)} \tag{4.18}$$

The units of ρ_{actual} (corresponding to the experimental synthesis plan) and ρ_{I-MCR} (corresponding to the hypothetical single step MCR using the I reactant inputs of the experimental synthesis plan) are number of reactant inputs per reaction stage. A completely convergent plan, such as a multi-component single stage synthesis, has $\rho_{rel} = 1$ and a completely linear plan as that described above has $\rho_{rel} = 0$. For the triclosan plan $\rho_{rel} = 0.485$.

An asymmetry or skewness parameter, β, ranging between 0 and 1 may be defined as

$$\beta = \frac{\bar{P}\bar{Q}_1}{\bar{Q}_1\bar{Q}_2} = \frac{\langle y \rangle - ((I-1)/2)}{((I-1)/2)} = \frac{2\langle y \rangle}{I-1} - 1 \tag{4.19}$$

where $\beta = 0$, $(\langle y \rangle = ((I-1)/2))$ represents a completely symmetric plan and $\beta = 1$, $(\langle y \rangle = I - 1$ represents a completely asymmetric plan. This parameter essentially tells us the degree of skewness of the triangle R_1R_2P compared to triangle $R_1R_2Q_1$ where both involve the same number of input reactants and number of reaction stages. Note that the hypothetical one-step MCR plan represents the most convergent plan possible for the I reactant inputs involved and is completely symmetric as expected. For the triclosan plan $\beta = 0.941$.

The three parameters δ, ρ_{rel} and β may be used to rank the performances of different synthetic plans to a common target structure rather succinctly. Figure 4.16 shows a comparison of these parameters for a number of combinations of synthesis plans involving four reactant input structures. It is observed that as the number of reaction stages increases the degree of convergence decreases, the relative rate of convergence decreases, and the degree of asymmetry increases. The key trends to increase convergence are to decrease the number of reaction stages, to increase the number of parallel reactions per stage, and to increase the number of reactant input components per reaction. The last strategy of increasing the frequency of multi-component reactions is particularly effective if such reactions gravitate toward the end stages in a synthesis plan.

A second evaluation that can be made, by analogy with plots that trace the building up of molecular complexity of intermediate products as the synthesis proceeds[92,94–98] is to

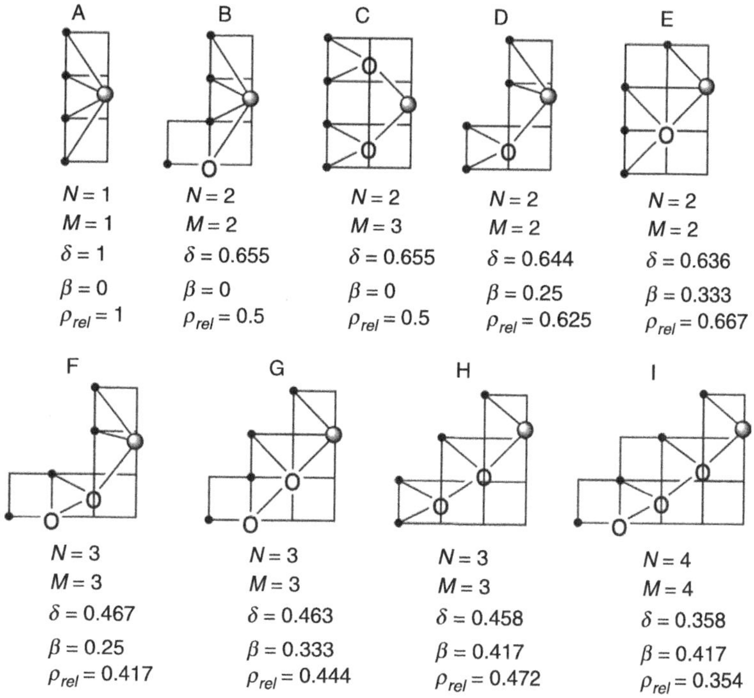

Figure 4.16 Summary of parameters for all combinations of synthesis trees involving four input reagents.

track the molecular weight of starting substrates and intermediate products as a function of reaction stage. We may derive using the method of moments a parameter that describes the net building up of structure from the set of initial input and intermediate structures toward the final target product. The molecular weight first moment per reaction stage about the target product molecular weight, p_n, in units of grams per mole per reaction stage is given by

$$\mu_1 = \frac{\Omega_1 - \Omega_2}{N + 1} \tag{4.20}$$

where

$\Omega_1 = \sum(MW_{intermediates} + MW_{starting\ material\ at\ beginning\ of\ each\ branch})$
$\Omega_2 =$ (total number of intermediate nodes + starting material nodes at beginning of each branch)p_n.

and N is the number of reaction stages in the synthesis plan and the starting materials correspond to those inputs at the beginning stage of each branch that actually get incorporated in the subsequent intermediate product. The zeroth stage representing the starting substrates for the longest branch or root of the synthesis tree is accounted for by the extra stage in the denominator. If a reaction stage has parallel reactions and therefore consists of more than one intermediate product being formed in that stage, then each of their respective molecular weights are included in the first summed term, Ω_1. The second term in Ω_1 accounts for molecular weights of input starting materials at the beginning of each branch provided they contribute to the structure of the immediately resulting product. A positive value for μ_1 indicates an overall net loss in MW per reaction stage (net degradation) and a negative value indicates an overall net gain in MW per reaction stage (net building up). The larger the magnitude of the first moment the greater is the effect of degradation or building up. Good synthesis plans are characterized by fewer reaction stages, the frequent occurrence of convergent reaction stages (i.e., parallel reactions), and large negative molecular weight first moments per reaction stage. For the triclosan plan shown in Figure 4.14, equation (4.20) results in $\mu_1 = (998.5 + 461.15 - 8(289.35))/(4 + 1) = -171.03$ g/mol/stage. This value is consistent with the observation that this plan begins with low molecular weight starting materials that are assembled via progressively heavier intermediates. The only drop in molecular weight appears in the final step where the acetyl group is eliminated. The sacrificial acetyl group introduced in reaction (i) serves to enhance the electrophilicity of the ortho chloro-substituted aromatic carbon atom in 2,5-dichloroacetophenone so that coupling with 2,4-dichlorophenol preferentially occurs at this site and not at the para position.

4.3.4 Energy efficiency

If each step in a synthesis plan has an associated input energy, Ψ_j, then the total energy input for the entire synthesis is

$$\Psi_{total} = \sum_{j}^{M} \Psi_j \tag{4.21}$$

We assume here that each Ψ_j is an *actual* input energy after subtracting all thermal losses. This input energy includes energy consumption during the reaction and all post-reaction phases such as heating, cooling, and operating under a pressure exceeding 1 atm. Basically, input energy will be required if any reaction operations are performed above or below standard temperature and pressure conditions of 25°C and 1 atm. It is important to note that input energy is additive. This means that regardless of the synthesis tree diagram, linear or convergent, one must add the input energies for each reaction to obtain the absolute overall input energy required for the plan. Clark[12] has described experimental determinations of energy inputs for chemical reactions using a domestic electricity meter set up in series with the laboratory power supply. Another useful energy metric that is characteristic of how well the input energy is utilized in a synthesis plan is the fraction of the total energy input that is directed to making the target product (FTE). For the jth reaction in a synthesis plan with input energy Ψ_j the amount of input energy directed toward product is

$$\Psi_{product,j} = (RME)_j \Psi_j \tag{4.22}$$

and the amount of input energy directed toward producing waste is

$$\Psi_{waste,j} = [1 - (RME)_j]\Psi_j \tag{4.23}$$

Therefore, the total amount of input energy for a synthesis plan that is directed to the target product is

$$\Psi_{product} = \sum_{j}^{M} (RME)_j \Psi_j \tag{4.24}$$

and that directed toward producing waste is

$$\Psi_{waste} = \sum_{j}^{M} [1 - (RME)_j] \Psi_j \tag{4.25}$$

The respective product forming and waste forming input energy fractions are given by

$$\Phi_{product} = FTE = \frac{\sum_{j}^{M} (RME)_j \Psi_j}{\sum_{j}^{M} \Psi_j} \tag{4.26a}$$

and

$$\Phi_{waste} = \frac{\sum_{j}^{M} [1 - (RME)_j] \Psi_j}{\sum_{j}^{M} \Psi_j} \tag{4.26b}$$

Equations (4.26a) and (4.26b) show that these parameters are weighted quantities and that the materials kernel green metric RME parameters for each reaction are the weighting factors. This provides a strong connection between the materials and energy usage metrics for a synthesis plan as it tells us how well the input energy is partitioned toward making a desired product. In accordance with Clark's recent analysis[12] in which he proposes an energy metric as the reaction input energy per mole of target product formed in units of kWh/mol we may rewrite equation (4.22) for the jth reaction in a synthesis plan as

$$\frac{\Psi_{product,j}}{(x/(\varepsilon_j \varepsilon_{j+1} \ldots \varepsilon_M))} = \left[\prod_{j}^{M} \varepsilon_j \right] (RME)_j \Psi_j / x \tag{4.27}$$

and equation (4.26a) for all reactions in the entire plan as

$$\frac{\Phi_{product}}{x} = \frac{\sum_{j}^{M} (RME)_j \Psi_j}{x \sum_{j}^{M} \Psi_j} \tag{4.28}$$

Unfortunately, the absolute energy input for a synthesis plan and the fraction of total input energy directed to final target product are not given as part of the standard protocol of reporting experimental results of individual chemical reaction or synthesis plan performances in scientific journals or patents. At this time there are no documented examples in the literature that can be used to illustrate the utility of the relationships shown above.

4.3.5 Hypsicity (oxidation level)

By analogy with the tracking of molecular weights of starting materials and intermediates leading to the final target product in a synthesis plan to gauge how efficiently a target structure is synthesized, it is possible to track the changes in oxidation state, or hypsicity, (Gk: *hypsos*, meaning level or height) of key atoms involved in bond-making and bond-breaking steps. This follows Hendrickson's proposal to aim for *isohypsic* synthesis plans which are characterized by a zero net change in oxidation state of all atoms of starting materials and intermediates involved until the target product is reached.[99] This can be achieved by designing synthesis plans which eliminate redox reactions entirely. This is consistent with the conclusions drawn in Section 4.2.6 regarding the minimization of such reactions because they are the most material inefficient and contribute to significant attenuations in RMEs. If these cannot be avoided due to practical considerations then the next best thing to achieve the isohypsic condition is to strategically sequence redox reactions in such a way that for every increase in oxidation level of an atom occurring in a step it is matched immediately by a decrease in oxidation level of equal magnitude in the next step, or vice versa. This cuts down on the accumulation of excess gains or losses in oxidation level of atoms, as the case may be, in starting materials and intermediates with respect to the oxidation levels of those atoms in the final target molecule over the course of the synthesis. Formally we may define an hypsicity index, HI, as given in equation (4.29).

$$HI = \frac{\sum_{stages,j}\left[\sum_{atoms,i}\left[(Ox)^{atom,i}_{stage,j} - (Ox)^{atom,i}_{stage,N}\right]\right]}{N+1} = \frac{\sum_{stages,j}\Delta_j}{N+1} \tag{4.29}$$

where (Ox) represents the relevant oxidation number of an atom. If HI is zero, then the synthesis is *isohypsic*. If HI is positive valued then to get to the target molecule, a net reduction is required over the course of the synthesis since an accumulated gain in oxidation level has resulted. Such a condition is termed *hyperhypsic*, by analogy with the term *hyperchromic* which describes increases in intensities of absorption bands in spectroscopy. Conversely, if HI is negative valued then to get to the target molecule a net oxidation is required over the course of the synthesis since an accumulated loss in oxidation level has resulted. Such a condition is termed *hypohypsic*, again by analogy with the term *hypochromic*. It is important to note that changes in oxidation number can occur for atoms in reactions that are not formally classified as reductions or oxidations with respect to the substrate of interest. A good example of this is the Grignard reaction with is classified as a carbon–carbon bond-forming reaction and yet involves a formal oxidation with respect to magnesium in the preparation of the Grignard reagent. Another is electrophilic aromatic substitution which begins with an oxidation state of -1 for the ArC-H carbon atom which then increases to $+1$ when hydrogen is substituted for chlorine. The HI therefore accounts for all such changes in oxidation numbers of atoms regardless of the reaction type. A thorough investigation of these phenomena will be reported on elsewhere.

The following sequence of steps may be followed to determine *HI* for a synthesis:

1. Enumerate atoms in the target structure that are only involved in the building up process from corresponding starting materials. This set of atoms defines those that are involved in bonding changes occurring in the relevant reaction steps.
2. Work backwards intermediate by intermediate to trace the oxidation numbers of the above set of atoms back to original starting materials as appropriate following the reaction stages back to the zeroth stage.

3. For each key atom, i, in each reaction stage, j, determine the difference in oxidation number of that atom with respect to what it is in the final target structure. Hence, $(Ox)^{atom,i}_{stage,j} - (Ox)^{atom,i}_{stage,N}$.
4. Sum the differences determined in step 3 over all key atoms in stage j. This yields the term

$$\sum_{atoms,i} \left[(Ox)^{atom,i}_{stage,j} - (Ox)^{atom,i}_{stage,N} \right] = \Delta_j$$

5. Finally, take the sum $\sum_{stages,j} \Delta_j$ over the number of stages and divide by $N+1$ accounting for the extra zeroth reaction stage.

An illustration of this sequence of steps is shown in Scheme 4.9 and Figure 4.17 for the hypsicity analysis of the synthesis of triclosan given originally in Scheme 4.8. For example, comparing the oxidation numbers of atoms a, b, c, d, e, f and f^* in the product structure for step 2 with those in the product structure for the final step 4 yields $\Delta_2 = (0-1)+(1-1)+(1-1)+(1-1)+(-1-(-1))(-1-(-1)) = -1$, where the first difference refers to atom b, the second to atom c, the third to atom d, the fourth to atom e, the fifth to atom f, and the sixth to atom f^*. Comparison of step 1* with step 4 only involves comparing oxidation numbers of atoms d, e, f and f^*. Hence, $\Delta_{1^*} = (1-1)+(1-1)+(-1-(-1))+(-1-(-1)) = 0$. Similarly, comparison of step 1 with step 4 only involves comparing oxidation numbers of atoms b and c. Hence, $\Delta_1 = (0-1)+(1-1) = -1$. Since we are comparing oxidation number changes over stages, for stage 1 we take the sum of Δ_{1^*} and Δ_1 which is -1. An overall HI of -1 indicates that the synthesis is hypohypsic and is consistent with the observation that the oxidation numbers of the key atoms in the starting materials and intermediates are below those in the target product, and that a net oxidation occurs over the course of the synthesis. Indeed from the graph in Figure 4.17 we can track the steady increase in oxidation number changes: -4 (stage 0) $\rightarrow$ -1 (stage 1) $\rightarrow$ 0 (stage 2) $\rightarrow$ 0 (stage 3) $\rightarrow$ 0 (stage 4). Atom b undergoes steady increase in oxidation number from -1

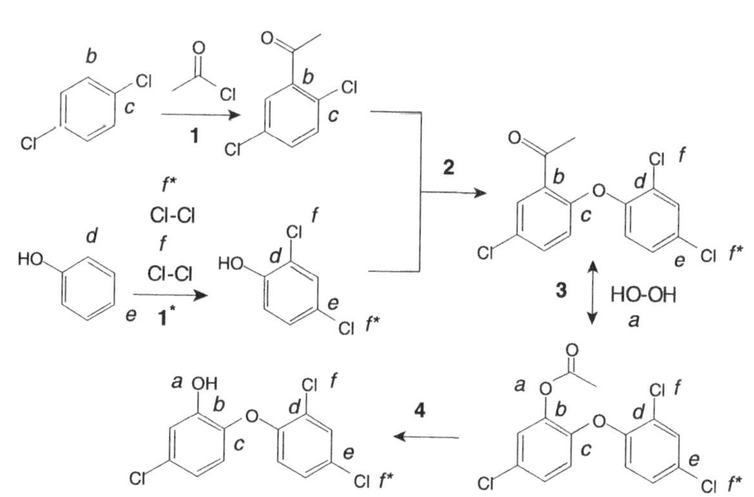

Scheme 4.9 Synthesis plan for triclosan showing labelled atoms involved in target bond forming steps.

Reaction	*a*	*b*	*c*	*d*	*e*	*f*	*f**	Δ_j	
4	−2	+1	+1	+1	+1	−1	−1	0	
3	−2	+1	+1	+1	+1	−1	−1	0	
2	−1	0	+1	+1	+1	−1	−1	0	
1*				+1	+1	−1	−1	0	−1
1		0	+1					−1	
0				−1	−1	0	0	−2	−4
0		−1	+1					−2	
								−5	

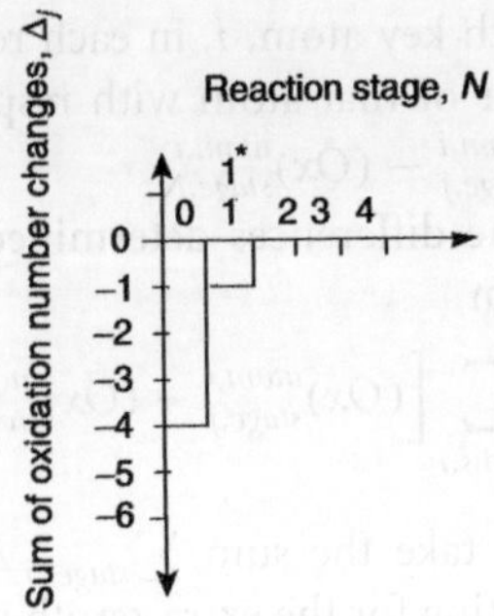

Figure 4.17 Summary of oxidation number changes by reaction stage for labelled atoms in Scheme 4.9 for triclosan synthesis.

in 1,4-dichlorobenzene to 0 in 2,5-dichloroacetophenone (via Friedel–Crafts acylation) to +1 in the target triclosan (via Baeyer–Villiger oxidation). Similarly, atoms *d* and *e* each increase in oxidation number from −1 in phenol to +1 in 2,4-dichlorophenol via the chlorine substitution reaction. Chlorine atoms *f* and *f** undergo reductions from 0 to −1 in the chlorination step 1*. Matching of oxidation numbers is achieved in stage 2 of the synthesis.

4.3.6 *Strategies for good syntheses*

We have shown up to this point how various plans to a common target may be characterized and assessed according to a number of green metrics which gauge the material efficiency, energy efficiency, and synthetic strategy and elegance. We may use these performance variables to effectively rank plans in an unbiased way to assess their 'greenness'. The following list summarizes the criteria for good and 'green' synthesis plans:

Metrics:

- As far a possible design reactions in a plan so that $AE \geq 62\%$ and $RME \geq 62\%$ with minimum use of reaction solvent and post-reaction materials, or at least incorporate their recovery.
- Maximize reaction yields per step to be no lower than 75%.
- As far as possible operate under conditions where no excess reagents are used keeping *SF* equal to 1 for each reaction.
- Minimize number of overall steps (M) and number of reaction stages (N).
- Maximize number of parallel steps/branches.
- Minimize overall number of input materials needed (I).
- Maximize number of input materials that end up in product in any given step (i.e., maximize atom economy (AE) for each step).
- Maximize overall kernel reaction mass efficiency (RME).
- Maximize degree of convergence (δ).
- Minimize degree of asymmetry (β).
- Maximize relative rate of convergence (ρ_{rel}).
- Maximize fraction of total input energy directed toward target product.

- Minimize overall raw materials cost (*RMC*).
- Design plans with large negative molecular weight first moments ($\mu 1$) (i.e., increase degree of building up from light starting materials via progressively heavier intermediates to final target product).
- Aim for the isohypsic condition of no net change in oxidation level for atoms involved in bond-making and bond-breaking reaction steps ($HI = 0$).

Design:

- Design convergent-reflexive plans from the outset that maximize the frequency of multicomponent, tandem,[100–104] domino,[105–109] and cascade reactions[110,111] and minimize the frequency of substitutions particularly protecting group strategies[112] and redox reactions. Reflexive plans[92] involve the production of a key source or 'hub' intermediate which is then shunted to the production of other intermediates down the line which in turn are combined in a late-stage convergent step toward the target. A key strategy in deciphering such a hub intermediate is to look for the largest repeated structural motif in the target product. This is where exploitation of structural symmetries in the target molecule becomes important.
- If redox reactions cannot be avoided aim for the isohypsic condition.
- For catalytic oxidations opt for hydrogen peroxide or molecular oxygen as oxidant; for catalytic reductions opt for molecular hydrogen as reductant.
- Minimize overall number of steps and stages preferably using nested MCRs as far as possible as the main strategy in a synthetic plan to a complex target. Overall convergence is maximized dramatically if a plan contains a late-stage MCR that involves components that are themselves terminal products of independent parallel linear sequences.
- Devise regio- and stereoselective synthetic strategies particularly regio- and stereoselective MCRs.
- Devise catalytic methods where catalysts are recycled and reused.
- Capitalize on opportunities to recycle byproducts back to starting materials.
- Opt for solventless reactions, recycle solvents, or use benign solvents (e.g., water, ionic liquids, supercritical media).
- Minimize energy demands such as heating, cooling and carrying out reactions under pressure exceeding 1 atm; aim for reactions run under standard temperature and pressure conditions (room temperature and 1 atm).

Many of these goals have been identified by others in qualitative[113–131] or semi-quantitative terms.[132–137] In this work they are couched more rigorously in quantitative terms. With respect to ranking synthesis plans, the key point is that we desire all of the metrics attributes to be in the *same* plan so that maximization and minimization of appropriate metrics occur synergistically. This is a good measure of knowing if iterations in syntheses are actually leading to an 'optimal' synthesis. If a comparison of a set of plans leads to a situation where there is a scattering of optimum metrics across many of them, then this is a sure sign that optimization has not been achieved. When only one plan exists for a given target the only possible comparison is the hypothetical 'ideal' synthesis involving a single step multicomponent reaction with $AE = 1$, $RME = 1$, $\varepsilon = 1$, $HI = 0$, μ_1 with a large negative value, $\beta = 0$, $\delta = 1$, and $\rho_{rel} = 1$. This situation is difficult to assess because this hypothetical scenario is often practically unattainable.

4.3.7 Illustrated examples

In this section several examples of syntheses to important classical and modern target molecules are assessed using the above new paradigm of measuring and ranking synthetic efficiency. These rankings help to identify the state of affairs of where we are at in the total synthesis of any given target so that further improvements can be made using the discussed metrics to guide future optimization efforts. As a key example, the quantitative analysis of 11 literature plans for the alkaloid nicotine is reported in detail including graphical outputs since this example highlights true progression in synthesis optimization. The remaining examples are dealt with briefly only highlighting key features of synthesis trees and metrics values. All plans are worked out under the assumptions of no excess reagents used and recovery of reaction solvents and post-reaction materials so that their intrinsic chemical performance may be compared. When several total syntheses to a common target are compared best values for the metrics parameters are bolded in the accompanying tables. Plans are ranked in descending order according to overall kernel RME.

4.3.7.1 Nicotine[138–151]

The structure of nicotine has one stereogenic centre and two rings that are connected by a single bond. Strategies employed in the literature revolve around three main themes: (a) begin with a pyridine derivative, extend the chain and cyclize to make the pyrrolidine ring and then methylate; (b) begin with a pyrrolidine ring precursor, extend the chain and cyclize to make the pyridine ring; and (c) begin with pyridine and pyrrolidine derivatives and couple these stereoselectively. Figure 4.18 summarizes various sets of starting materials for each route and Figures 4.19–4.29 illustrate the specific tree diagrams. All plans are of the linear type. Figures 4.19, 4.20, 4.27, and 4.29 show the synthesis of the racemic product, Figures 4.21–4.26 show the stereoselective synthesis of the natural (+)(S) enantiomer, and Figure 4.28 shows the stereoselective synthesis of the (−)(R) enantiomer. Figures 4.19, 4.21, 4.23, 4.24, 4.25, and 4.27 follow the first strategy beginning with pyridine, 3-pyridinecarboxyaldehyde, a 3-substituted pyridine carbonate derivative, 3-bromopyridine, 3-pyridinecarboxyaldehyde and nicotinic acid, respectively. Figure 4.22 follows the second strategy beginning with L-proline. Figures 4.20, 4.26 and 4.28 follow the third beginning with *N*-methylpyrrolidone and ethyl nicotinate, ethyl 5-bromonicotinate and *N*-vinylpyrrolidone, and 3-bromopyridine and pyrrolidine, respectively. There are some key features in each of these coupling strategies. The Späth plan involves extrusion of carbon dioxide in the second step which actually destroys the pyrrolidone ring joined to the nicotinate moiety that was made in the first coupling step. The choice of a condensation step with a nicotinate in the coupling strategy yields an adduct with an extra carbon atom that must be removed. The extrusion and subsequent recyclization step to restore the pyrrolidine ring create a redundancy in the synthesis plan. The Jacob plan involves a resolution step with a chiral acid, (+)-α-methoxy-α-(trifluoromethyl)phenylacetic acid. The Campos plan is a direct stereoselective coupling that employs a Negishi strategy with (−)sparteine as chiral auxiliary. It should be pointed out that the Campos synthesis was actually for an N-Boc protected derivative of (−)(R) nornicotine; however, we have adapted it for the present metrics analysis to yield (−)(R)nicotine assuming reaction yields remain the same. The earliest reported synthesis of nicotine by Pictet shown in Figure 4.29 is an

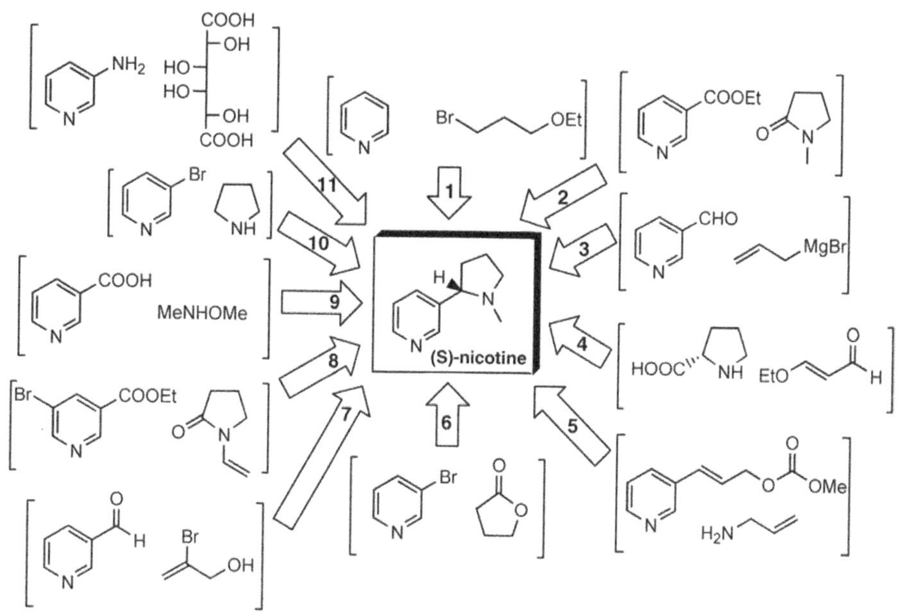

Figure 4.18 Synthesis map showing starting materials used for the synthesis of nicotine.

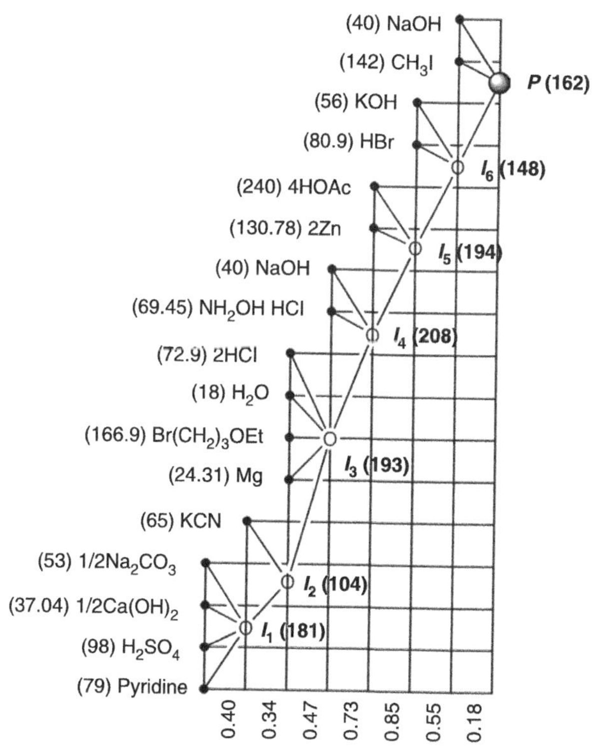

Figure 4.19 Synthesis tree for nicotine plan 1 (Craig, 1933).

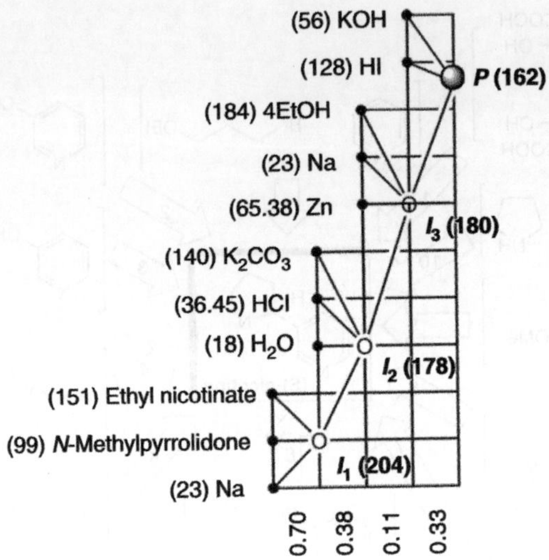

Figure 4.20 Synthesis tree for nicotine plan 2 (Späth, 1928).

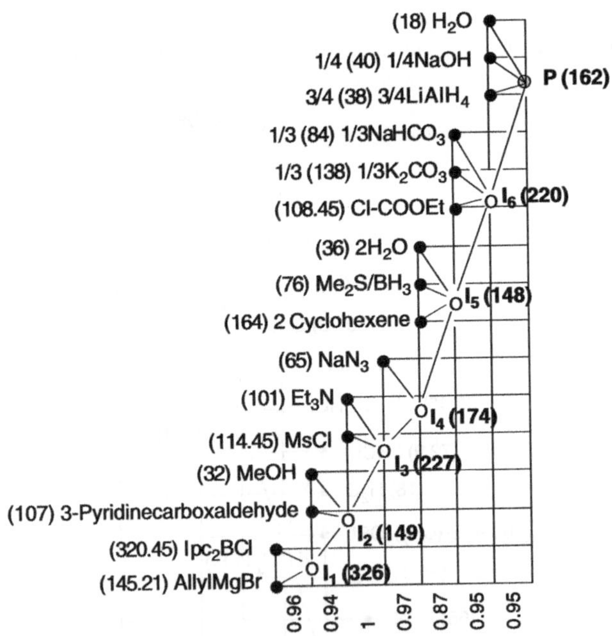

Figure 4.21 Synthesis tree for nicotine plan 3 (Lebreton, 2001).

unusual coupling strategy that involves first making *N*-β-pyridylpyrrole by heating mucic acid with 3-aminopyridine. In the second step this rearranges to 2-(3′-pyridyl)pyrrole by further heating in alkaline solution containing iron(III)chloride as noted by the horizontal line connecting nodes of identical molecular weight in the synthesis tree diagram.

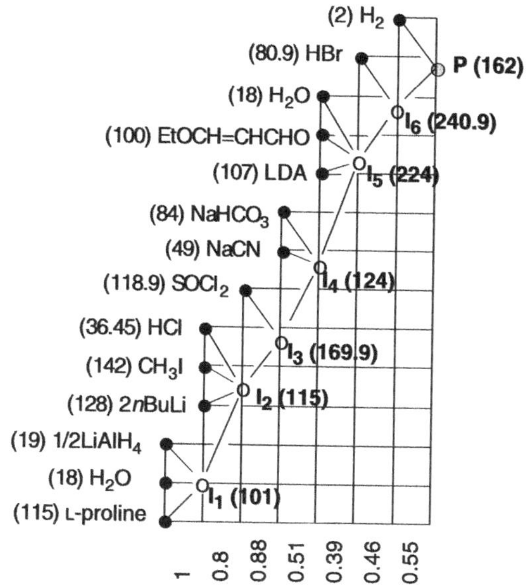

Figure 4.22 Synthesis tree for nicotine plan 4 (Chavdarian, 1982).

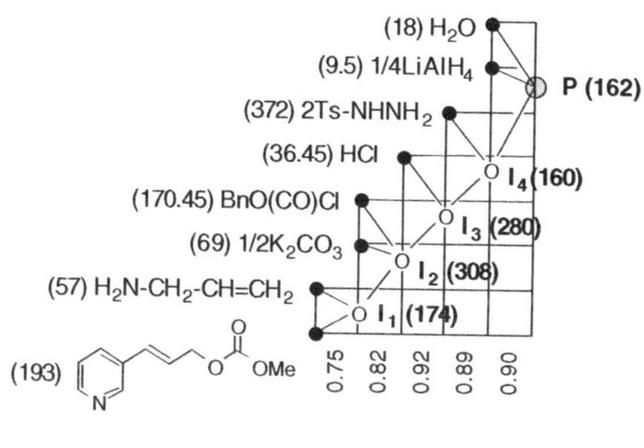

Figure 4.23 Synthesis tree for nicotine plan 5 (Helmchen, 2005).

The complete numerical analysis of all plans is summarized in Table 4.7. Since the Ley plan (see Figure 4.27) employed solid-supported reagents maximum values for AE and RME and a minimum value for μ_1 were determined. The overall kernel RME and overall yield could not be determined for the Pictet plan since reaction yields were not reported except for the first step which was 42% with respect to mucic acid. There have also been reported criticisms of this plan with respect to product identification and purity as noted by Späth.[139] Owing these shortcomings further graphical analysis of this plan was not done. Figures 4.30–4.31 show the individual and cumulative reaction performances according to AE, reaction yield, and RME for the remaining plans. Figure 4.32 showcases the waste distribution profiles as a function of reaction stage. Figure 4.33 shows the molecular weight building up profile for each plan. It can be readily seen that the Campos two-step plan, based

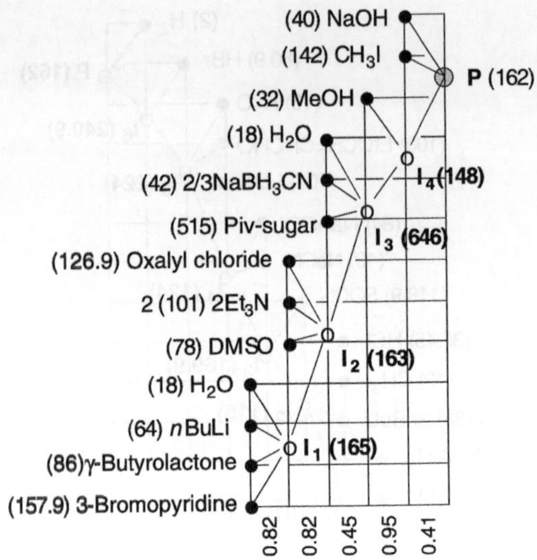

Figure 4.24 Synthesis tree for nicotine plan 6 (Loh, 1999).

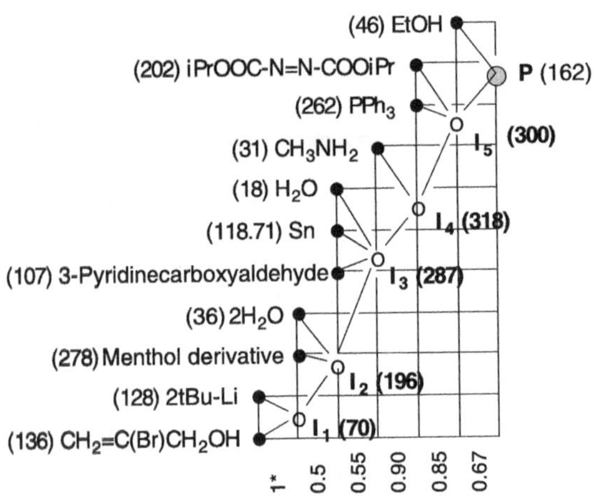

Figure 4.25 Synthesis tree for nicotine plan 7 (Delgado, 2001).

on final product material recovered regardless of stereochemistry, has all the key attributes described for an optimal synthesis in the *same* plan; that is, it has the highest overall RME, the fewest number of input materials, the highest degree of building up in the synthesis, and the highest degree of convergence. These criteria suggest that this plan is the 'greenest' on the basis of material efficiency and input material cost alone. The consideration of any other factors within the framework of green chemistry, most notably energy consumption, toxicity of materials, safety of materials and procedures particularly on scale-up, and waste treatment, may change this conclusion. However, it should be noted that plans with high material efficiency and which utilize cheap, readily available and safe starting materials are strong contenders for overall 'green' plans.

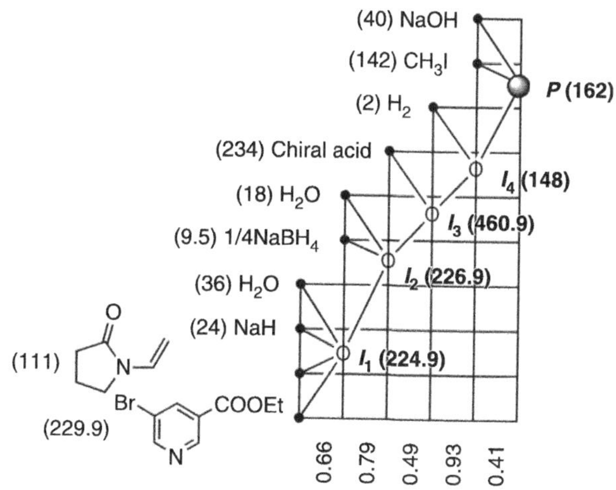

Figure 4.26 Synthesis tree for nicotine plan 8 (Jacob, 1982).

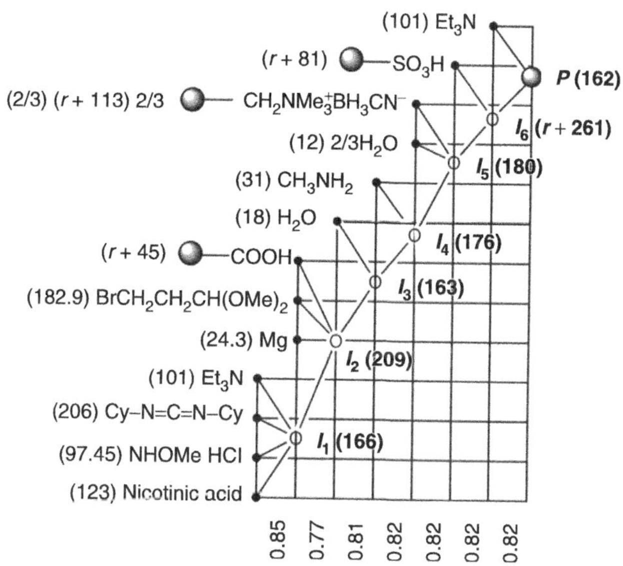

Figure 4.27 Synthesis tree for nicotine plan 9 (Ley, 2002).

As an example, Scheme 4.10 and Table 4.8 summarize the hypsicity analysis for the Craig plan. Hypsicity indices for all plans are given in column 8 of Table 4.7. Figure 4.34 illustrates the change in oxidation level profiles as a function of reaction stage. It turns out that the Campos plan with the highest RME is also the one with an *HI* closest to the isohypsic condition of $HI = 0$. In Table 4.8 a check on the computation of $\sum \Delta_j$ was made by taking the sum, over all stages, of the oxidation number differences for a given atom in a given stage and the final stage. These are represented by the integers in the last row of the table. Both the vertical and horizontal columns sum to 14.

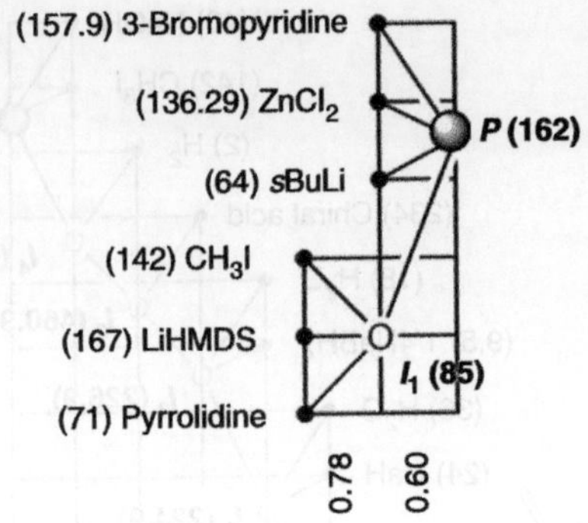

Figure 4.28 Synthesis tree for nicotine plan 10 (Campos, 2006).

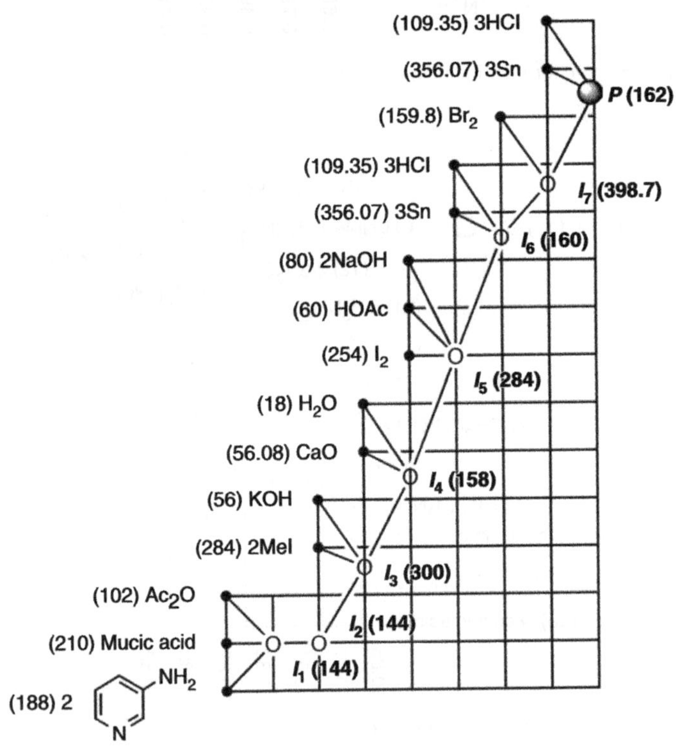

Figure 4.29 Synthesis tree for nicotine plan 11 (Pictet, 1895–1904).

4.3.7.2 *Ascorbic acid (vitamin C)*[152–159]

Table 4.9 summarizes the materials metrics analysis for the Reichstein and Haworth plans to ascorbic acid. Scheme 4.11 shows the industrial synthesis plan according to the Reichstein

Table 4.7 Summary of reaction metrics and synthesis tree parameters for nicotine synthesis plans ranked according to overall kernel (maximum) *RME*[a].

Plan	*N*	*M*	*I*	δ	β	μ_1	*HI*	% *AE*	% $\langle\varepsilon_T\rangle$	% *RME*
Campos (2006)	**2**	**2**	**6**	**0.722**	**0.35**	**−62.67**	**−0.33**	**22.0**	**46.8**	**11.5**
Helmchen (2005)	5	5	8	0.426	0.641	+33.33	+4.5	17.5	45.3	10.9
Lebreton (2001)	7	7	16	0.461	0.772	+51.71	+1.75	11.6	67.1	8.6
Ley (2002)	7	7	13	0.398	0.805	+10.06[b]	+2.38	14.7[c]	24.4	4.5[c]
Delgado (2001)	6	6	11	0.424	0.767	+43.00	+2.67	11.9	14.1	2.7
Jacob (1982)	5	5	10	0.461	0.717	+72.94	+2.17	19.1	9.7	2.7
Loh (1999)	5	5	13	0.496	0.777	+65.65	+1.5	11.4	11.7	1.9
Chavdarian (1982)	7	7	14	0.424	0.821	−5.53	+1	15.8	3.5	0.8
Späth (1928)	4	4	12	0.541	0.753	+0.4	+1.6	16.1	1.0	0.3
Craig (1933)	7	7	16	0.408	0.955	−25.00	+1.75	11.5	0.4	0.02
Pictet (1895–1904)	8	8	15	0.451	0.815	+58.74	+3.78	7.1	d	d

[a] See Appendix for definitions of terms.
[b] Minimum value.
[c] Maximum value.
[d] No reaction yields reported.

procedure which has a kernel overall RME nearly two orders of magnitude larger than the Haworth sequence.

4.3.7.3 Co-enzyme A[160–169]

The Khorana–Todd synthesis of co-enzyme A is a highly convergent route shown as a reduced tree diagram in Figure 4.35. The reduced tree shows the main branches and convergent steps omitting all other input materials for clarity. It can be seen that the plan has 5 points of convergence with 6 tiers of reaction yields equivalent to the number of starting material nodes. The 'overall yield' along the main lower branch beginning from uric acid is 1.6%. Cumulative yields from ribose, 2 $(BnO)_2P(OH)$, pantolactone, benzylthiol, and $(BnO)_2P(OH)$ are $(0.28 \times 0.50 \times 0.70 \times 0.66 \times 0.70 \times 0.90 \times 0.65) \times 100 = 2.6$, $(0.80 \times 0.70 \times 0.90 \times 0.65) \times 100 = 32.8$, $(0.72 \times 0.76 \times 0.60 \times 0.88 \times 0.88 \times 1 \times 0.65) \times 100 = 16.5$, $(0.87 \times 0.60 \times 0.88 \times 0.88 \times 1 \times 0.65) \times 100 = 26.3$, and $(0.80 \times 0.88 \times 0.88 \times 1 \times 0.65) \times 100 = 40.3\%$ respectively. Table 4.10 summarizes the metrics parameters. Generally, the number of points of convergence is one less than the number of paths (branches) from starting material nodes to final target node, which in turn is equal to the number of reaction yield tiers.

4.3.7.4 Colchicine[170–185]

Various routes to the total synthesis of colchicine are summarized in Figure 4.36. So far the plan with the highest RME is that of Evans as shown by the data in Table 4.11. However, the Schmalz plan has the highest degree of convergence and the Scott plan has the highest degree of building up in the synthesis. This molecule remains a synthetic challenge because of the serious problem of isomerization occurring in the tropolone ring as shown below

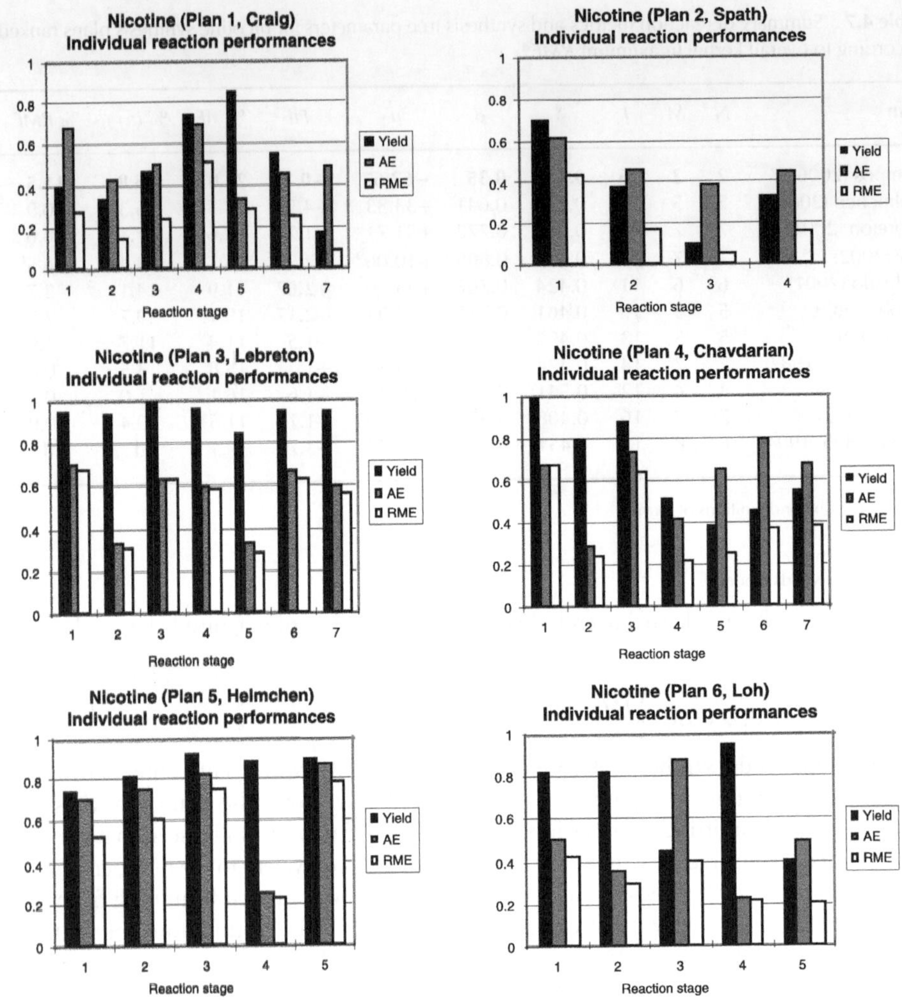

Figure 4.30 Histograms showing individual reaction performances for nicotine synthesis plans.

where a significant proportion of material is lost in the methylation step to an unwanted geometric isomeric product via the left-hand structure.

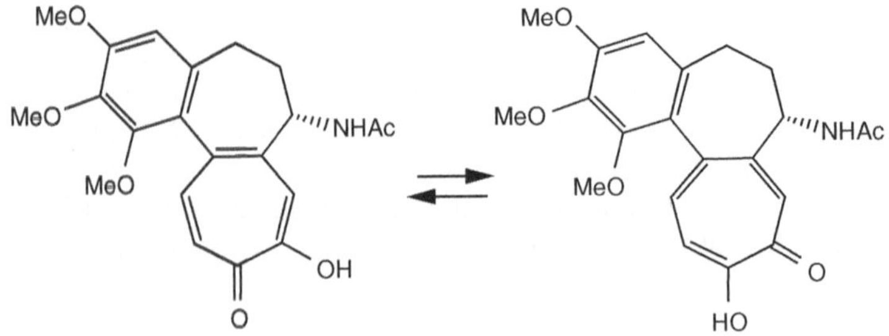

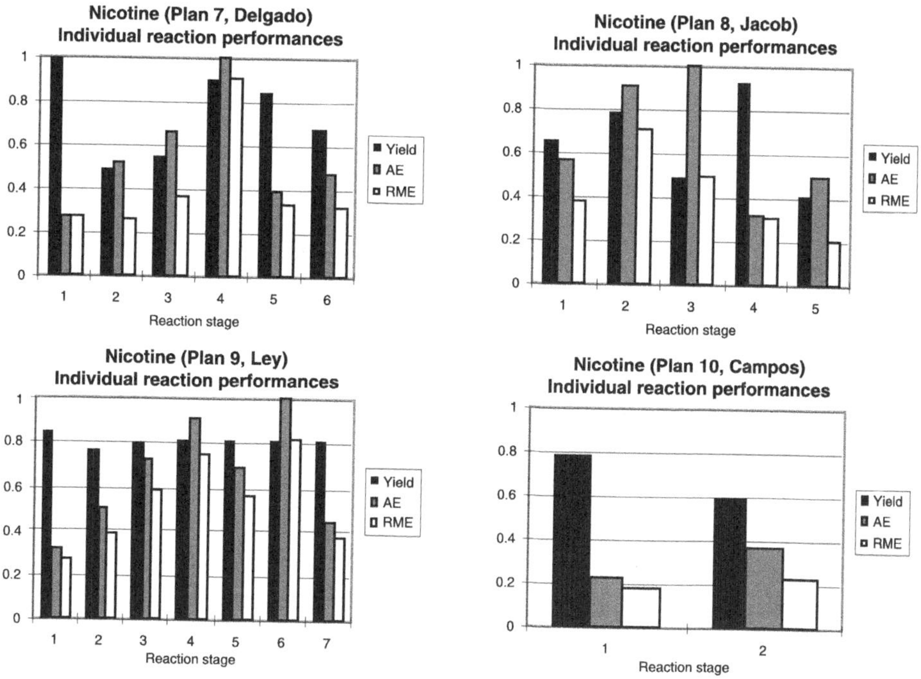

Figure 4.30 Continued.

4.3.7.5 Cyclooctatetraene[186–189]

Cyclooctatetraene is among the set of classic molecules to have been synthesized. The celebrated synthesis plans of Willstätter and Reppe are shown in Schemes 4.12 and 4.13. The Reppe plan is a single step 4 component coupling that has an AE and kernel RME 16 and 450 times better than the Willstätter plan, respectively as shown by the results given in Table 4.12. This reaction may be considered a pseudo-multicomponent reaction because the four constituent acetylene components are identical. Figure 4.37 shows two ways of drawing out the corresponding tree diagram depending on whether we treat the acetylene as a single node or 4 separate nodes. These two possibilities lead to two different conclusions about the synthesis. When it is drawn with a single input node the molecular weight first moment is zero indicating that essentially no net synthesis has really taken place; whereas, when it is drawn with 4 input nodes it is strongly negative indicating a significant degree of synthesis has indeed taken place.

4.3.7.6 Discodermolide[190–202]

The synthesis of discodermolide, a marine natural product found in the marine sponge *Discoderma dissoluta* is an excellent example of illustrating the progress to optimization of its total synthesis. Figure 4.38 shows a synthesis map for the various starting materials used to get to this target and Figure 4.39 shows a reduced synthesis tree for the Novartis plan using a convergent strategy employing a key 'hub' intermediate that is used as a linchpin compound

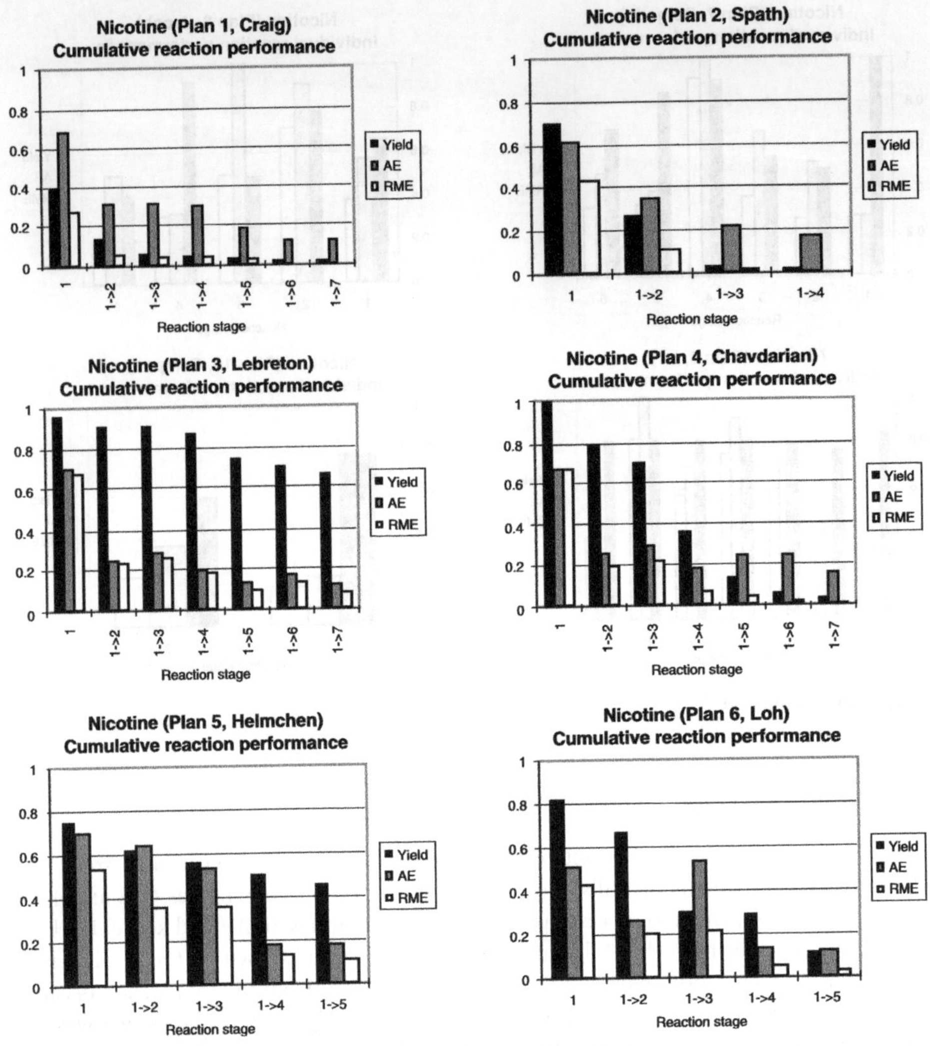

Figure 4.31 Histograms showing cumulative reaction performances for nicotine synthesis plans.

for each of the three branches of the synthesis. Synthetic plans exhibiting this strategy have been termed 'reflexive'[92,95] and are often employed in industrial scale syntheses. Such a strategy is viable only if there is a common structural motif that is repeated in the target structure as shown by the circled moieties in Figure 4.38. The more a hub intermediate resembles the final target structure, this means that its node co-ordinates in the tree diagram are pushed toward the final target node co-ordinate. This will improve both the degree of convergence and degree of asymmetry parameters. A highly material efficient sequence to this hub intermediate with a high overall AE and high reaction yields is essential to ensure a high overall kernel RME since this path or branch in the tree diagram will be repeated several times. In the present case there are three repeated identical sequences from

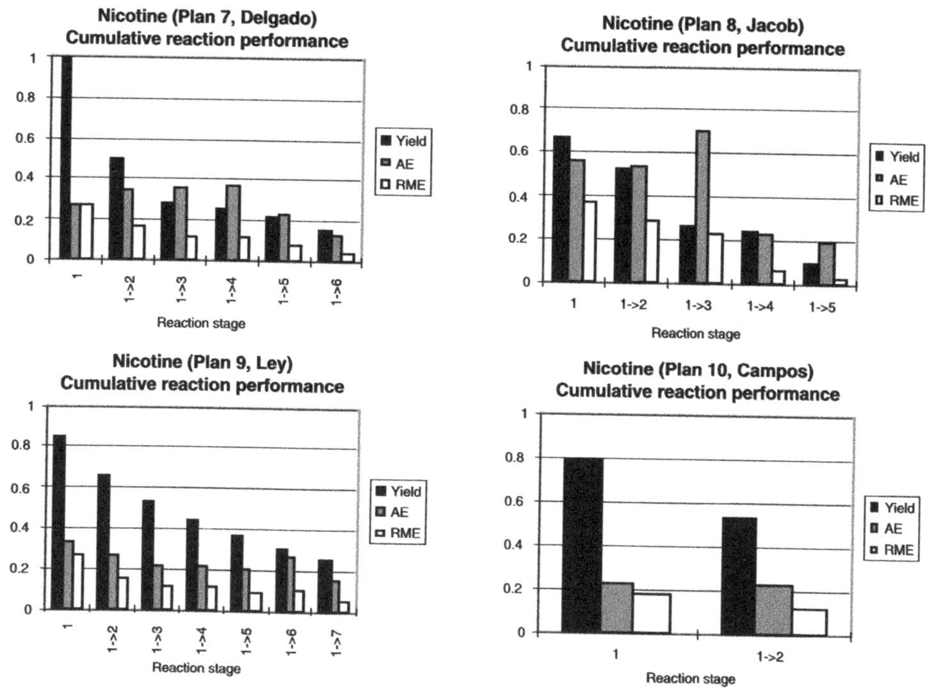

Figure 4.31 Continued

S to H. The ultimate situation is to begin the synthesis with such a hub intermediate if it is an available starting material. Hence, the best scenario is to begin a synthesis with a readily available and cheap starting material whose structure corresponds to the highest common structural motif found in the target structure. From the results shown in Table 4.13 several comparisons can be made. The Smith 3rd generation synthesis and the Novartis synthesis begin with the same starting materials and go through the same hub intermediate. The Smith 4th generation plan begins with two advanced starting materials and so its metrics performance is understandably better than that of the 3rd generation. Note that the 4th generation plan has an overall kernel RME 10 times higher than the 3rd while its molecular weight building up parameter is more positive indicating that the degree of synthesis in the 4th generation plan is less than the 3rd. Similar conclusions can be drawn when comparing two versions of the Novartis plan depending on whether the metrics analysis begins directly with the hub intermediate H. For the Novartis and Smith 3rd generation plans it is possible to determine the mass split ratio for the hub intermediate H for the branches beginning with H and ending up at the target node *P* by multiplying the respective chain of reaction yield parameters starting from P and working backwards guided by the tree diagrams for each plan. For the Novartis plan shown in Figure 4.39 the mass split ratio for H is 47.4%:48.7%:3.9% for the lower, middle and upper branches, respectively and the corresponding ratio of overall yields for these branches is 0.91%:0.89%:11.1%. For the Smith 3rd generation plan the corresponding mass split ratio is 35.8%:52.8%:11.4% and the overall yield ratio is 1.6%:1.1%:5.0% for the lower, middle and upper branches.

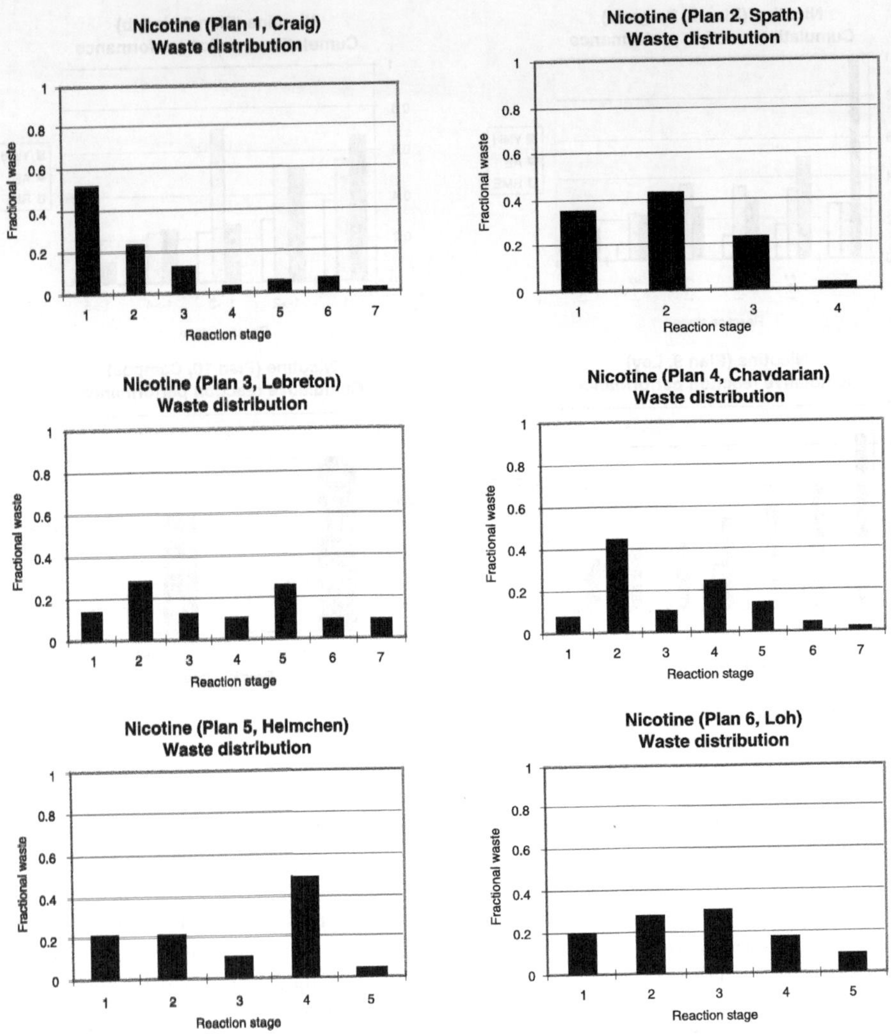

Figure 4.32 Histograms showing waste distribution profiles as a function of reaction stage for nicotine syntheses.

These kinds of calculations are easily done when a tree diagram is constructed for the plan. The Schreiber plan which proved the structure of the final target molecule has the highest degree of convergence since it has 3 points of convergence and 4 branches or paths in its tree diagram.

4.3.7.7 Diversity-oriented synthesis sequence[203]

An example diversity-oriented synthesis strategy as shown in Scheme 4.14 illustrates how complex targets can be quickly synthesized from smaller molecules via combinations of

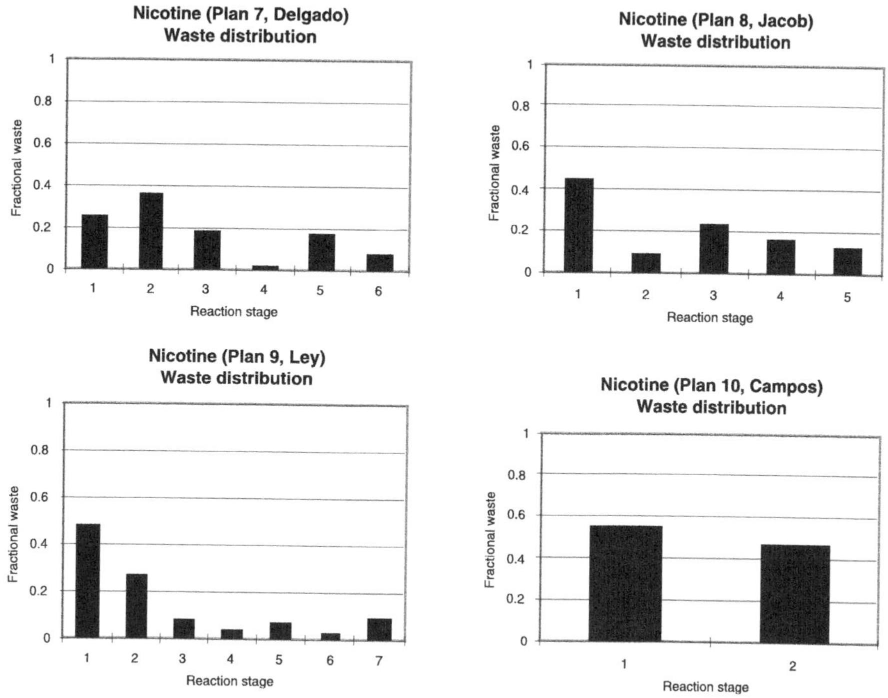

Figure 4.32 Continued.

multicomponent reactions and strategic design of cascade or domino reactions. The metrics summary for this plan given in Table 4.14 nicely corroborates the advantages of such approaches in total synthesis as was discussed in Section 4.3.6.

4.3.7.8 Epibatidine[204–222]

There has been a great deal of interest in the synthesis of the alkaloid (−)-epibatidine found in the skin of the Ecuadorian poison tree frog *Epipedobates tricolor* particularly because it is about 200 times more powerful than morphine as an analgesic. Eighteen total synthesis plans were analysed and the metrics are summarized in Table 4.15. Figure 4.40 shows a synthesis map of starting materials for the better performing plans. The plan with the highest overall kernel RME is the Giblin plan with a value of 6.6%. The Carroll plan is particularly noteworthy for discussion since as shown in Scheme 4.15 it incorporates a recycling loop for the epimerization of a key intermediate, it has overall the fewest number of steps and uses the fewest number of input materials, and it is the most symmetric. The third step in this plan produces this intermediate in a 1:2 ratio for the exo to endo forms. Separation of the exo product and continuing with this material to the target product leads to an overall RME of 2.5% for the production of (−)-epibatidine. Since the exo form of the intermediate is required for getting to this target and it is formed as a minor product it is imperative to introduce a recycling loop into the synthesis to improve the RME performance. Incorporation of an epimerization reaction that converts the key intermediate

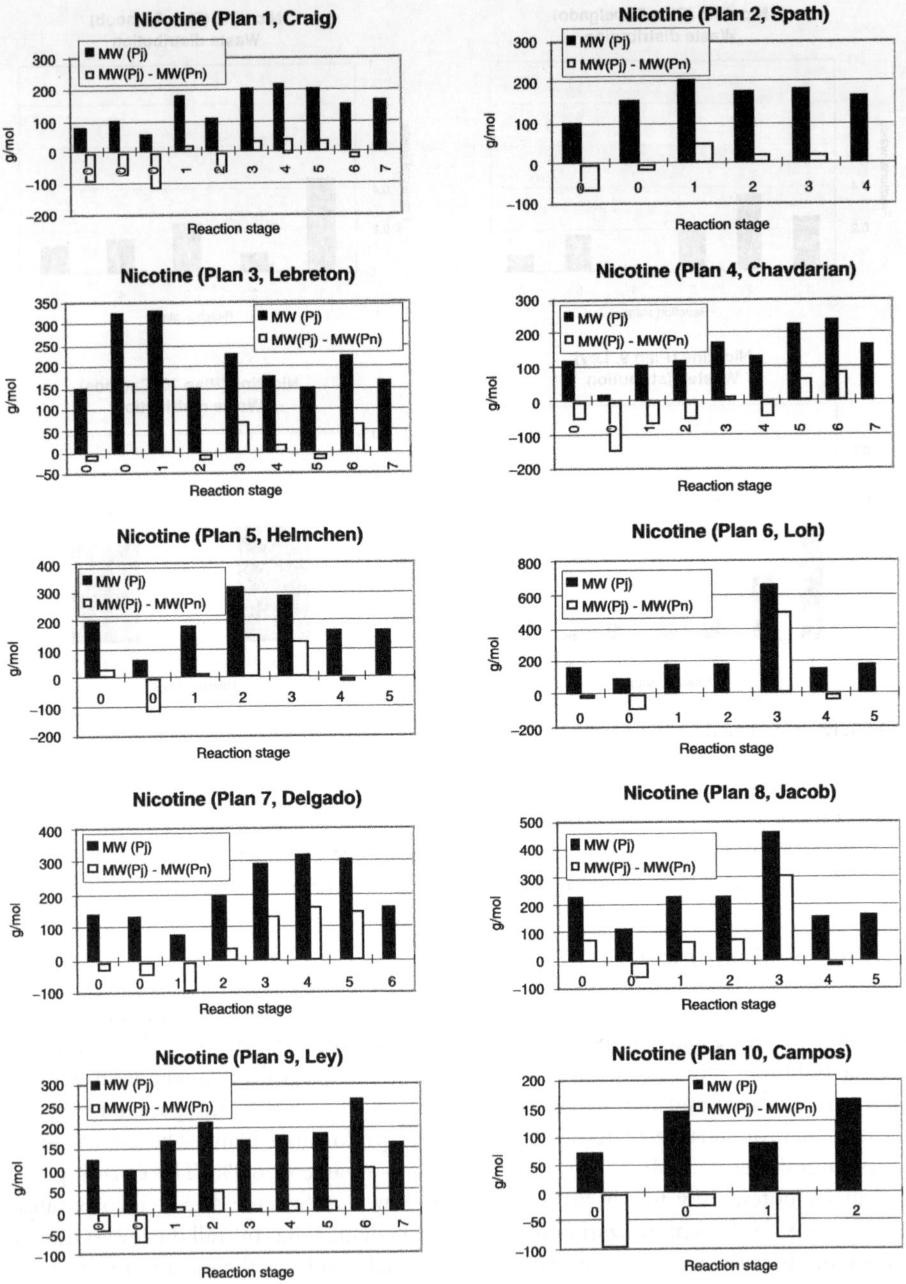

Figure 4.33 Histograms showing molecular weight distribution profiles as a function of reaction stage for nicotine syntheses.

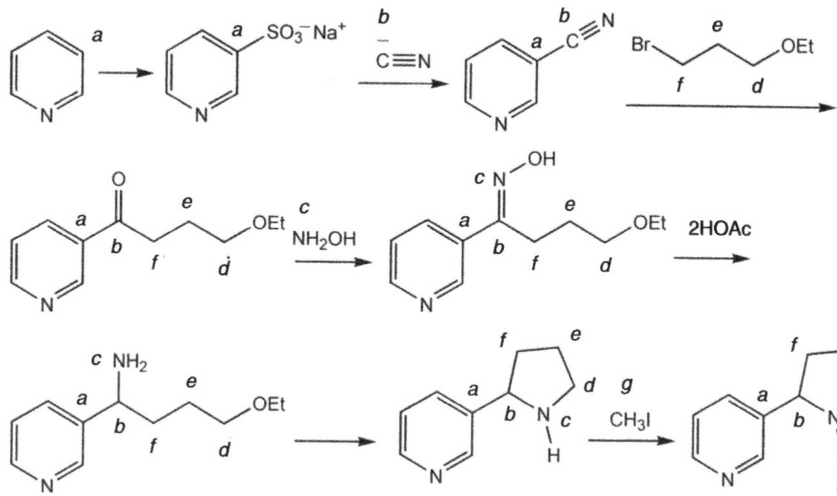

Scheme 4.10 Craig synthesis of nicotine showing key atoms for hypsicity (oxidation level) analysis.

Table 4.8 Oxidation number analysis of Craig plan for the synthesis of nicotine.

Reaction stage	Atom *a*	Atom *b*	Atom *c*	Atom *d*	Atom *e*	Atom *f*	Atom *g*	Sum of oxidation number changes for row
7	0	0	−3	−1	−2	−2	−2	0
6	0	0	−3	−1	−2	−2	−2	0
5	0	0	−3	−1	−2	−2		0
4	0	2	−1	−1	−2	−2		4
3	0	2	−1	−1	−2	−2		4
2	0	3		−1	−2	−1		4
1	1	2						3
0	−1							−1
								14
Sum of oxidation number changes for column	0	9	4	0	0	1	0	14

produced in step 3 from the endo into exo form adds an extra 1.7% to the overall kernel RME value for the production of product with the correct stereochemistry. Figure 4.41 shows the accompanying synthesis trees for the Carroll plan with and without this base catalysed epimerization step. Note that the overall AE for each tree is identical at 13.3%. Addition of a horizontal line in the tree diagram does not change the value of the ordinate for P and hence its degree of asymmetry, but as expected it results in a decrease in the degree of convergence for the plan. The major weakness of the Carroll plan that brings down the

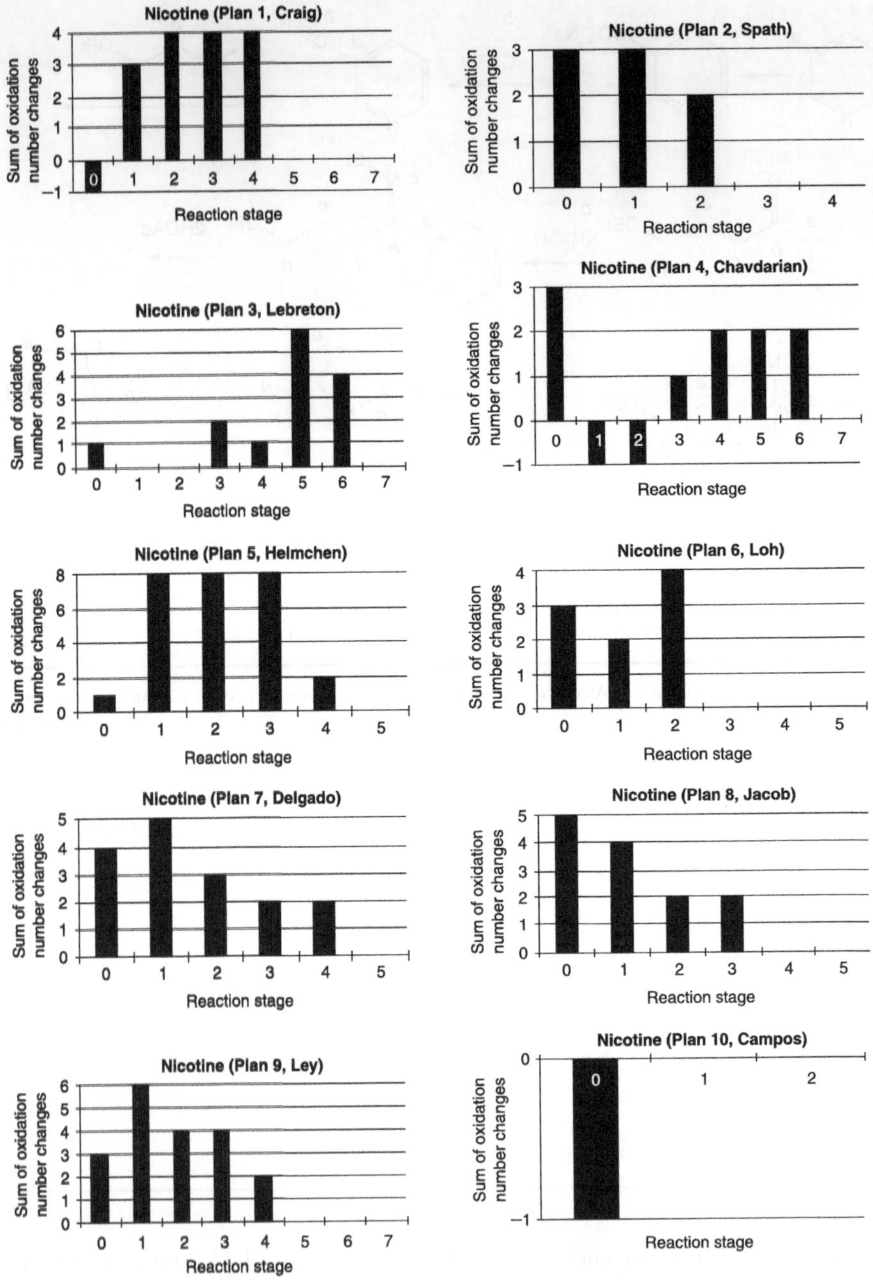

Figure 4.34 Histograms showing oxidation level difference distribution profiles as a function of reaction stage for nicotine syntheses.

Table 4.9 Summary of reaction metrics and synthesis tree parameters for ascorbic acid (vitamin C) synthesis plans ranked according to overall kernel (maximum) RME[a].

Plan	N	M	I	δ	β	μ_1	% AE	% $\langle \varepsilon_T \rangle$	% RME
Reichstein (1934)	8	8	12	0.377	0.749	+25.11	23.1	11.2	5.7
Haworth (1933)	12	12	27	0.412	0.895	+42.12	5.1	0.3	0.06

[a] See Appendix for definitions of terms.

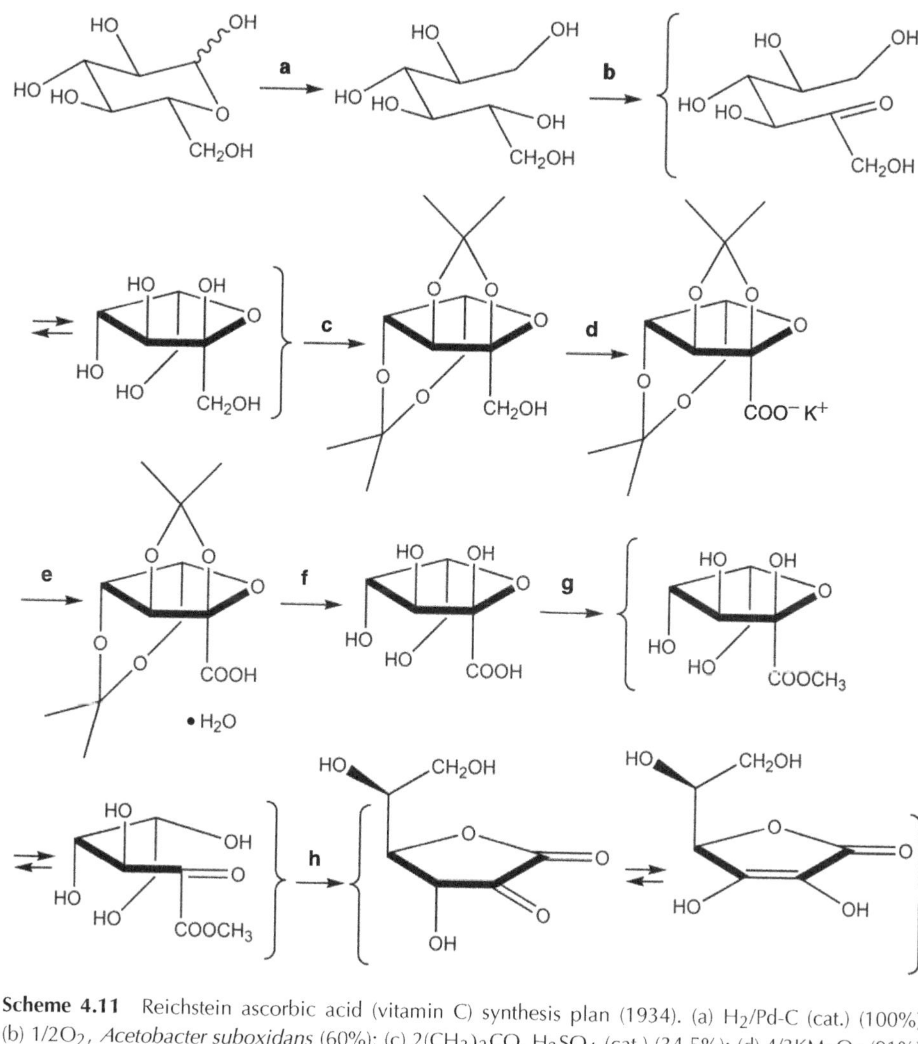

Scheme 4.11 Reichstein ascorbic acid (vitamin C) synthesis plan (1934). (a) H_2/Pd-C (cat.) (100%); (b) $1/2O_2$, *Acetobacter suboxidans* (60%); (c) $2(CH_3)_2CO$, H_2SO_4 (cat.) (34.5%); (d) $4/3KMnO_4$ (91%); (e) H_2O, HCl (90%); (f) $2H_2O$, H_2SO_4 (cat.) (82%); (g) $1/2(CH_3O)_2SO_2$, (84%); (h) Na, CH_3OH, then HCl (96%).

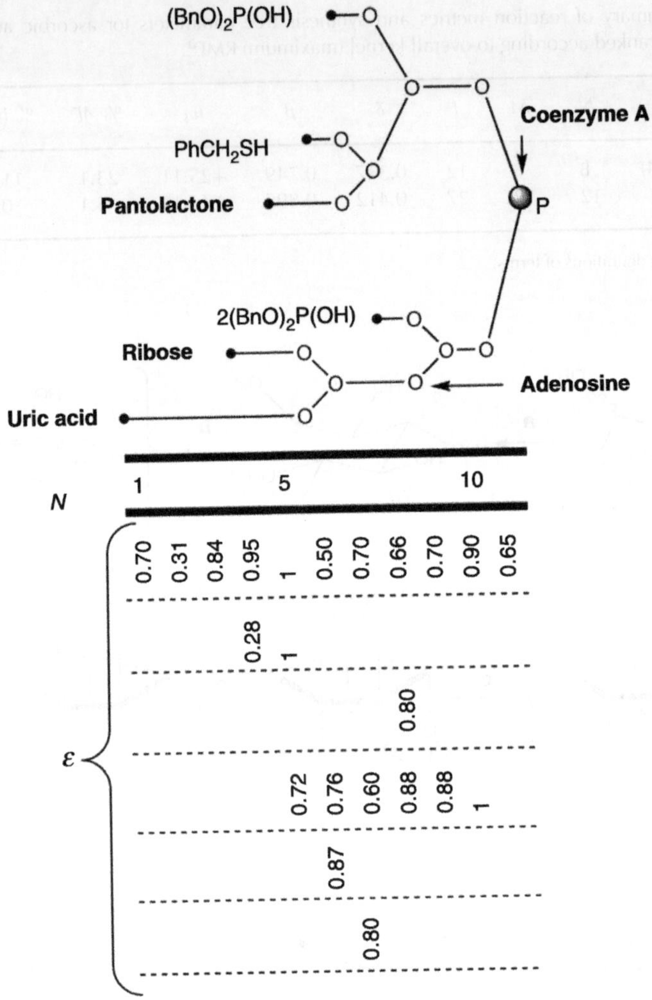

Figure 4.35 Reduced synthesis tree for Khorana–Todd synthesis of co-enzyme A. Step counts are shown in parentheses.

Table 4.10 Summary of reaction metrics and synthesis tree parameters for coenzyme-A synthesis plans ranked according to overall kernel (maximum) RME[a].

Plan	N	M	I	δ	β	μ_1	% AE	% $\langle \varepsilon_T \rangle$	% RME
Khorana-Todd (1961)	11	22	45	0.593	0.712	−1383.72	13.1	1.6	1.1

[a] See Appendix for definitions of terms.

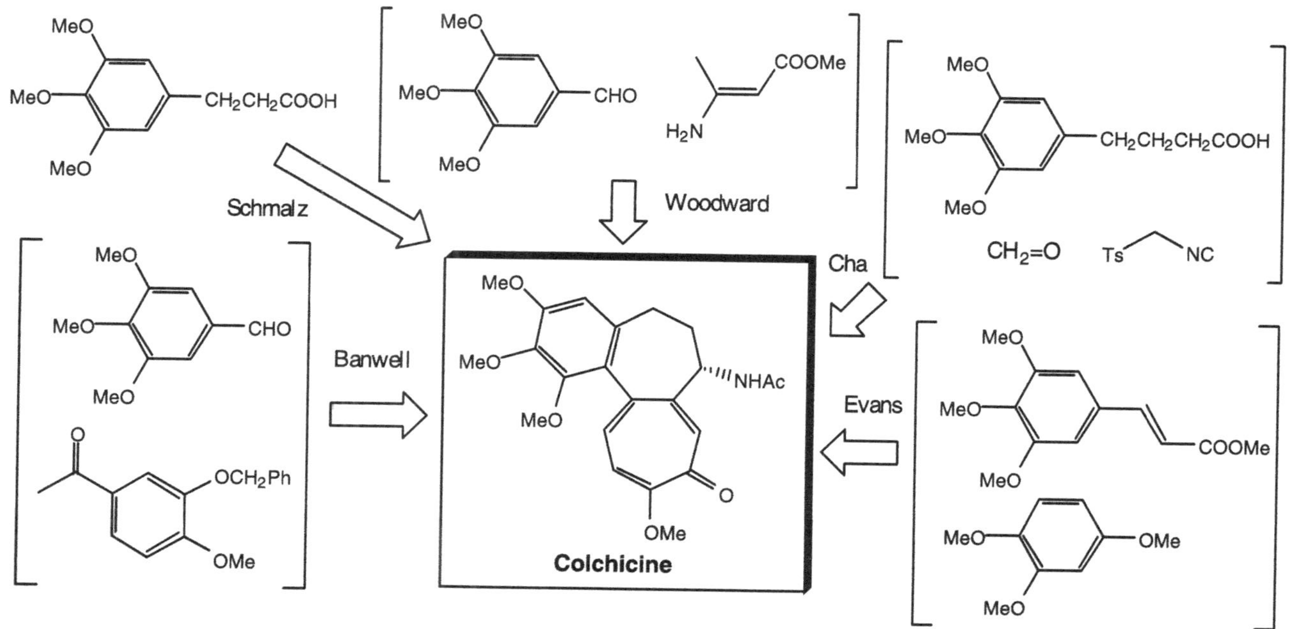

Figure 4.36 Synthesis map showing starting materials used for the synthesis of colchicine.

Table 4.11 Summary of reaction metrics and synthesis tree parameters for colchicine synthesis plans ranked according to overall kernel (maximum) RME[a].

Plan	*N*	*M*	*I*	δ	β	μ_1	% *AE*	% $\langle\varepsilon_T\rangle$	% *RME*
Evans (1984)	**12**	16	**23**	0.391	**0.867**	−147.54	**17.0**	**1.8**	**0.6**
Cha (1998)	17	17	28	0.355	0.904	−35.89	11.7	0.8	0.4
Banwell (1996)	15	**15**	25	0.361	0.888	−33.50	11.2	0.9	0.2
Schmalz (2005)	16	18	37	**0.402**	0.929	−54.82	9.4	0.9	0.2
Toromanoff (1965)	23	26	34	0.332	0.912	−127.40	11.9	0.1	0.03
van Tamelen (1959)	19	22	29	0.342	0.897	−167.20	13.4	0.008	0.005
Scott (1965)	16	19	22	0.328	0.880	**−196.48**	12.4	<2.6	<0.6
Eschenmoser (1959)	25	28	35	0.322	0.915	−114.30	11.7	<0.006	<0.003
Woodward (1963)	26	28	41	0.344	0.904	−50.09	9.1	b	b

[a] See Appendix for definitions of terms.
[b] No reaction yields given for synthesis plan in original paper (see reference [173]).

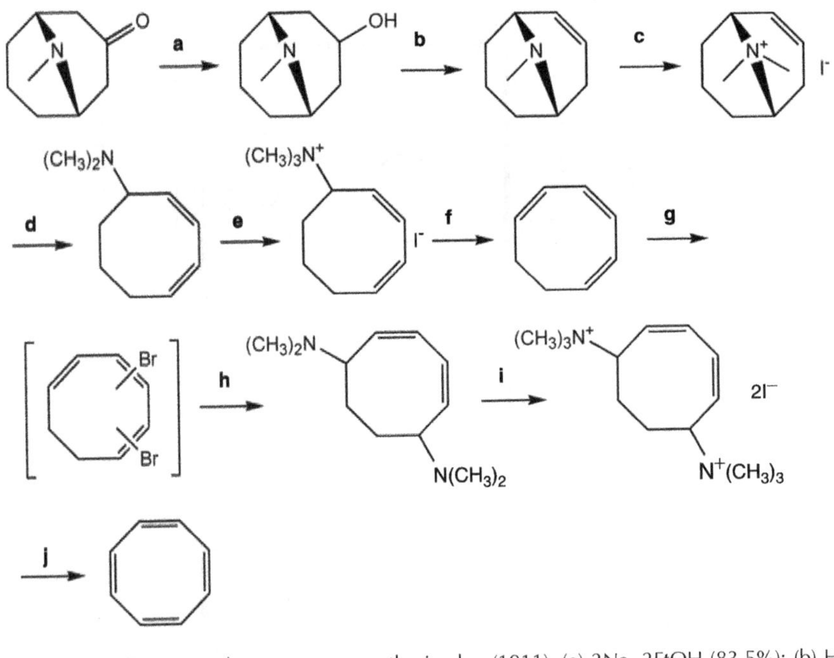

Scheme 4.12 Willstätter cyclooctatetraene synthesis plan (1911). (a) 2Na, 2EtOH (83.5%); (b) H_2SO_4 (cat.); (c) CH_3I (96%); (d) $1/2Ag_2O$, heat (83%); (e) CH_3I, (94.5%); (f) $1/2Ag_2O$, heat (70%); (g) Br_2; (h) $4(CH_3)_2NH$ (47%, 2 steps); (i) $2CH_3I$ (97.5%); (j) Ag_2O (8.6%).

overall kernel RME is the reduction step in the second stage which involves producing nickel boride catalyst *in situ* and then destroying it upon reaction completion. The generation and destruction components when added to the net hydrogenation transformation result in a low atom economy of 34% for this step. If regioselective hydrogenation could be carried out using hydrogen gas with a recoverable catalyst this would yield an atom economy of 100%

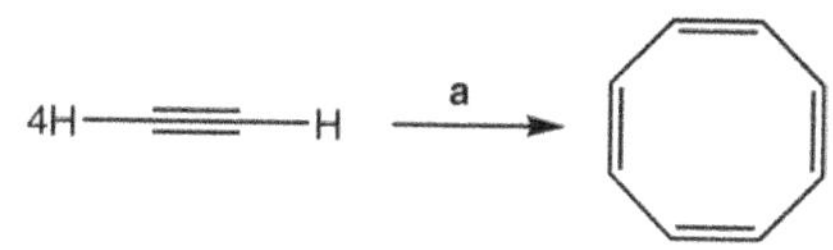

Scheme 4.13 Reppe cyclooctatetraene synthesis plan (1948). (a) $Ni(CN)_2$ (cat.), 15 atm, 80–120°C (90%).

Table 4.12 Summary of reaction metrics and synthesis tree parameters for cyclooctatetraene synthesis plans ranked according to overall kernel (maximum) RME[a].

Plan	N	M	I	δ	β	μ_1	% AE	% $\langle\varepsilon_T\rangle$	% RME
Willstätter (1911)	10	10	11	0.303	0.800	+105.44	6.3	1.5	0.2
Reppe (1948)	1	1	1	1	0	0	100	90.0	90.0
Reppe (1948)	1	1	4	1	0	−156.00	100	90.0	90.0

[a] See Appendix for definitions of terms.

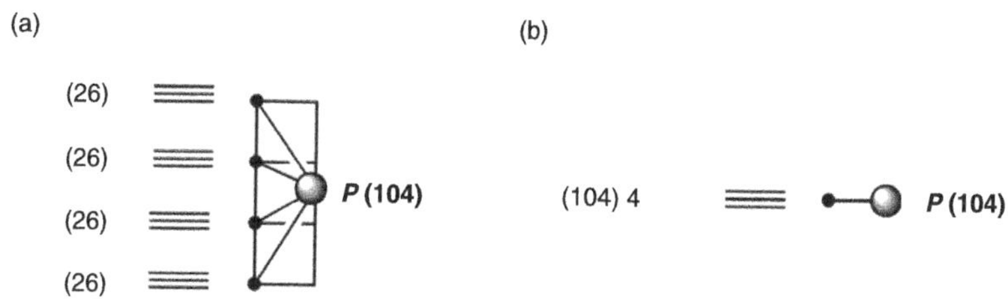

Figure 4.37 Synthesis trees for Reppe synthesis of cyclooctatetraene: (a) treating each molecule of acetylene as a separate input; (b) treating all 4 acetylene molecules as a single input.

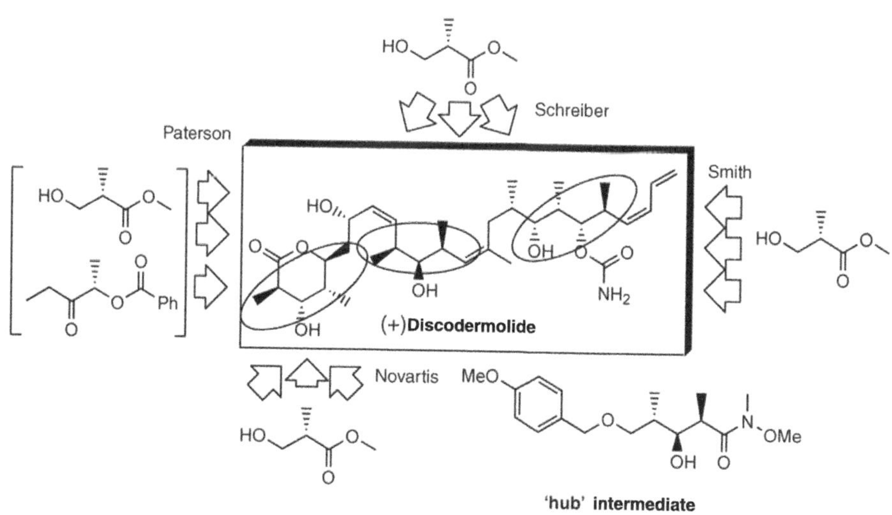

Figure 4.38 Synthesis map showing starting materials used for the synthesis of discodermolide.

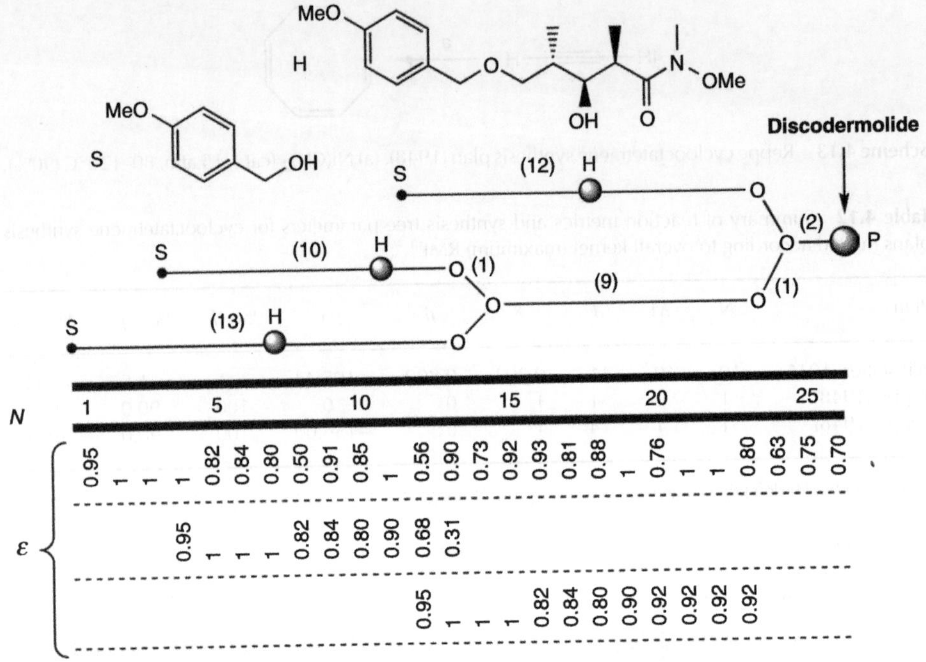

Figure 4.39 Reduced synthesis tree for Novartis synthesis of discodermolide. Step counts are shown within parentheses.

Table 4.13 Summary of reaction metrics and synthesis tree parameters for discodermolide synthesis plans ranked according to overall kernel (maximum) RME[a].

Plan	N	M	I	δ	β	μ_1	% AE	% $\langle\varepsilon_T\rangle$	% RME
Smith, 4th generation (2005)	21	**26**	**49**	**0.395**	**0.947**	−232.00	**8.1**	10.8	**2.0**
Paterson (2004)	22	41	89	0.435	0.988	−290.15	5.8	**15.4**	1.7
Mickel (Novartis) (2004)[b]	**19**	27	56	0.424	0.951	−99.90	6.4	1.7	0.3
Mickel (Novartis) (2004)	26	48	103	0.446	0.967	−347.33	4.3	0.9	0.1
Smith, 3rd generation (2003)	24	44	90	0.439	0.969	−283.20	5.1	1.6	0.2
Schreiber (1996)	25	49	112	0.449	0.979	−352.70	4.3	0.9	0.1

[a] See Appendix for definitions of terms.
[b] Analysis repeated with the assumption that the hub intermediate H (see Figure 4.36) is an available starting material.

for this step and thus make this plan the leader in RME performance when the recycling loop is taken into account. Without the recycling loop the overall kernel RME would be 5.2% and with one recycle included it would be 7.7% assuming that the yield performance for the reduction step remains at 96%. The Corey plan has the highest overall yield and the Regan plan has both the highest atom economy and the greatest degree of building up in the synthesis.

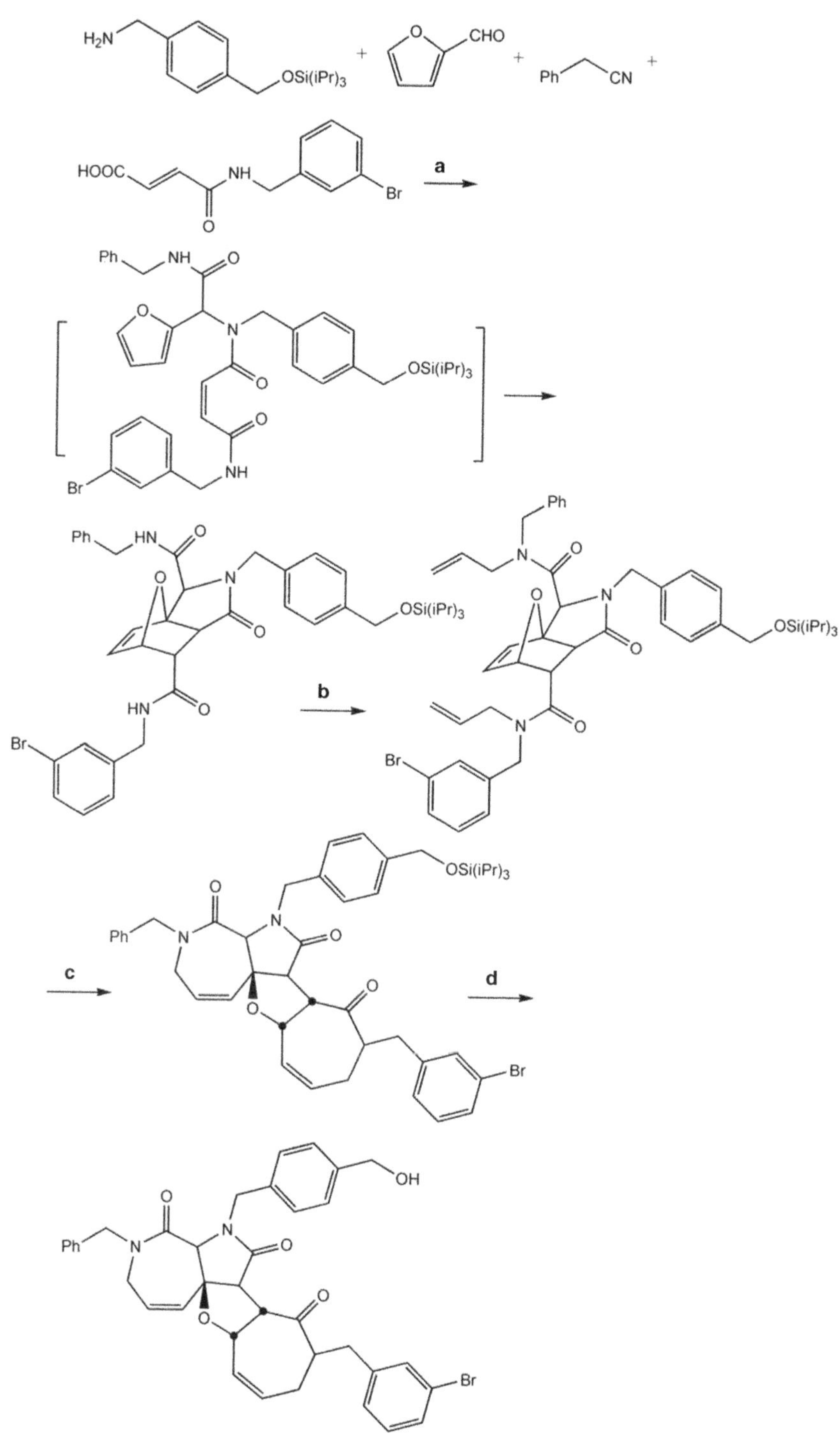

Scheme 4.14 Schreiber diversity-oriented synthesis plan (2000). (a) MeOH/THF, heat (67%); (b) $2KN(SiMe_3)_2$, 2 $CH_2{=}CH_2Br$ (89%); (c) Grubbs catalyst (69%); (d) HF, pyridine (95%).

Table 4.14 Summary of reaction metrics and synthesis tree parameters for diversity-oriented synthesis (DOS) plan example ranked according to overall kernel (maximum) *RME*[a].

Plan	N	M	I	δ	β	μ_1	% *AE*	% $\langle\varepsilon_T\rangle$	% *RME*
Schreiber (2000)	4	4	7	0.467	0.604	−289.94	46.2	39.1	21.3

[a] See Appendix for definitions of terms.

Table 4.15 Summary of reaction metrics and synthesis tree parameters for epibatidine synthesis plans ranked according to overall kernel (maximum) *RME*[a].

Plan	N	M	I	δ	β	μ_1	% *AE*	% $\langle\varepsilon_T\rangle$	% *RME*
Giblin (1997)	6	6	12	0.438	0.780	+83.19	19.9	24.2	**6.6**
Corey (1993)	9	9	13	0.355	0.830	+84.63	9.7	**35.5**	4.6
Hoashi (2004)	10	10	14	0.351	0.795	+50.88	12.5	18.6	4.4
Trudell (1996)	9	9	15	0.362	0.850	+67.41	13.5	17.6	4.3
Carroll (1995)	**4**	**4**	**9**	**0.500**	**0.706**	+120.31	13.3	15.7	2.5
				0.441[b]	0.706[b]	+116.9[b]	13.3[b]	14.5[b]	2.3[b]
Aggarwal (2005)	7	7	11	0.388	0.767	+112.28	13.1	4.3	2.0
Olivo (1999)	10	10	14	0.346	0.833	+43.57	13.2	8.1	2.0
Bai (1996)	9	9	17	0.404	0.831	+39.92	9.1	10.4	1.9
Regan (1993)	**5**	7	11	0.486	0.692	**+9.23**	**22.6**	3.7	1.6
Evans (2001)	13	14	21	0.357	0.892	+162.93	6.7	11.7	1.2
Pandey (1998)	10	13	21	0.409	0.877	+59.46	8.2	8.9	1.1
Shen (1993)	6	6	10	0.407	0.762	+54.43	20.6	2.5	0.9
Szántay (1996)	10	10	11	0.310	0.710	+56.95	11.3	2.6	0.9
Loh (2005)	12	13	20	0.367	0.877	+64.60	6.5	7.0	0.8
Fletcher (1994)	13	13	18	0.333	0.873	+41.78	11.6	1.2	0.3
Broka (1993)	20	20	33	0.351	0.918	+122.79	4.2	1.4	0.3
Kibayashi (1998)	11	12	23	0.412	0.903	+111.83	7.4	1.0	0.3
Kosugi (1997)	16	16	30	0.376	0.908	+87.50	6.4	<12.0	<1.2

[a] See Appendix for definitions of terms.
[b] Takes into account racemization recycling step.

4.3.7.9 Hemibrevetoxin B intermediate[223,224]

Two syntheses for a key intermediate in the total synthesis of hemibrevetoxin B as shown in Figure 4.42 are compared in Table 4.16. The Nakata plan is a long linear sequence involving a number of protection–deprotection steps. The Nelson plan, on the other hand, though it is linear, is based on a strategy that exploits the inherent symmetry in the molecule. This redesign shortens the step count considerably from 29 to 8 and results in a seven-fold increase in overall kernel RME. This two-directional approach is accounted for in the corresponding tree diagram by doubling the stoichiometric coefficients for all input materials from step 1 to step 6 leading to a highly symmetrical intermediate product. The

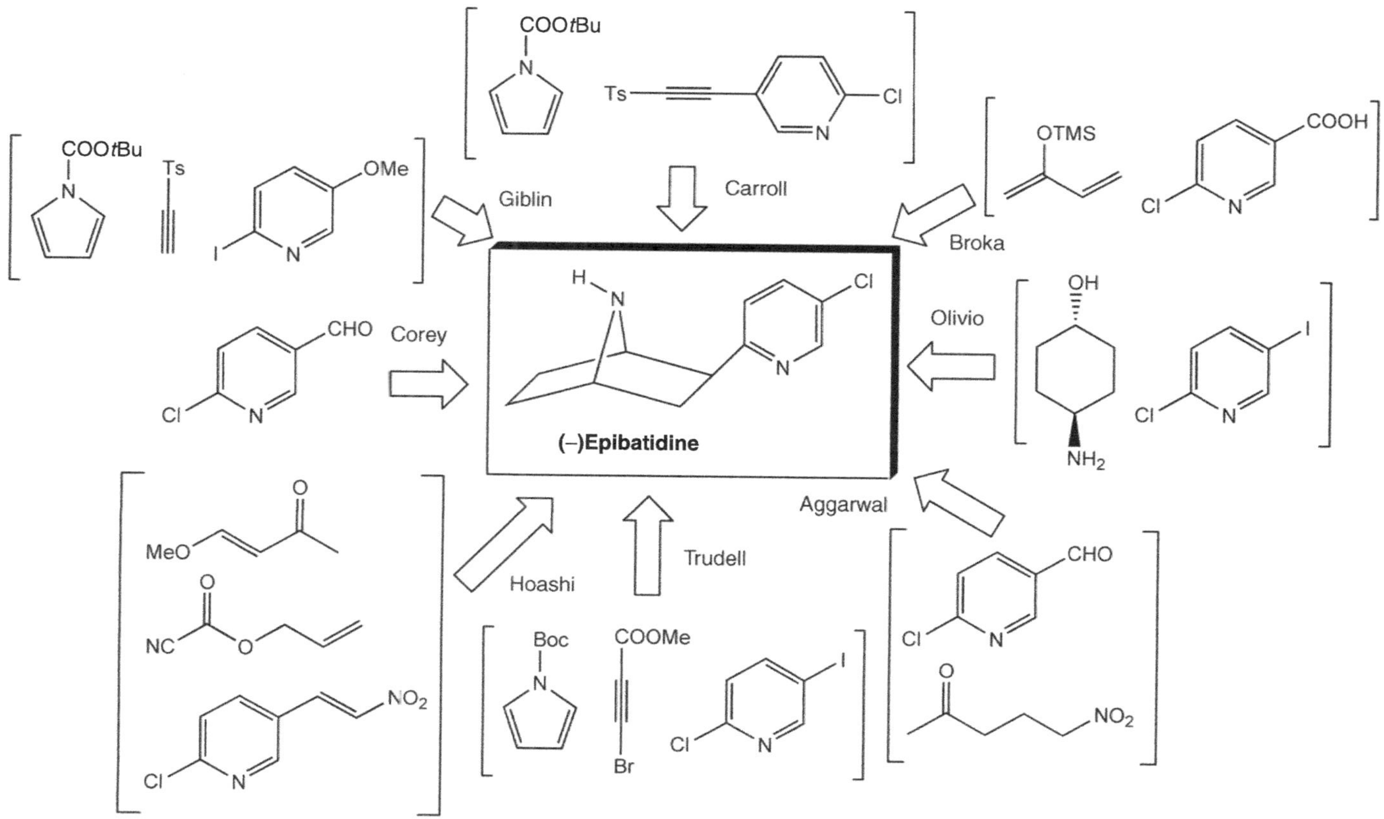

Figure 4.40 Synthesis map showing starting materials used for the synthesis of (−)-epibatidine.

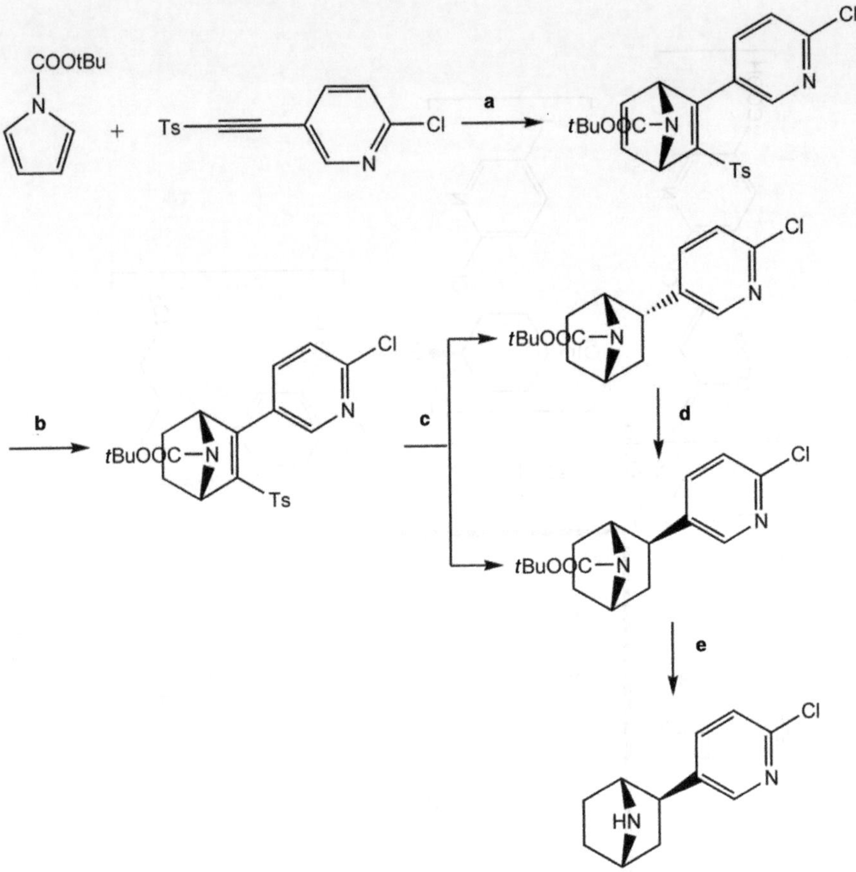

Scheme 4.15 Carroll (−)-epibatidine synthesis plan (1995). (a) Heat (78%); (b) EtOH, 2[Ni(OAc)$_2$ · 4H$_2$O], 4NaBH$_4$, then 4HCl, 4H$_2$O (96%); (c) 5Na, 4MeOH (65%); (d) heat, KO*t*Bu, *t*BuOH (46%); (e) CH$_2$Cl$_2$, CF$_3$COOH (cat.), then H$_2$O (97%).

next two steps involve selective hydrolysis of one of the two epoxide moieties followed by its protection as a dimethylketal.

4.3.7.10 *Ibogamine*[225–228]

Figure 4.43 shows a synthesis map for various routes to the alkaloid ibogamine and Scheme 4.16 highlights the Trost plan to this target which turns out to be the most material efficient to date according to the results reported in Table 4.17. This highly atom economical ($AE = 49\%$) and short synthesis ($N = 4$) is designed using a novel silver–palladium catalysed olefin arylation reaction in the final cyclization step. This plan illustrates how the discovery of a new kind of reaction can greatly fast track the synthesis to a complex target. Though it is older than the other two reported in the literature it stands head and shoulders above the newer plans in all of the metrics parameters. Its only drawback is the poor yields for the final two steps at 45% each.

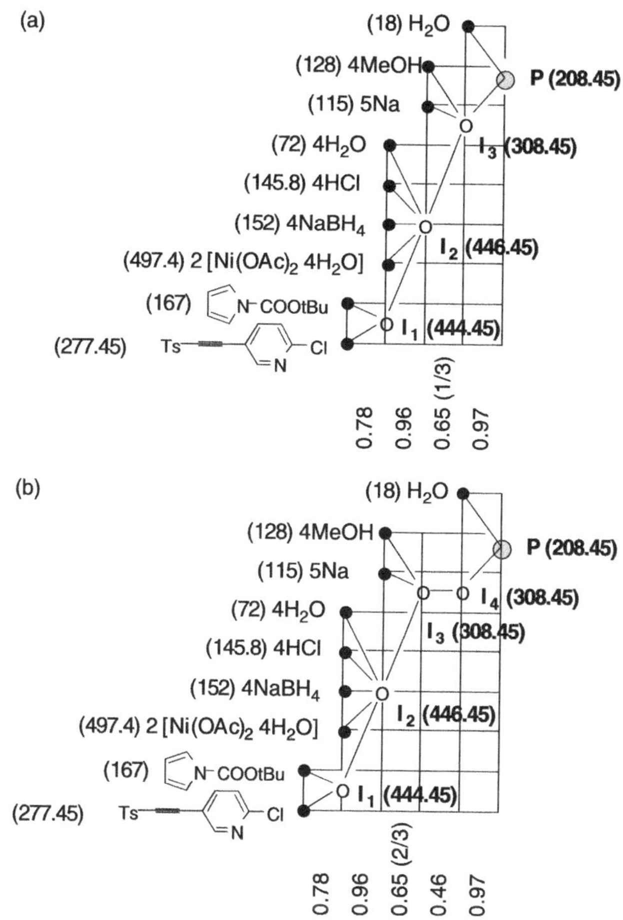

Figure 4.41 Synthesis trees for synthesis of (−)-epibatidine by the Carroll method: (a) direct synthesis; (b) includes epimerization recycling step (see Scheme 4.15).

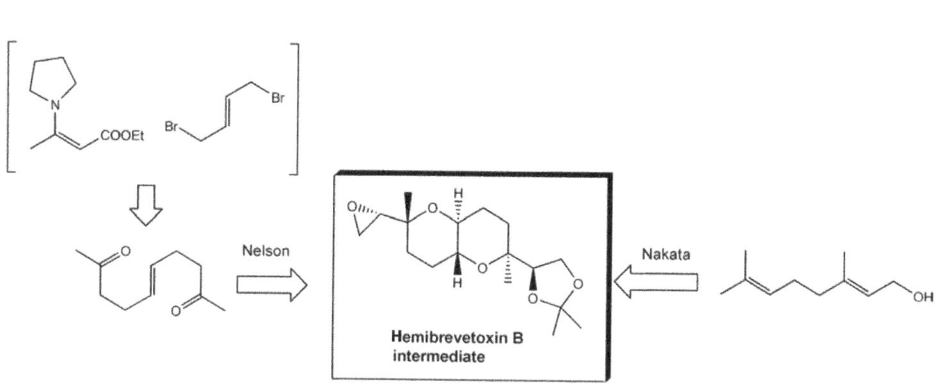

Figure 4.42 Synthesis map showing starting materials used for the synthesis of hemibrevetoxin B intermediate.

Table 4.16 Summary of reaction metrics and synthesis tree parameters for synthesis plan to hemibrevetoxin B intermediate ranked according to overall kernel (maximum) *RME*[a].

Plan	*N*	*M*	*I*	δ	β	μ_1	% *AE*	% $\langle \varepsilon_T \rangle$	% *RME*
Nelson (2003)	8	8	14	0.392	0.820	−128.91	12.8	31.8	6.0
Nakata (1996)	29	29	41	0.319	0.943	−2.17	6.8	4.3	0.8

[a] See Appendix for definitions of terms.

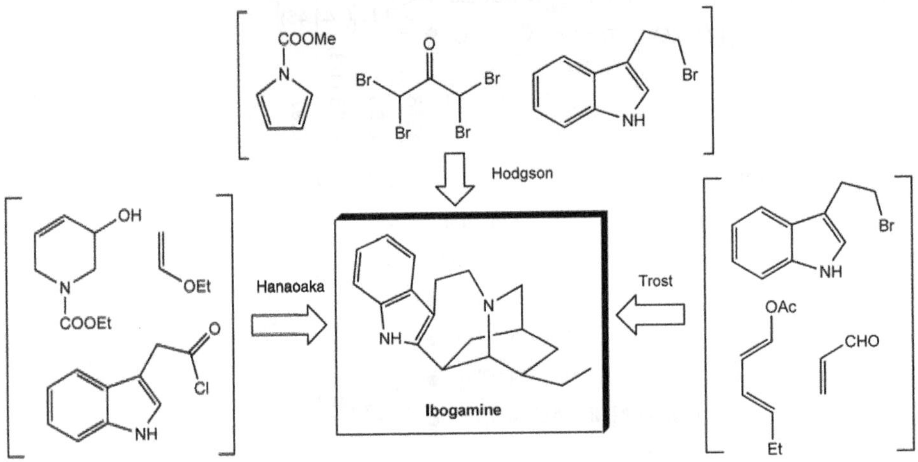

Figure 4.43 Synthesis map showing starting materials used for the synthesis of ibogamine.

4.3.7.11 *Lysergic acid*[229–239]

Table 4.18 summarizes the metrics parameters for 9 total synthesis plans to lysergic acid. Figures 4.44 and 4.45 show the synthesis map for various starting materials to this target and the synthesis tree for the most material efficient Hendrickson plan so far achieved. This plan also has the lowest molecular weight first moment indicating that the synthesis has the highest degree of building up occurring. However, complete optimization has yet to be achieved since other plans have merit. The Rebek plan has the least number of input materials and the highest atom economy. The Caccia plan has the highest degree of convergence and the original Woodward plan is the most symmetric. Though the Hendrickson plan has the highest overall kernel RME the separation between the plans is small. Any further improvements on the total synthesis of this compound will need to focus on bringing all of the metrics parameters working together in the same direction preferably by some combination of tandem, domino, or multicomponent strategies.

4.3.7.12 *Morphine*[240–259]

Table 4.19 summarizes the metrics parameters for 10 total synthesis plans to morphine. Figure 4.46 shows a synthesis map of various starting materials for some of the better performing plans to this target. The Rice plan remains the most material efficient plan

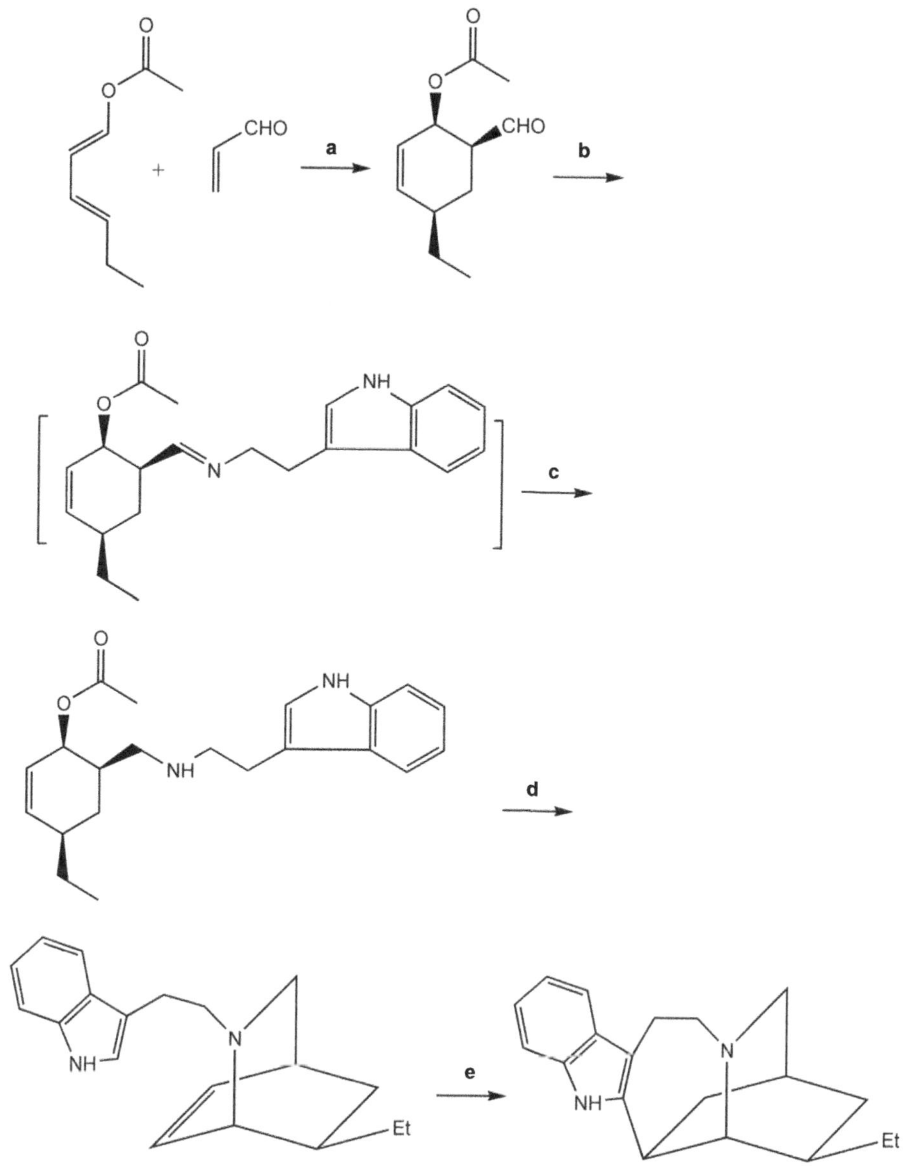

Scheme 4.16 Trost (+)-ibogamine synthesis plan (1978). (a) Heat (90%); (b) tryptamine; (c) 1/4$NaBH_4$, (93%, 2 steps); (d) $Pd(PPh_3)_4$ (cat.), (45%); (e) $PdCl_2(CH_3CN)_2$ (cat.), $AgBF_4$, then $NaBH_4$ (45%).

despite recent reports of newer syntheses. The data in Table 4.19 once again show that complete optimization has yet to be achieved for this challenging and important target molecule. The Trost plan has the only negative valued molecular weight first moment and reaches the target in the fewest reaction steps and reaction stages. The Overman plan has the highest degree of convergence.

Table 4.17 Summary of reaction metrics and synthesis tree parameters for ibogamine synthesis plans ranked according to overall kernel (maximum) *RME*[a].

Plan	*N*	*M*	*I*	δ	β	μ_1	% *AE*	% $\langle\varepsilon_T\rangle$	% *RME*
Trost (1978)	**4**	**4**	7	**0.474**	**0.557**	**−62.82**	**49.3**	**16.9**	**10.0**
Hodgson (2005)	11	11	25	0.416	0.893	−15.56	7.5	2.0	0.4
Hanaoka (1981)	19	19	32	0.355	0.919	+11.60	10.1	0.5	0.2

[a] See Appendix for definitions of terms.

Table 4.18 Summary of reaction metrics and synthesis tree parameters for lysergic acid synthesis plans ranked according to overall kernel (maximum) *RME*[a].

Plan	*N*	*M*	*I*	δ	β	μ_1	% *AE*	% $\langle\varepsilon_T\rangle$	% *RME*
Hendrickson (2004)	**10**	12	21	0.411	0.868	**−68.6**	12.0	**5.4**	**1.0**
Rebek (1984)	**10**	**10**	**15**	0.356	0.851	+32.2	**14.5**	1.2	0.6
Oppolzer (1981)	13	14	22	0.369	0.874	+41.2	12.2	3.3	0.6
Kurihara (1987)	16	16	27	0.356	0.923	+45.0	10.8	2.2	0.5
Ramage (1981)	17	17	29	0.360	0.910	+59.3	8.3	1.5	0.3
Cacchia (1988)	**10**	14	22	**0.428**	0.862	−6.7	11.1	1.2	0.2
Woodward (1956)	15	17	27	0.389	**0.815**	−19.7	12.5	0.7	0.2
Szántay (2004)	15	17	27	0.374	0.891	+3.8	10.5	0.5	0.2
Ninomiya (1985)	18	18	34	0.370	0.934	+3.7	9.2	0.1	0.04

[a] See Appendix for definitions of terms.

4.3.7.13 *Oxomaritidine*[260,261]

Ley and co-workers have introduced a new paradigm for molecular assembly which uses a flow process with polymer supported reagents. This was applied to the total synthesis of the *Amaryllidaceae* alkaloid (±)-oxomaritidine. Figure 4.47 shows two ways of representing the corresponding synthesis tree diagram. The left one assumes that intermediates are actually isolated at each step and the right one depicts the complete telescoped sequence in a single multicomponent stage. The overall kernel RME metrics results shown in Table 4.20 represent maximum values since the molecular weight of the polymer support is ignored. These estimates are valid so long as the solid support is recovered and not destined for waste.

4.3.7.14 *Paclitaxel (taxol) synthesis*[262–274] *and semi-synthesis*[275–284]

The total synthesis of paclitaxel (taxol), the cytotoxic constituent found in the bark of the Pacific Yew tree, *Taxus brevifolia*, is arguably the most significant triumph of modern organic synthesis. The structure of this molecule and its analogues has attracted the attention of most of the top synthetic groups in the world. Figure 4.48 shows a synthesis map of various sets of starting materials to this target. Figure 4.49 shows the reduced tree diagram for the

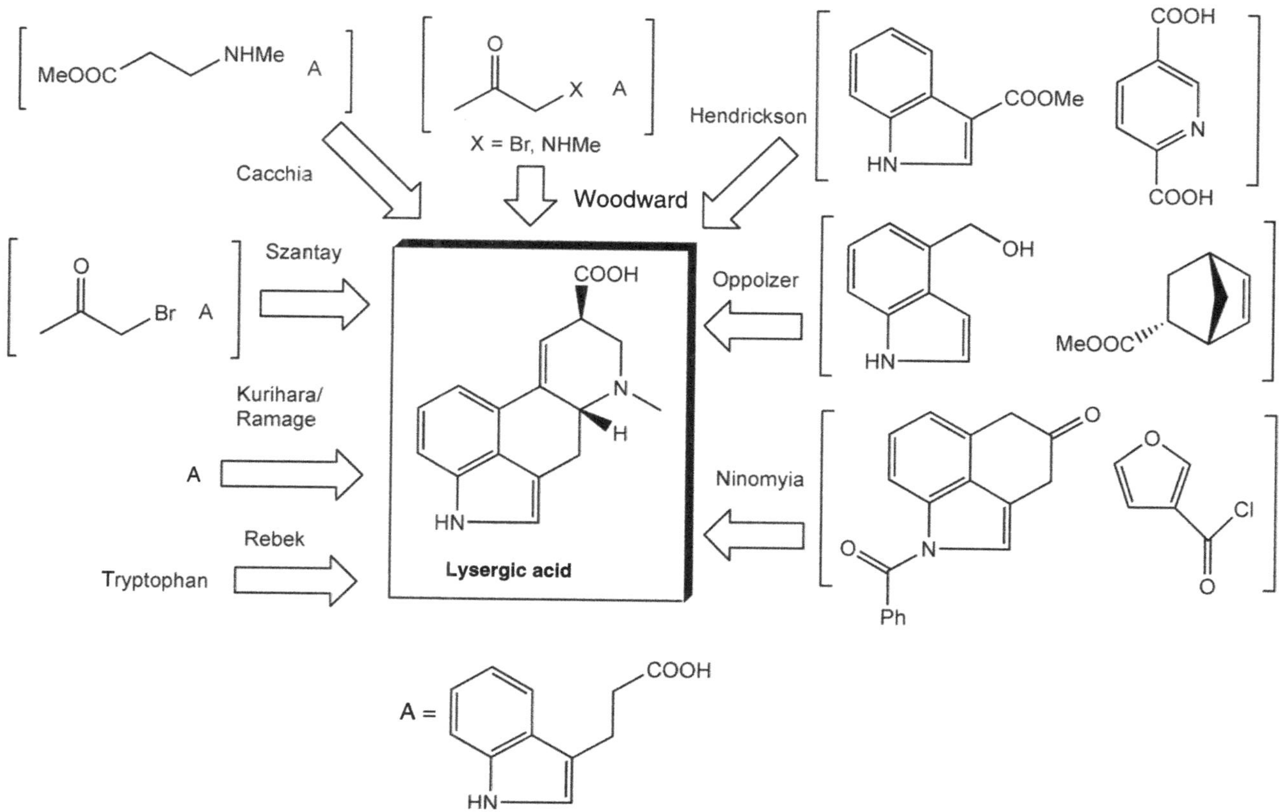

Figure 4.44 Synthesis map showing starting materials used for the synthesis of lysergic acid.

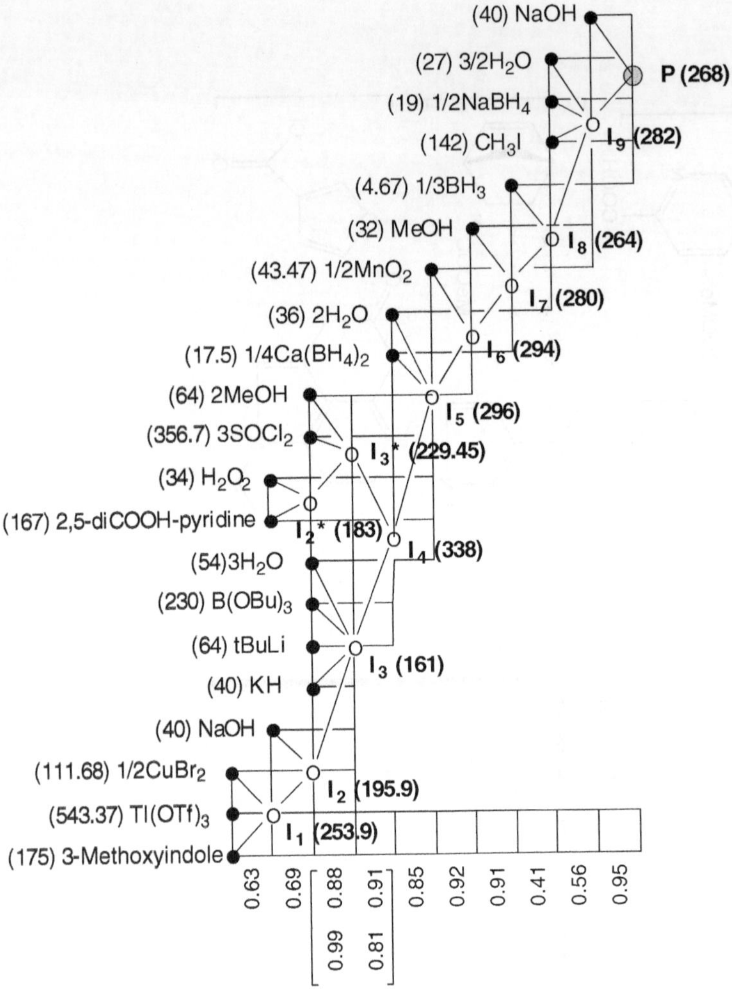

Figure 4.45 Synthesis tree for synthesis of lysergic acid by the Hendrickson method.

Holton plan starting from homocamphor. Table 4.21 summarizes the metrics parameters for the total synthesis routes reported in the literature. The Nicolaou route has most of the best performing metrics except for the overall kernel RME. The Holton plan beginning with homocamphor nudges ahead because two of the steps in the Nicolaou plan in stages 20 and 25 have poor reaction yields of 25 and 42% respectively. The worst two steps in the Holton plan occurring in stages 7 and 8 have reaction yields of 41 and 52% respectively. Slightly lower values for overall yield and overall kernel RME are found for the Holton plan if the synthesis is evaluated from camphor as starting material. Table 4.22 summarizes the metrics results for semi-syntheses of taxol beginning with the key starting material 10-deacetylbaccatin III obtained from the dried needles of *Taxus baccata*. The dramatic reduction in number of steps and increase in overall yields propel the overall kernel RMEs to values two to three orders of magnitude better than those for the total syntheses. What is

Table 4.19 Summary of reaction metrics and synthesis tree parameters for morphine synthesis plans ranked according to overall kernel (maximum) *RME*[a].

Plan	N	M	I	δ	β	μ_1	% AE	% $\langle\varepsilon_T\rangle$	% RME
Rice (1980)	18	18	**26**	0.333	**0.894**	+38.39	**15.1**	**10.1**	**3.3**
Trost (2005)	**16**	**17**	28	0.367	0.899	**−0.52**	7.0	2.8	0.7
Mulzer (1998)	22	22	42	0.370	0.935	+39.66	8.0	2.7	0.6
Overman (1993)	19	24	41	**0.389**	0.933	+53.65	7.1	3.9	0.4
Parker (2006)	21	21	39	0.367	0.930	+71.16	7.1	2.4	0.4
Fukuyama (2006)	26	32	58	0.386	0.954	+87.97	3.9	1.4	0.3
White (1997)	30	32	39	0.306	0.931	+34.25	7.3	<1.6	<0.3
Ogasawara (2001)	30	30	53	0.353	0.949	+14.96	5.4	0.5	0.07
Taber (2002)	28	28	49	0.352	0.945	+67.00	4.9	0.2	0.05
Gates (1956)	26	26	47	0.357	0.950	+33.89	5.6	0.004	0.0008

[a] See Appendix for definitions of terms.

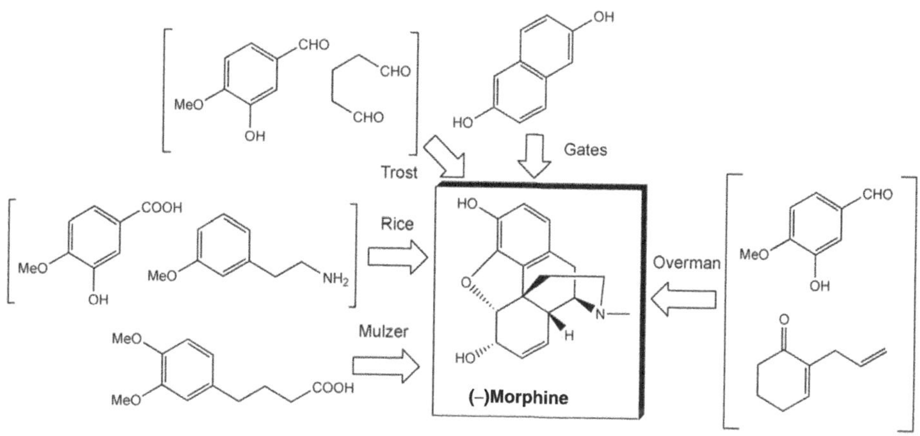

Figure 4.46 Synthesis map showing starting materials used for the synthesis of morphine.

interesting is that if the yield of 10-deacetylbaccatin III is taken into account in the metrics evaluation for each case, the overall kernel RMEs drop to values comparable to those of the total syntheses! This really speaks both to the triumph of the total syntheses efforts and the obvious advantages of the semi-syntheses. The Greene and Kingston plans are the leading material efficient semi-syntheses plans for taxol.

4.3.7.15 Papaverine[285–293]

The opiate alkaloid, papaverine, from *Papaver somniferum* is an anti-spasmodic, vasodilator, and smooth muscle relaxant. Its total synthesis has been studied since Pictet and Gams early work in 1909 and has since been followed up by various industrial syntheses up till the early 1950s using important industrial commodities as vanillin, acetovanillone, veratraldehyde (methylvanillin), and homoveratric acid as starting materials (see Figure 4.50). Table 4.23 summarizes the results of the five synthetic plans for this natural product. All are convergent

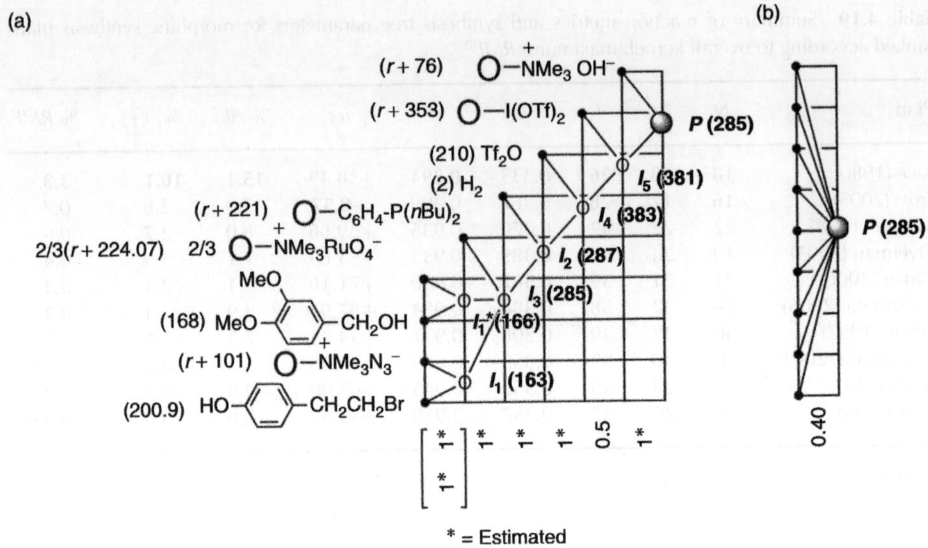

Figure 4.47 Synthesis trees for (±)-oxomaritidine by Ley method.

Table 4.20 Summary of reaction metrics and synthesis tree parameters for oxomaritidine synthesis plans ranked according to overall kernel (maximum) *RME*[a].

Plan	*N*	*M*	*I*	δ	β	μ_1	% *AE*	% $\langle \varepsilon_T \rangle$	% *RME*
Ley (2006)	6	7	9	0.389	0.740	−61.30	19.2	50.0	9.9
Ley (2006)	1	1	9	1	0	−437.55	19.2	40.0	7.7

[a] See Appendix for definitions of terms.

plans except for the linear Redel–Bouteville plan. The Kindler–Peschke–Pal plan is the most efficient material performer with the highest kernel RME of 15% and pseudo-overall yield of 55% in 8 stages, and greatest building up of 280 g/mol per reaction stage. It begins with very cheap starting materials and has very high reaction yields per step with no step below 79% yield. The runner-up in material performance is the Dean method with a kernel RME of 8% and pseudo-overall yield of 40% also in 8 stages. The Redel–Bouteville plan has the highest atom economy at 32% but since its pseudo-overall reaction yield is lower at 13% its resulting kernel RME drops to 4%. The Dean method utilizes an expensive and toxic reagent early in the second stage, thallium(III)nitrate, to transform an acetophenone to a methyl arylacetate in a redox reaction. In order for this method to be competitive with the Kindler–Peschke–Pal method it becomes imperative to either replace it with a cheaper and less hazardous reagent to effect the same unique transformation or, failing that, to find an appropriate cheap and efficient recycling reaction that recycles the thallium(I) byproduct back to thallium(III) reagent.

Tree diagrams for all five plans are shown in Figures 4.51–4.55. The tree analyses of the Kindler–Peschke–Pal and Dean methods highlight a useful material efficient synthetic

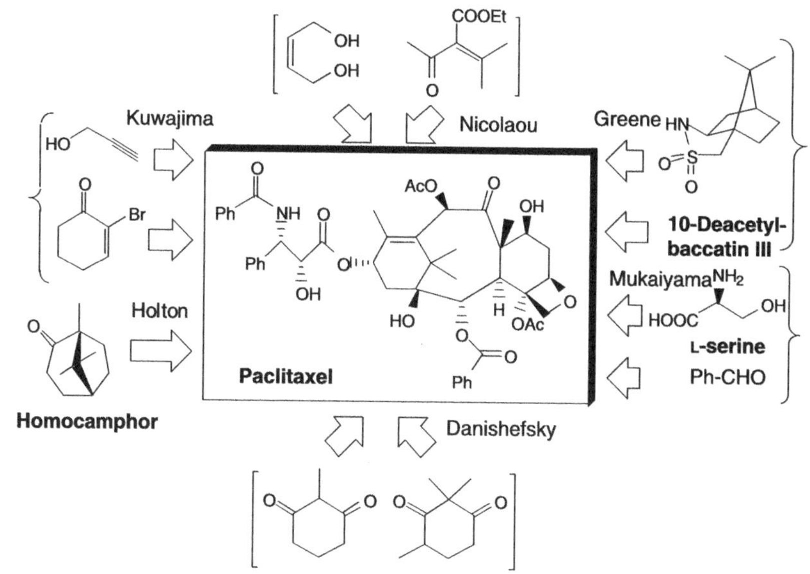

Figure 4.48 Synthesis map showing starting materials used for the synthesis of paclitaxel (taxol).

strategy used in both plans; namely, the synthesis of a key intermediate which is partitioned into two pathways which in turn converge at a later stage in the synthesis. As in the case of the Novartis synthesis of discodermolide these plans are reflexive. In the Kindler–Peschke–Pal plan the key intermediate is homoveratronitrile which is split into two paths, one part is reduced to the corresponding amine and the other part is hydrolysed to the corresponding acid. Both of these transformations occur in parallel steps in the 5th reaction stage. These products are then recombined in the 6th stage to produce homoveratrylhomoveratramide. From the synthesis tree it is possible to determine the partitioning ratio for the transformations homoveratronitrile to homoveratric acid and homoveratronitrile to homoveratramine to be 1:0.79 or 56% and 44%, respectively. In the Dean plan the key intermediate is methyl 3,4-dimethoxyphenylacetate. Rather than partitioning this material into two different paths where both parts are transformed to intermediates which are later recombined, one portion is saved and left untransformed until a later reaction stage. With this modification, about 64% of the arylacetate produced in the second stage is committed to the next steps leading to homoveratramine and 36% is saved for reaction with homoveratramine in the 6th reaction stage to produce homoveratrylhomoveratramide.

With respect to the tree parameters, as expected the only linear plan (Redel–Bouteville) has the lowest degree of convergence. The most material efficient Kindler–Peschke–Pal plan is also the most convergent with a degree of convergence of 45% and is the most symmetric. Interestingly, the original Pictet–Gams synthesis has the highest relative rate of convergence at 0.23 and second highest degree of convergence at 44% despite its lowest kernel RME rating of 0.5% and 4% pseudo-overall yield among all the plans.

The Decker–Wahl plan has two identical branches that converge in the 4th stage in a pseudo-dimerization of 3,4-dimethoxyphenylpyruvic acid in the presence of ammonia. The linearized Decker–Wahl plan in Figure 4.52 shows a net degradation of 86 g/mol per

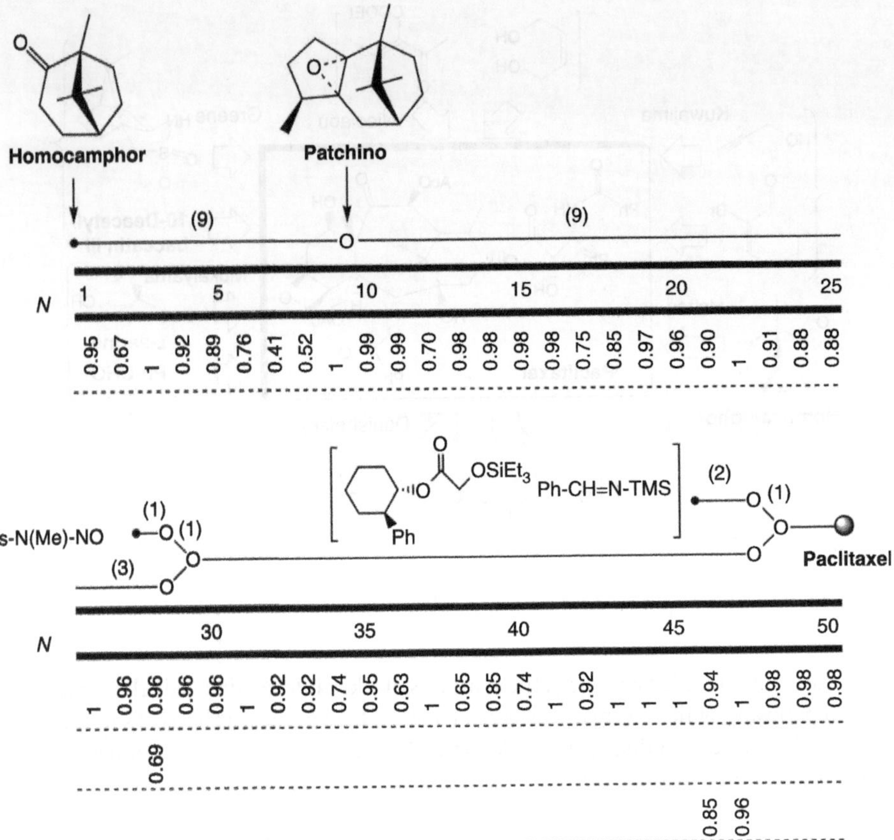

Figure 4.49 Reduced synthesis tree for Holton synthesis of paclitaxel (taxol). Step counts are shown within parentheses.

Table 4.21 Summary of reaction metrics and synthesis tree parameters for paclitaxel (taxol) synthesis plans ranked according to overall kernel (maximum) *RME*[a].

Plan	*N*	*M*	*I*	δ	β	μ_1	% *AE*	% $\langle \varepsilon_T \rangle$	% *RME*
Holton (1994)[b]	50	53	106	0.371	0.972	−366.95	8.6	**0.2**	**0.2**
Holton (1994)[c]	56	60	113	0.360	**0.987**	−421.33	8.3	0.1	0.09
Nicolaou (1995)	**38**	**47**	**85**	**0.384**	0.955	**−486.27**	**8.8**	0.08	0.02
Mukaiyama (1999)	59	67	125	0.373	0.961	−299.60	5.9	0.02	0.01
Danishefsky (1996)	49	55	98	0.366	0.961	−472.64	6.5	0.01	0.008
Kuwajima (2000)	57	63	107	0.353	0.975	−352.21	5.9	0.010	0.004

[a] See Appendix for definitions of terms.
[b] From homocamphor.
[c] From camphor.

Table 4.22 Summary of reaction metrics and synthesis tree parameters for paclitaxel (taxol) semi-synthesis plans ranked according to overall kernel (maximum) *RME*[a].

Plan	N	M	I	δ	β	μ_1	% *AE*	% $\langle \varepsilon_T \rangle$	% *RME*
Greene (1994)	**6**	**8**[b] 9[c]	**16**	**0.507**	0.735	−561.73	35.4	**45.2**	**21.3**[b] 0.02[c]
Kingston (1994)	8	10[b] 11[c]	**16**	0.486	**0.483**	−664.95	**44.7**	32.5	20.8[b] 0.02[c]
Gennari (1997)	7	9[b] 10[c]	20	0.496	0.795	−642.51	30.2	29.4	12.2[b] 0.01[c]
Commercon (1992)	11	15[b] 16[c]	27	0.427	0.893	−531.95	32.3	18.5	7.5[b] 0.003[c]
Greene-Gueritte-Voegelein (1988)	6	8[b] 9[c]	23	0.541	0.819	**−828.51**	31.1	21.6	12.3[b] 0.01[c]

[a] See Appendix for definitions of terms.
[b] Starting from 10-deacetylbaccatin III.
[c] Starting from needles of *Taxus baccata* (1 g 10-deacetylbaccatin III obtained from 1000 g dried needles).

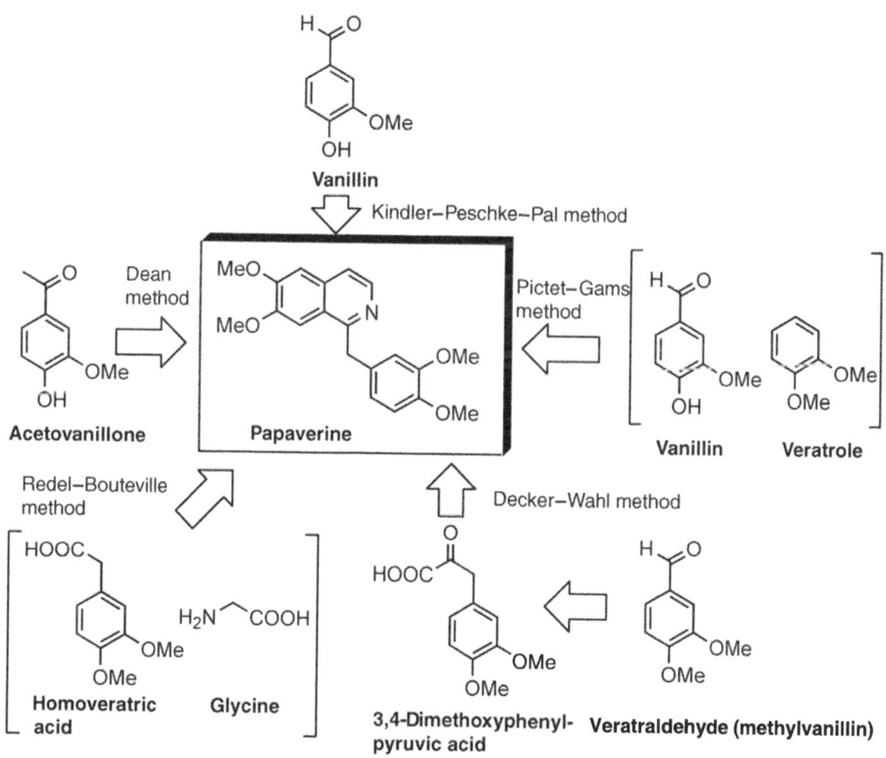

Figure 4.50 Synthesis map showing starting materials used for the synthesis of papaverine.

Table 4.23 Summary of reaction metrics and synthesis tree parameters for papaverine synthesis plans ranked according to overall kernel (maximum) *RME*[a].

Plan	*N*	*M*	*I*	*δ*	*β*	μ_1	% *AE*	% $\langle \varepsilon_T \rangle$	% *RME*
Kindler–Pescke–Pal (1958)	**8**	13	15	**0.450**	**0.604**	**−280.23**	**−27.4**	**42.2**	**15.1**
Dean (1960s)	**8**	10	12	0.392	0.630	−212.89	19.9	24.9	8.0
Redel–Bouteville (1949)	**8**	**8**	**11**	0.359	0.746	−29.96	31.7	7.6	4.0
Decker–Wahl (1950)	9	9	13	0.360	0.791	+86.40	19.7	8.3	2.8
Pictet–Gams (1909)	**8**	11	18	0.443	0.813	−223.18	13.6	5.0	0.5

[a] See Appendix for definitions of terms.

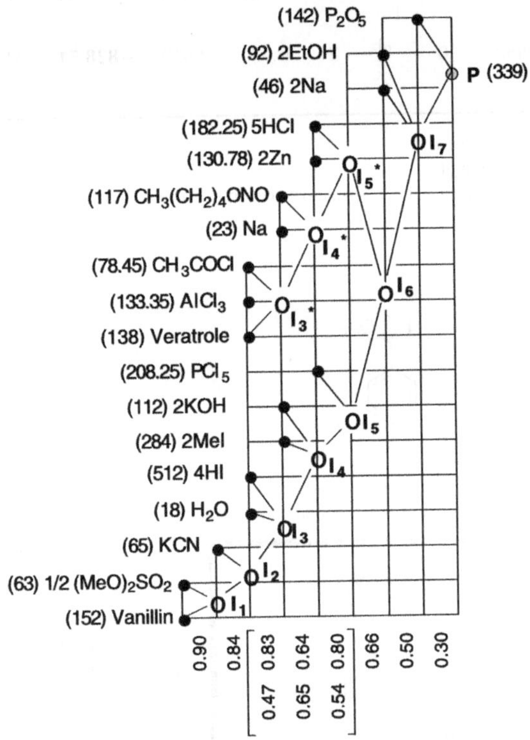

Figure 4.51 Synthesis tree for synthesis of papaverine by the Pictet–Gams method.

reaction stage compared to a net building up of 83 g/mol per reaction stage in the convergent version; whereas, the degree of convergence is 36% and 43%, respectively. Atom economies and reaction mass efficiencies are identical for both.

4.3.7.16 Penicillin[294–296]

Reduced synthesis tree diagrams for the Sheehan plan for penicillin are shown in Figure 4.56 and their metrics summary is shown in Table 4.24. The two plans differ in an extra racemization step between the two intermediate structures shown. The α-isomer is the

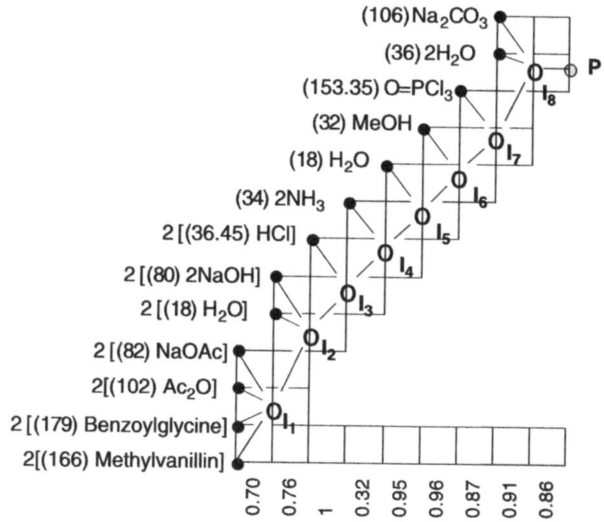

Figure 4.52 Synthesis tree for synthesis of papaverine by the Decker–Wahl method.

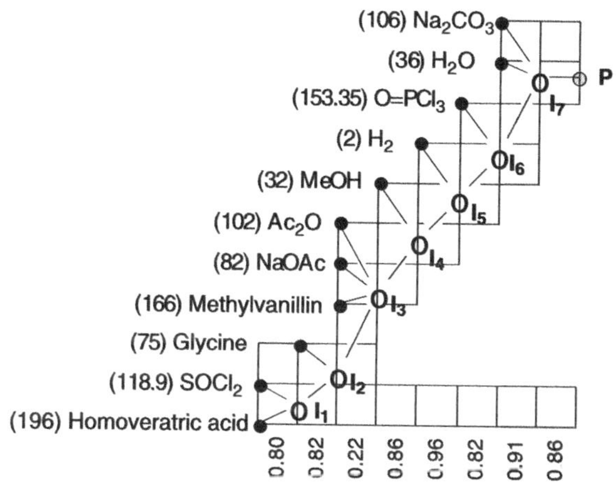

Figure 4.53 Synthesis tree for synthesis of papaverine by the Redel–Bouteville method.

required one to reach the target with the correct stereochemistry. Just like the case with the Carroll synthesis of (−)-epibatidine, this extra step results in a combined overall kernel RME of 0.14%. The fairly high overall atom economy over 10 stages is undercut due to the poor performing reaction yields in stages 4*, 5 and 10.

4.3.7.17 Physostigmine[297–311]

The alkaloid physostigmine found in the calabar bean is another classic milestone molecule in total synthesis. Metrics for thirteen plans to this molecule are summarized in Table 4.25.

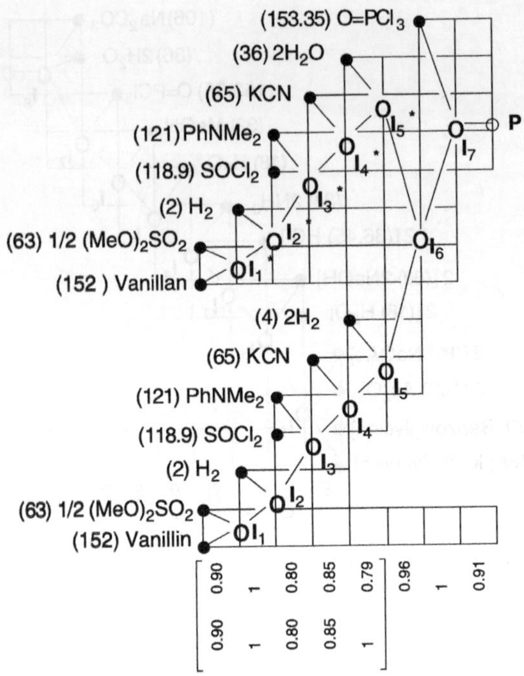

Figure 4.54 Synthesis tree for synthesis of papaverine by the Kindler–Peschke–Pal method.

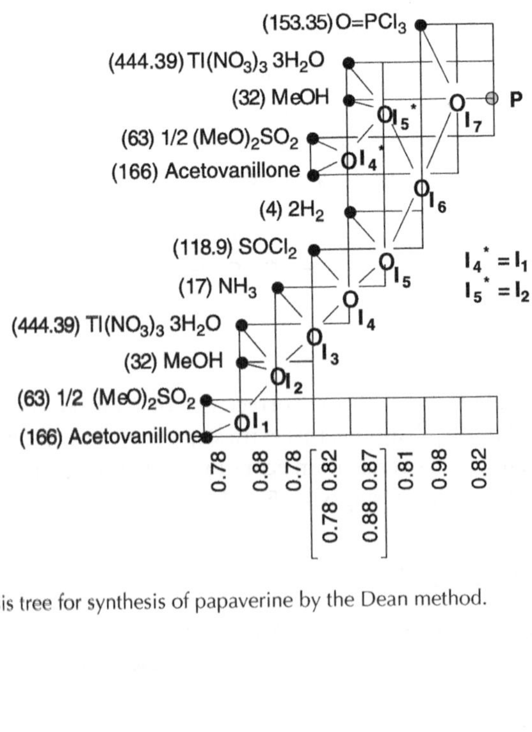

Figure 4.55 Synthesis tree for synthesis of papaverine by the Dean method.

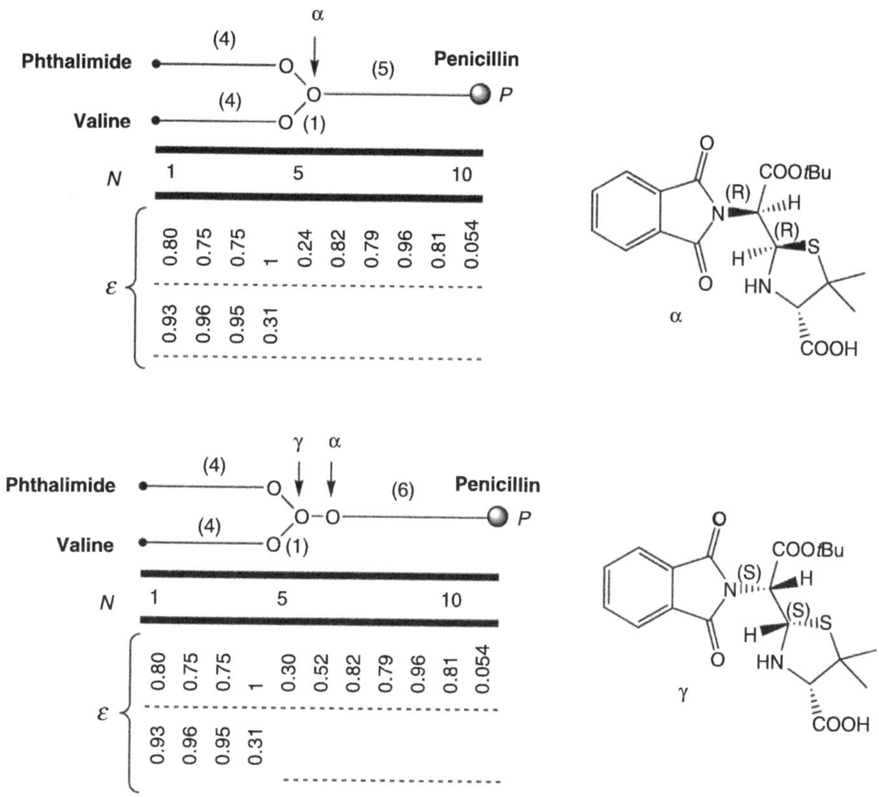

Figure 4.56 Reduced synthesis trees for Sheehan synthesis of penicillin. Step counts are shown within parentheses.

Table 4.24 Summary of reaction metrics and synthesis tree parameters for penicillin synthesis plans ranked according to overall kernel (maximum) *RME*[a].

Plan	N	M	I	δ	β	μ_1	% *AE*	% $\langle\varepsilon_T\rangle$	% *RME*
Sheehan (1959)[b]	10	14	26	0.437	0.894	−234.76	19.1	0.3	0.08
Sheehan (1959)[c]	11	15	26	0.421	0.894	−212.53	19.1	0.2	0.06

[a] See Appendix for definitions of terms.
[b] Synthesis plan via α-isomer.
[c] Synthesis plan via isomerization of γ-isomer to α-isomer.

Figure 4.57 summarizes various starting materials to this target for some of the better performing plans. The classic Julian and recent Yu–Lu and Trost plans are shown in Schemes 4.17, 4.18 and 4.19 respectively. Figure 4.58 shows the synthesis tree for the most material efficient Yu–Lu plan. Remarkably, the Julian plan ranks second among the list yet it was the first reported synthesis for this compound that also proved its correct structure. The

Table 4.25 Summary of reaction metrics and synthesis tree parameters for physostigmine synthesis plans ranked according to overall kernel (maximum) *RME*[a].

Plan	*N*	*M*	*I*	δ	β	μ_1	% *AE*	% (ε_T)	% *RME*
Yu–Lu (1994)	**7**	**7**	**12**	0.410	**0.741**	−44.91	**36.9**	**64.2**	**26.2**
Julian (1935)	9	9	20	0.428	0.851	−67.10	17.9	27.0	7.1
Trost (2006)	8	8	17	0.430	0.823	**−81.39**	18.2	22.6	6.6
Nakagawa (2000)	9	9	19	0.421	0.843	−54.16	16.4	13.9	4.8
Overman (1998)	10	10	22	0.420	0.865	+7.36	11.1	19.7	3.2
Johnson (2003)	8	8	19	0.444	0.842	−75.62	17.3	10.1	2.6
Mukai (2006)	9	9	24	**0.449**	0.877	−22.56	8.8	11.8	1.7
Speckamp (1978)	11	12	26	0.423	0.887	−41.75	16.3	4.8	1.5
Marino (1992)	11	12	23	0.408	0.873	−6.51	10.2	4.6	1.1
Takano (1991)	14	14	26	0.381	0.887	−49.40	11.7	3.0	0.8
Fuji (1991)	16	16	34	0.394	0.914	−53.41	7.4	4.1	0.7
Fukumoto (1986)	22	22	42	0.370	0.931	−43.85	7.3	1.3	0.3
Takano (1982)	12	12	28	0.418	0.895	−37.54	14.4	0.4	0.1

[a] See Appendix for definitions of terms.

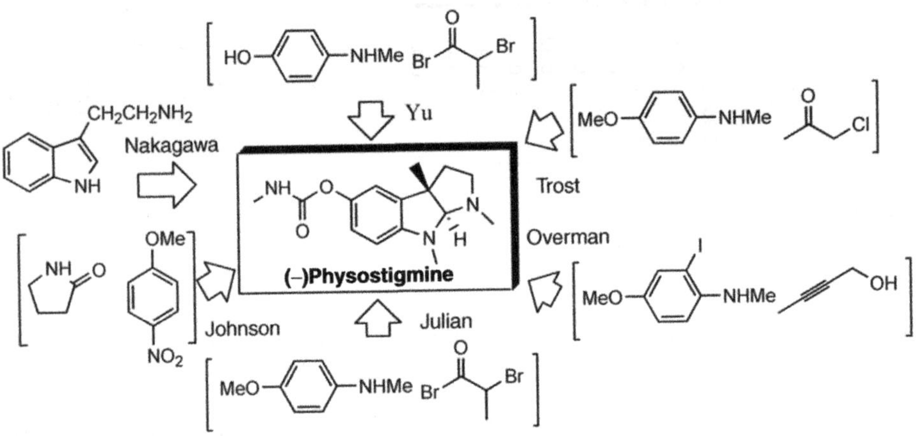

Figure 4.57 Synthesis map showing starting materials used for the synthesis of physostigmine.

Yu–Lu plan is a variant of the Julian one and its remarkable material efficiency is primarily due to a combined high atom economy and high reaction yields. The Trost plan has the best performing molecular weight first moment parameter. The Mukai plan has the highest degree of convergence.

4.3.7.18 *Quinine*[312–318]

The synthesis of quinine is considered the quintessential classic in the art of total synthesis. Figure 4.59 summarizes the starting materials for four routes beginning with the historical Woodward–Doering–Rabe plan to the racemic mixture and including three recent modern

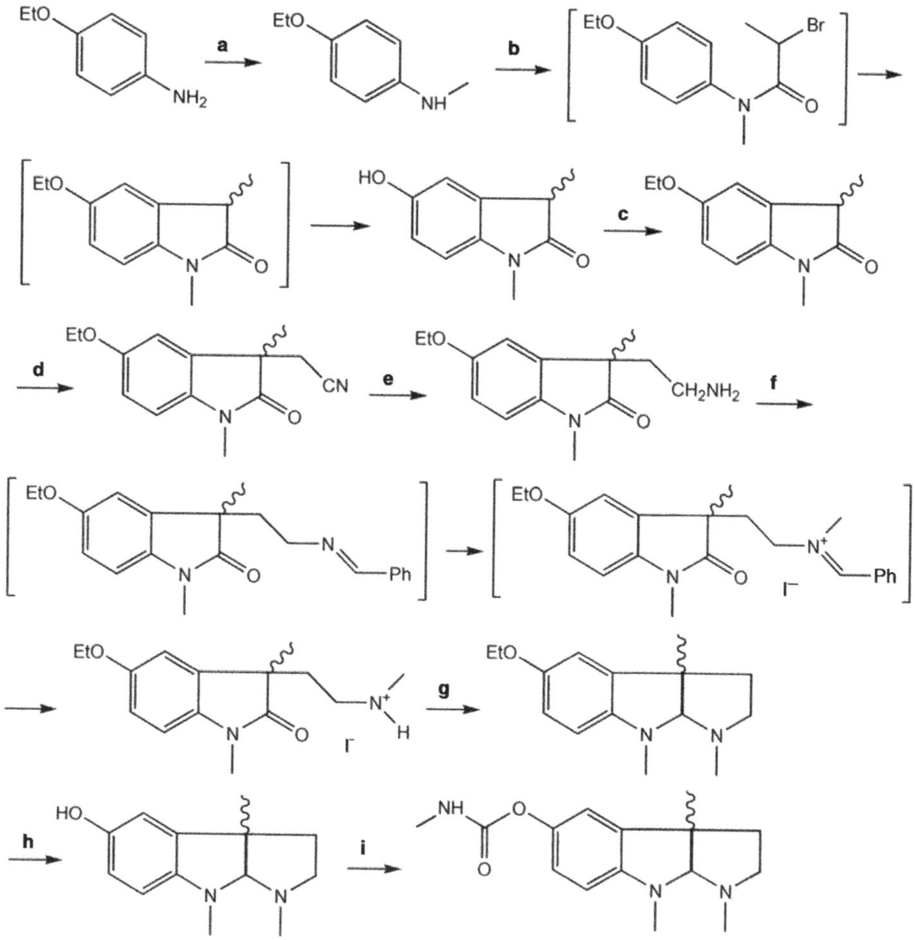

Scheme 4.17 Julian (±)-physostigmine synthesis plan (1935). (a) Na, $1/2(CH_3O)_2SO_2$ (96%); (b) $CH_3{-}CHBr{-}COBr$, $AlCl_3$, then H_2O (92%); (c) $1/2(EtO)_2SO_2$ (68%); (d) NaOEt, $ClCH_2CN$ (84%); (e) $2H_2$, Pd—C (cat.) (91%); (f) PhCHO, CH_3I, H_2O (86%); (g) 2Na, 2EtOH (99%); (h) $AlCl_3$, then $3H_2O$ (91%); (i) Na, $CH_3{-}N{=}C{-}O$, then H_2O (76%).

stereoselective routes to the (−) isomer. As can be seen from Table 4.26 the most material efficient synthesis is that of Stork with an overall kernel RME of 1.2% in 17 reaction stages despite the occurrence of a high number of protection–deprotection steps that produce triphenylphosphine as a byproduct. It prevails with the highest overall kernel RME mainly because it has the fewest reaction inputs, reaction stages, and reaction steps, and has the highest average reaction yield per step of 86%. The kernel overall RME values for the Woodward–Doering–Rabe and Stork plans of 0.0063% and 1.2% parallel previous determinations[25] of overall RME values of 0.0032% and 0.16%, respectively, when stoichiometric factors are taken into account for each step. The Jacobsen plan happens to be the most convergent plan with a 41% degree of convergence though all of the routes are for

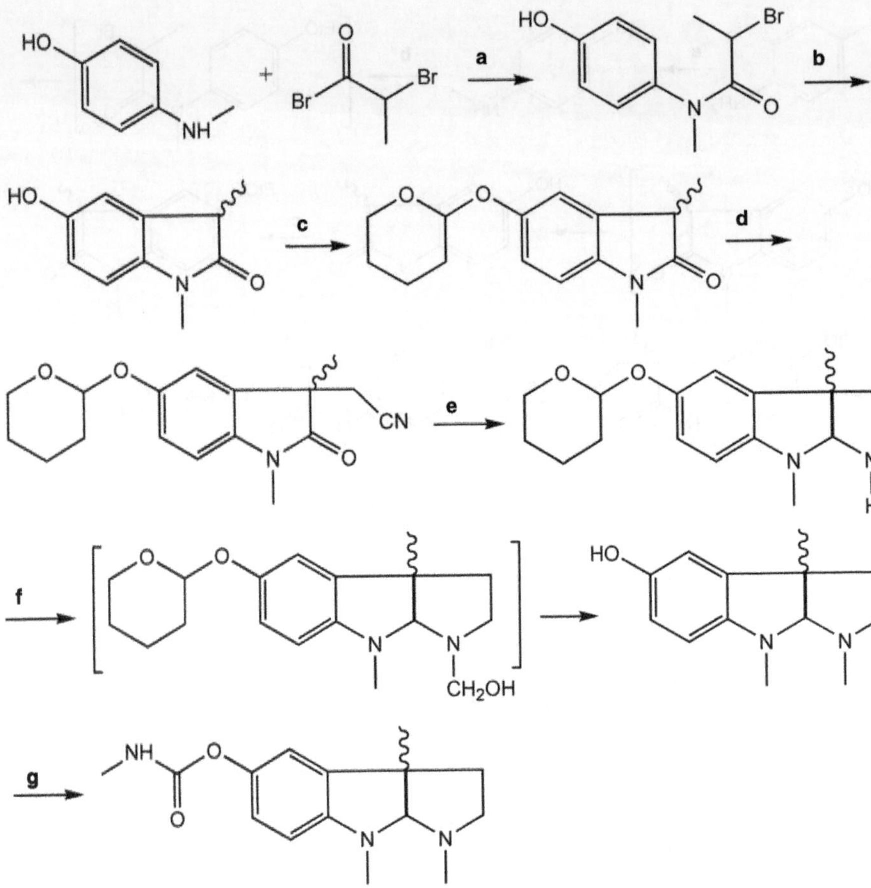

Scheme 4.18 Yu–Lu (±)-physostigmine synthesis plan (1994). (a) (95%); (b) $AlCl_3$ (cat.), (97%); (c) *p*-TsOH (cat.), 3,4-dihydro-2H-pyran (90.5%); (d) $ClCH_2CN$ (100%); (e) 3/4$LiAlH_4$, $2H_2O$ (100%); (f) CH_2O, then 1/2$NaBH_4$, $2H_2O$ (85%); (g) Na, $CH_3{-}N{=}C{=}O$, then H_2O (90.6%).

the most part dominated by long linear sequences as shown by their similar low relative rate of convergence of about 0.1 and high asymmetry. There are three points of convergence in the Jacobsen plan compared with one each for the Stork and Acharya–Kobayashi plans and none for the Woodward–Doering–Rabe plan. The Acharya–Kobayashi plan has by far the longest route (21 reaction stages) and begins with a very expensive reagent, *cis*-4-acetoxy-2-cyclo-penten-1-ol. Since it appears at the beginning of the synthesis tree, amplification of raw material cost is inevitable. In terms of tracking the building up character of the syntheses to the target quinine, the Jacobsen plan comes out again on top with a net building of 115 g/mol per reaction stage. Though the Stork plan is the most material efficient overall, this trend is offset by it having a net *loss* of 47 g/mol per reaction stage over the entire synthesis. This is due to the frequent use of bulky protecting groups which momentarily increase the molecular weights of intermediates along the way by a significant margin over the target threshold of 324 g/mol. Though this target compound has been used as a

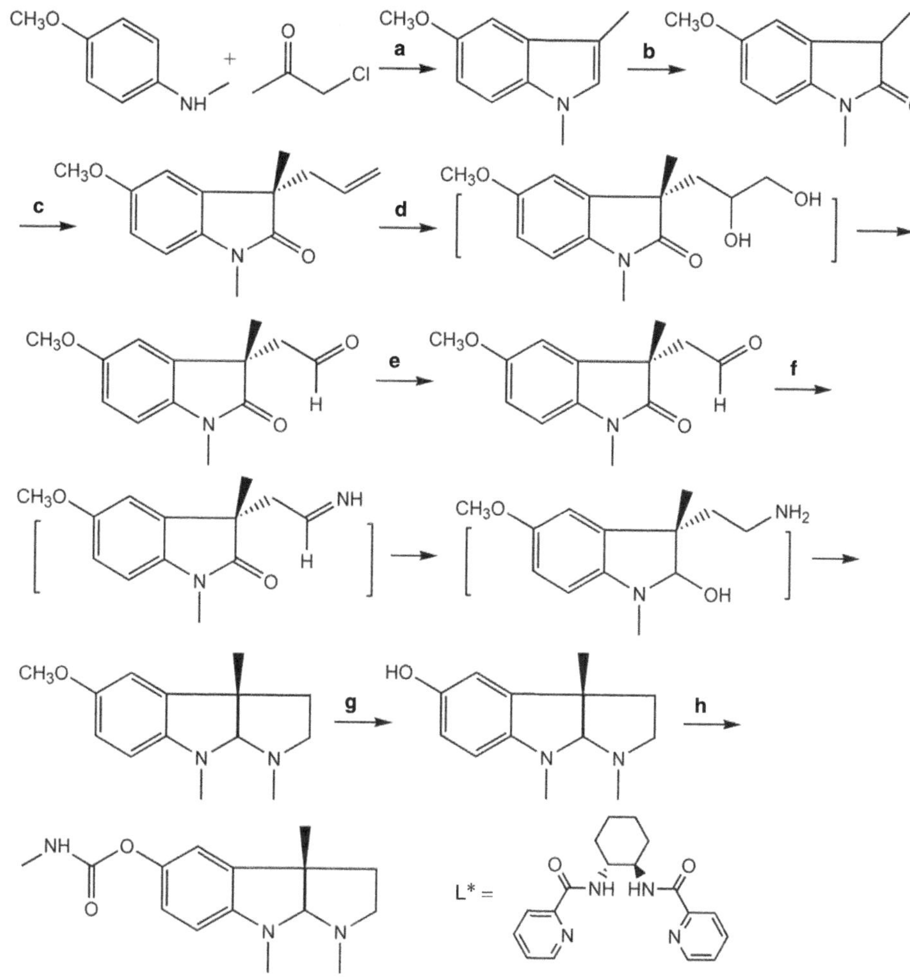

Scheme 4.19 Trost (−)-physostigmine synthesis plan (2006). (a) Et_3N (72%); (b) $(CH_3)_2SO$ (88%); (c) diallyl carbonate, LiO*t*Bu, Mo, L* (98%); (d) $1/2OsO_4$, then $NaIO_4$ (92%); (e) recrystallization from iPrOH-cyclohexane (72%); (f) $CH_3NH_2 \cdot HCl$, then $1/2LiAlH_4$, $2H_2O$ (88%); (g) BBr_3, $3H_2O$ (79%); (h) Na, $CH_3N{=}C{=}O$, then H_2O (79%).

benchmark for gauging synthetic prowess, the best total syntheses so far still pale in comparison to the observation that up to 13% of the dry weight of the bark of *Cinchona ledgeriana* trees is quinine, a value exceeding the overall kernel RMEs of all four total syntheses and represents the major commercial source of this material.

4.3.7.19 Reserpine[319–333]

Figure 4.60 shows a synthesis map for various starting materials for the synthesis of the alkaloid reserpine. Metrics for 10 plans are summarized in Table 4.27. The Liao and Stork plans are the best performers overall, however the Pearlman plan has the highest atom

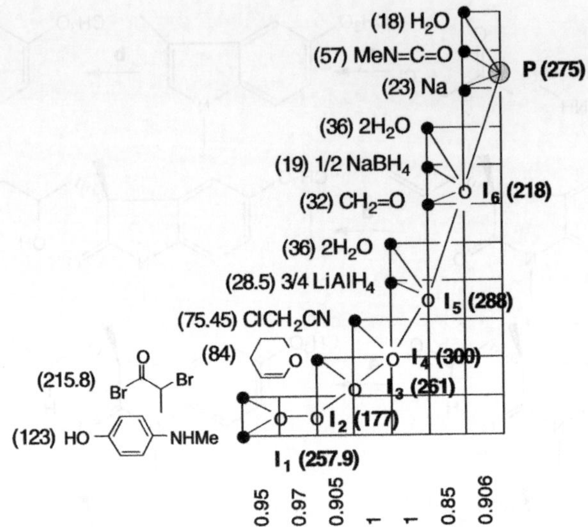

Figure 4.58 Synthesis tree for synthesis of physostigmine by the Yu method.

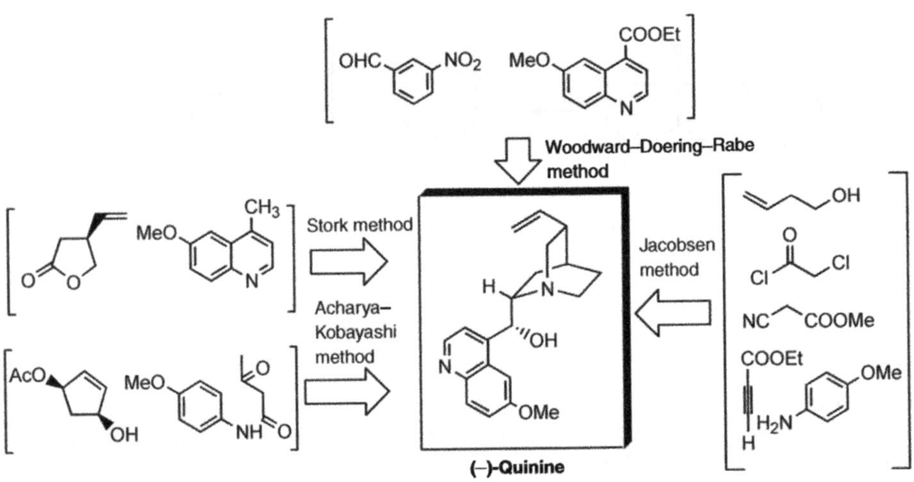

Figure 4.59 Synthesis map showing starting materials used for the synthesis of quinine.

economy and the original Woodward plan has both the highest degree of convergence and best performing molecular weight first moment parameter.

4.3.7.20 *Tetracycline*[334,335]

The recent synthesis of (−)-tetracycline by Myers and co-workers incorporates a biocatalytic step in the first stage which oxidizes benzoic acid aerobically to an α,β-dihydroxy derivative in the presence of a whole-cell mutant strain of *Alcaligenes eutrophus*. Figure 4.61 shows a reduced tree diagram for the synthesis and Table 4.28 summarizes the metrics parameters.

Table 4.26 Summary of reaction metrics and synthesis tree parameters for quinine synthesis plans ranked according to overall kernel (maximum) RME^a.

Plan	N	M	I	δ	β	μ_1	% AE	% $\langle\varepsilon_T\rangle$	% RME
Stork (2001)	**17**	**18**	**33**	0.384	**0.896**	+47.11	**8.6**	**7.1**	**1.2**
Jacobsen (2004)	16	21	38	**0.413**	0.907	**−114.71**	6.7	3.2	0.8
Acharya–Kobayashi (2004)	21	28	51	0.402	0.942	−57.14	5.3	3.3	0.6
Woodward–Doering–Rabe	19	19	36	0.372	0.924	−60.51	8.4	0.03	0.006

[a] See Appendix for definitions of terms.

There are three reaction paths to the final target via two points of convergence. The overall yield from the bromooxime starting material is 1.3% and from benzoic acid is 1.2%. The overall kernel RME is 0.4% over 18 reaction stages and 27 reaction steps.

4.3.7.21 *Thienamycin*[337–341]

Table 4.29 summarizes the metrics for 4 plans to the antibiotic thienamycin and Figure 4.62 shows the corresponding synthesis map with starting materials used. The two Melillo and Reider variants by Merck co-workers are the better performers however each plan has certain attributes that are not found in the other. The slightly shorter Reider plan has a higher overall reaction yield though it has a similar atom economy and overall kernel RME to the Melillo plan. Further optimization would be required to result in all optimum metrics occurring in the same plan.

4.3.7.22 *Tropinone*[342–350]

Tropinone is another classic compound in the history of total synthesis. The celebrated plans of Willstätter and Robinson are shown in Schemes 4.20 and 4.21 and Figure 4.63 shows a synthesis map for different ways this compound has been made. The synthesis tree for the three-component Robinson plan is shown in Figure 4.64. Calcium carbonate and hydrochloric acid are added as inputs to complete the balanced chemical equation since these are involved in a neutralization reaction. It is clear from the results summarized in Table 4.30 that the Robinson plan is the clear front-runner because the synthesis is achieved in a single step even though the reaction yield is modest. Any further improvements to this method would be directed to improving this parameter.

4.3.7.23 *Two-directional tandem sequence*[351]

Stockman and co-workers have demonstrated a two-directional tandem approach to a target structure as shown in Scheme 4.22. Again, the results given in Table 4.31 are consistent with the attributes for good material efficient syntheses as given in Section 4.3.6. Two representations of the tree diagram are given in Figure 4.65 depending on whether the last three intermediates are isolated or not. Just like the case of the extra epimerization step in the Carroll plan for the synthesis of (−)-epibatidine represented by a horizontal line in the tree diagram, the extra reaction stages here implied by isolation of the last three intermediates serve to decrease the degree of convergence while maintaining a constant

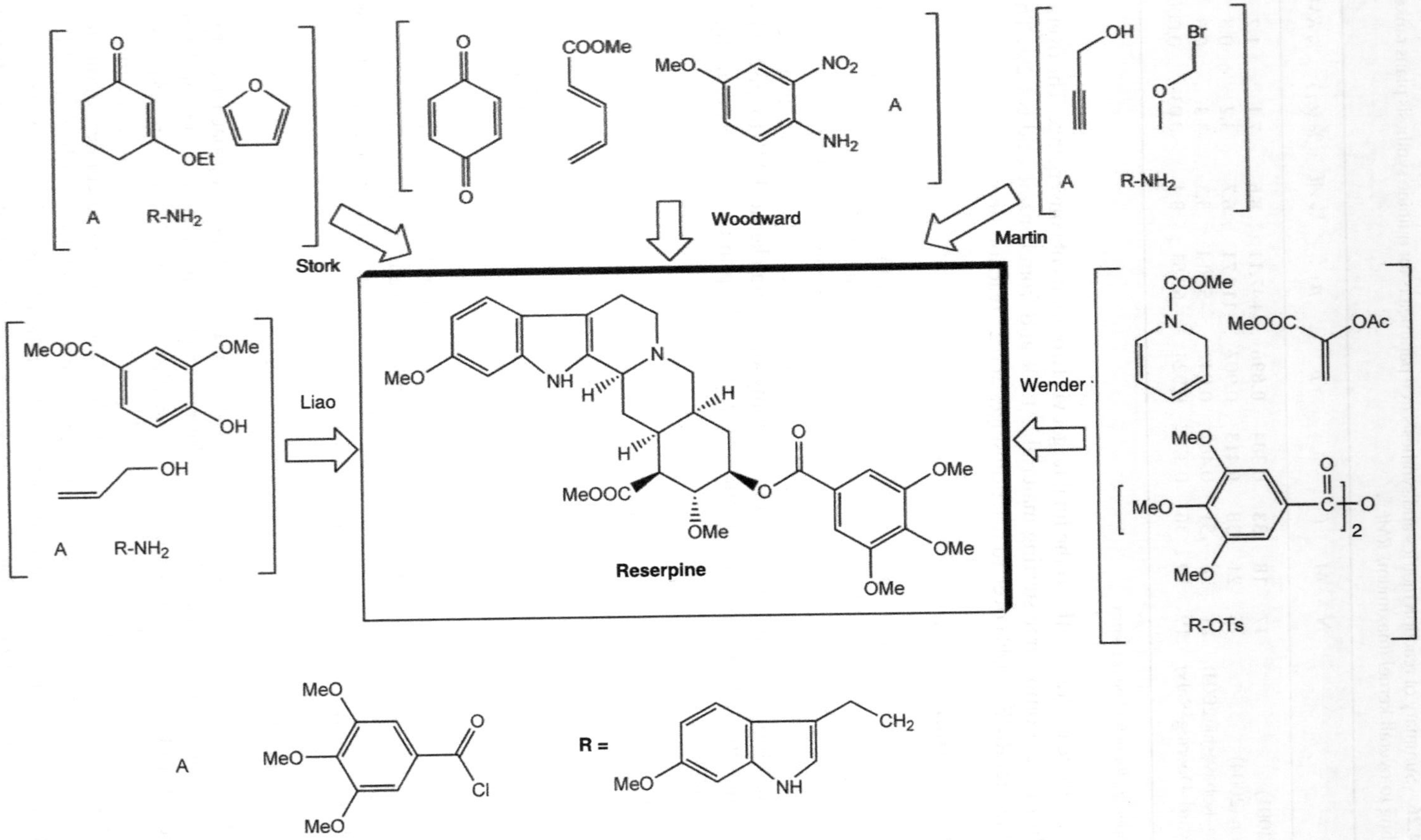

Figure 4.60 Synthesis map showing starting materials used for the synthesis of reserpine.

Table 4.27 Summary of reaction metrics and synthesis tree parameters for reserpine synthesis plans ranked according to overall kernel (maximum) *RME*[a].

Plan	*N*	*M*	*I*	δ	β	μ_1	% *AE*	% $\langle\varepsilon_T\rangle$	% *RME*
Liao (1996)	**11**	**11**	**19**	0.376	**0.871**	−309.75	21.5	10.2	**5.5**
Stork (2005)	12	14	25	0.398	0.903	−419.39	23.1	**15.2**	4.5
Martin (1985)	21	22	31	0.333	0.914	−289.69	18.0	3.2	1.5
Wender (1980)	17	19	29	0.360	0.908	−397.48	16.6	2.0	1.4
Hanessian (1997)	23	24	34	0.331	0.919	−162.40	12.0	3.1	1.0
Shea (2003)	21	22	34	0.346	0.923	−227.22	13.3	2.0	0.8
Pearlman (1979)	14	15	23	0.361	0.882	−399.66	**33.8**	0.1	0.3
Woodward (1958)	19	25	49	**0.410**	0.944	**−564.55**	15.2	0.09	0.1
Fraser–Reid (1995)	39	39	72	0.360	0.939	−236.22	8.2	0.5	0.1
Mehta (1997–2000)	28	29	47	0.345	0.949	−314.23	12.7	0.01	0.008

[a] See Appendix for definitions of terms.

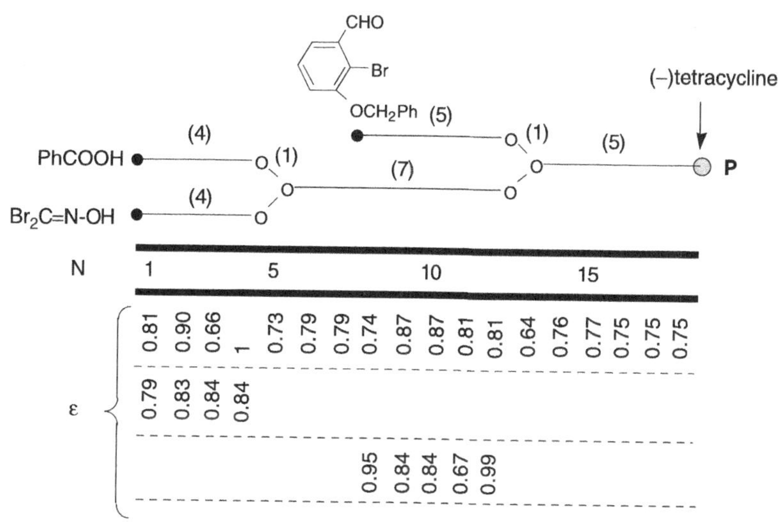

Figure 4.61 Reduced synthesis tree for Myers synthesis of tetracycline. Step counts are shown within parentheses.

degree of asymmetry. The two-directional approach is a key strategy that is amenable when the target structure has symmetric features as was seen earlier for the synthesis of hemibrevetoxin B intermediate.

4.3.7.24 Vinblastine[352–355] and vincristine[356]

The two alkaloids vinblastine and vincristine found in *Catharanthus roseus* have been recent targets of total synthesis because of their potency in cancer chemotherapy. The reduced tree diagram for the Fukuyama plan to vincristine is shown in Figure 4.66. There are three points of convergence, four branches leading to the target product and four tiers of reaction yields.

Table 4.28 Summary of reaction metrics and synthesis tree parameters for tetracycline synthesis plan ranked according to overall kernel (maximum) *RME*[a].

Plan	*N*	*M*	*I*	δ	β	μ_1	% *AE*	% $\langle \varepsilon_T \rangle$	% *RME*
Myers (2005)	18	27	45	0.408	0.939	−151.74	8.5	1.3	0.4

[a] See Appendix for definitions of terms.

Table 4.29 Summary of reaction metrics and synthesis tree parameters for thienamycin synthesis plans ranked according to overall kernel (maximum) *RME*[a].

Plan	*N*	*M*	*I*	δ	β	μ_1	% *AE*	% $\langle \varepsilon_T \rangle$	% *RME*
Melillo (1986)	18	19	**27**	**0.338**	**0.908**	+61.33	**8.0**	9.8	**2.4**
Reider (1982)	**13**	**13**	28	0.400	0.911	**+23.49**	7.8	**17.4**	2.3
Salzmann (1980)	19	19	37	0.374	0.933	+55.25	6.1	8.3	1.2
Hanessian (1990)	18	18	35	0.375	0.931	+192.84	6.9	0.6	0.2

[a] See Appendix for definitions of terms.

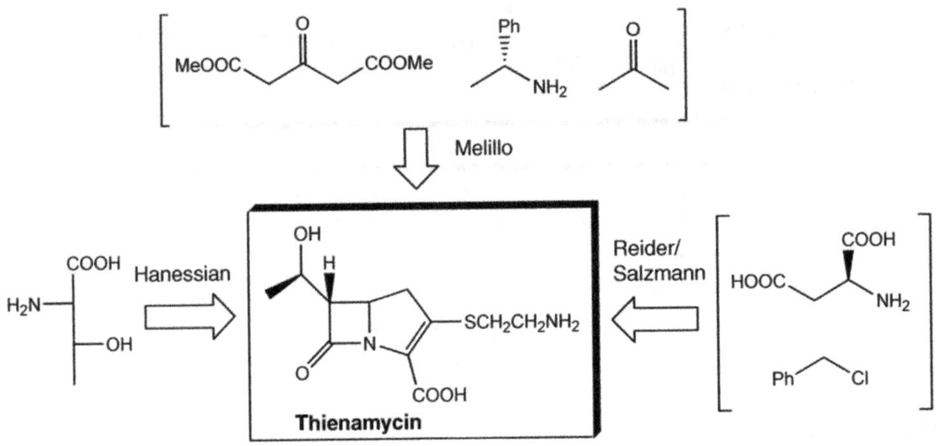

Figure 4.62 Synthesis map showing starting materials used for the synthesis of thienamycin.

Tables 4.32 and 4.33 summarize the metrics for the synthesis plans for both products. The Fukuyama plans to both targets are very similar differing only in the very late stages of each plan. The Kuehne plan to vinblastine is considerably shorter than the Fukuyama one since it uses (−)-vindoline as an available starting material in stage 11. This explains why its overall kernel RME is 17 times larger than that of the Fukuyama plan. For a more fair comparison, if the upper two branches leading to (−)-vindoline are omitted from the Fukuyama plan, the number of stages remain the same at 27 but the number of reactions and inputs decreases to 29 and 47, respectively. These changes result in an increase in overall kernel RME from 0.3% to 0.5% but it is still an order of magnitude less than that determined for the Kuehne

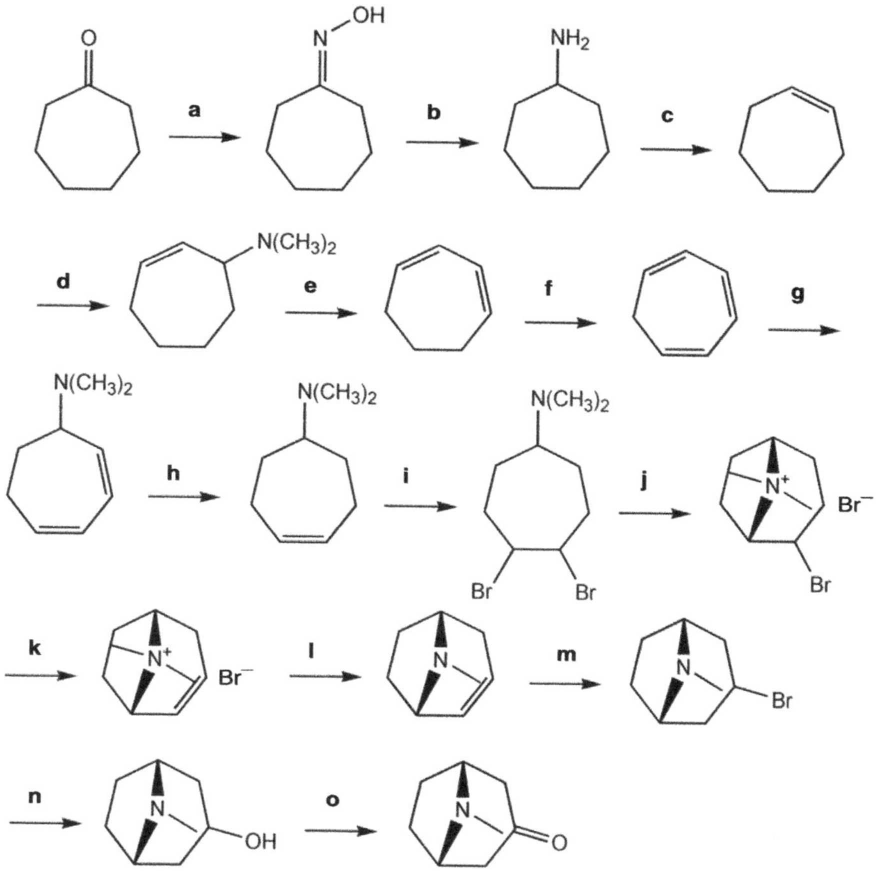

Scheme 4.20 Willstätter tropinone synthesis plan (1901). (a) NH_2OH (80%); (b) 4Na, 4EtOH (25%); (c) $3CH_3I$, $3/2Ag_2O$ (85.5%); (d) Br_2, $3CH_3NH_2$ (75%); (e) CH_3I, $1/2Ag_2O$ (80%); (f) Br_2, 2quinoline (77.5%); (g) $2CH_3NH_2$, HBr (60%); (h) 2Na, 2EtOH (83.5%); (i) Br_2 (100%, assumed); (j) heat (30%); (k) NaOH (100%, assumed); (l) heat (54%); (m) HBr (85%); (n) H_2O, H_2SO_4 (cat.) (24%); (o) $2/3CrO_3$ (89%).

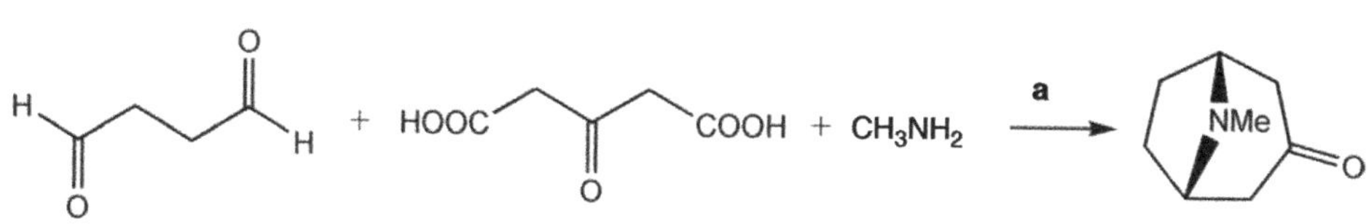

Scheme 4.21 Robinson Tropinone Synthesis Plan (1917). (a) $CaCO_3$, then 2HCl (42%).

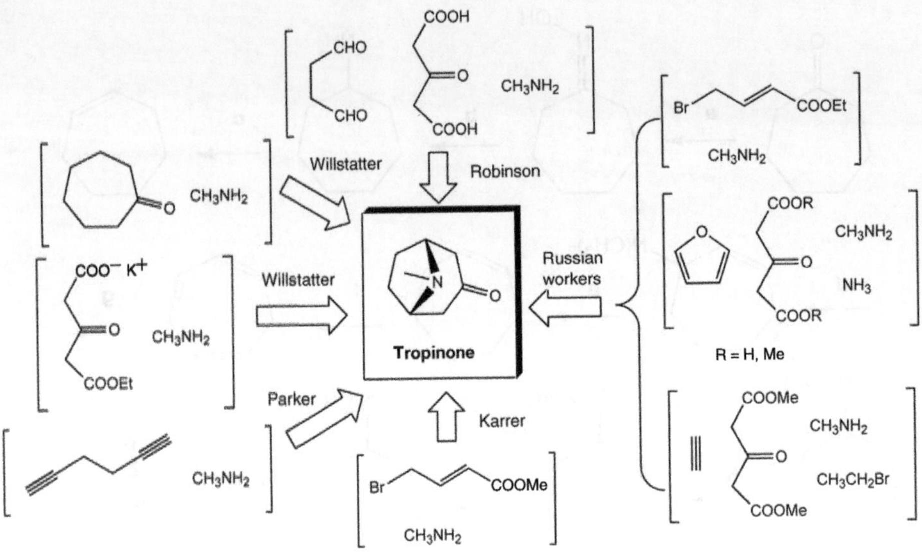

Figure 4.63 Synthesis map showing starting materials used for the synthesis of tropinone.

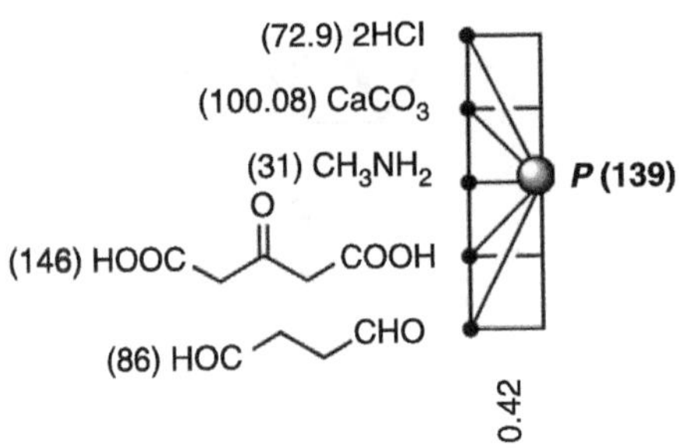

Figure 4.64 Synthesis tree for Robinson synthesis of tropinone.

plan. A more drastic change is the increase in the molecular weight first moment parameter from −844 to −282 g/mol/stage which is consistent with the decreased degree of synthesis occurring in the truncated tree if vindoline is an available starting material.

4.3.7.25 Welwitindolinone A[357–359]

The example of the synthesis of welwitindolinone A, an alkaloid found in marine blue-green algae, is an excellent one to illustrate the point that shorter syntheses do not always automatically result in higher material efficiencies. Figure 4.67 shows the two sets of starting materials that have been used to accomplish the total synthesis of this compound. From the results of Table 4.34 it can be seen that the Baran plan is the superior plan based on step count, atom economy, degree of convergence, degree of asymmetry and the molecular weight first moment for building up in the synthesis. However, all of these gains evaporate when reaction yields are taken into account and this gives the Wood plan a slight edge in

Table 4.30 Summary of reaction metrics and synthesis tree parameters for tropinone synthesis plans ranked according to overall kernel (maximum) *RME*[a].

Plan	*N*	*M*	*I*	δ	β	μ_1	% *AE*	% $\langle\varepsilon_T\rangle$	% *RME*
Robinson (1912)	**1**	**1**	5	**1**	**0**	**−77.00**	**31.9**	**42**	**13.4**
Parker (1959)	5	5	9	0.456	0.617	+38.33	20.2	6.2	1.7
Russian (1958)	4	4	**4**	0.358	0.417	+104.76	27.8	4.0	1.2
Willstätter (1921)	5	5	7	0.397	0.630	+100.83	25.1	2.5	0.7
Willstätter (1901)	15	15	21	0.330	0.899	+8.84	4.9	0.1	0.02
Russian (1958)	4	5	10	0.523	0.670	+0.20	13.5	<40.4	<6.2
Russian (1958)	3	3	9	0.600	0.637	−40.50	11.4	<41.1	<5.6
Karrer (1947)	6	6	9	0.396	0.687	+108.94	13.0	<4.9	<0.5
Russian (1958)	5	6	15	0.526	0.766	+19.5	9.3	<2.1	<0.4

[a] See Appendix for definitions of terms.

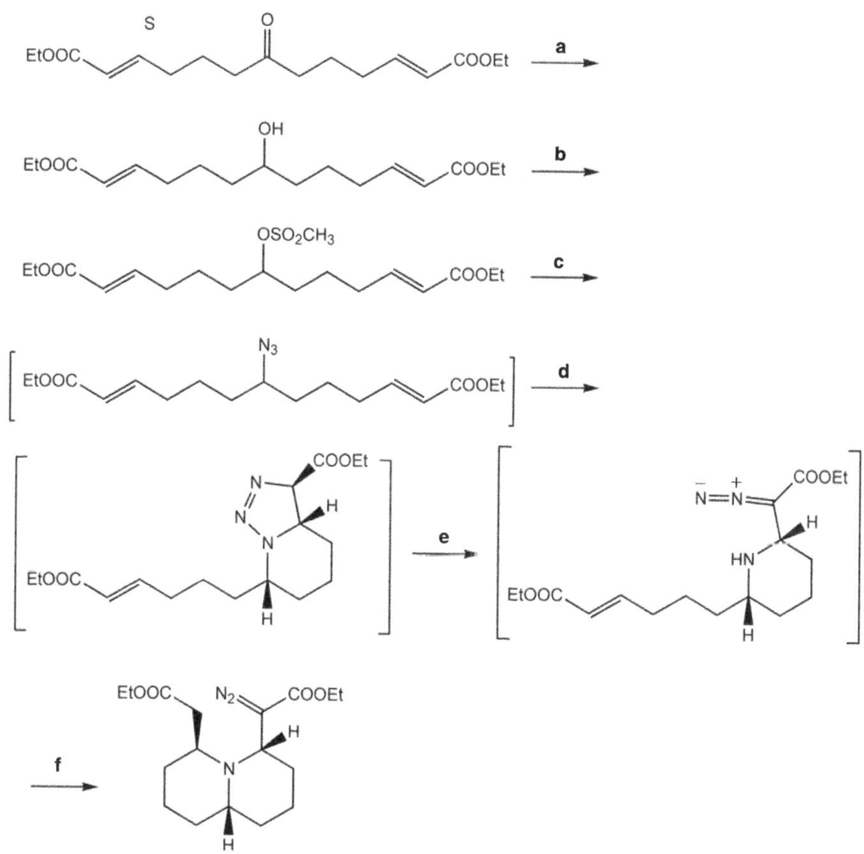

Scheme 4.22 Stockman two-directional tandem synthesis sequence (2005). (a) 1/4NaBH$_4$, H$_2$O (96%); (b) CH$_3$SO$_2$Cl, Et$_3$N (100%); (c) NaN$_3$; (d) [3 + 2]; (e) [1,4]-H; (f) thermal cyclization (52%, 4 steps).

Table 4.31 Summary of reaction metrics and synthesis tree parameters for two-directional tandem synthesis plan example ranked according to overall kernel (maximum) *RME*[a].

Plan	*N*	*M*	*I*	δ	β	μ_1	% *AE*	% $\langle\varepsilon_T\rangle$	% *RME*
Stockman (2005)	3	3	6	0.534	0.550	−81.63	54.5	49.9	27.7
Stockman (2005)	6	6	6	0.319	0.550	−46.64	54.5	49.9	27.7

[a] See Appendix for definitions of terms.

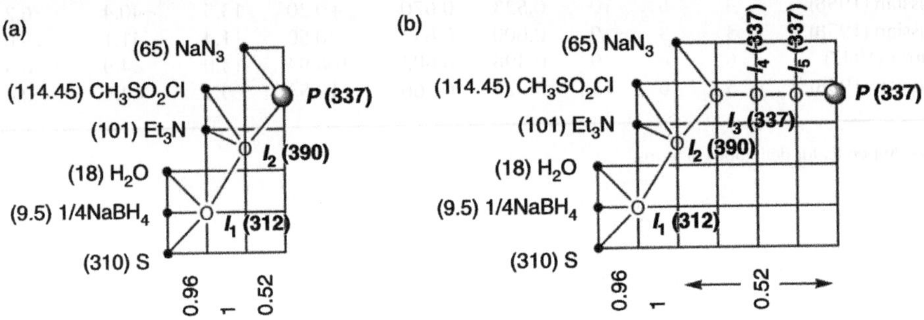

Figure 4.65 Synthesis trees for Stockman two-directional tandem synthesis plan: (a) assuming last three intermediates in Scheme 4.22 are not isolated. (b) Assuming last three intermediates in Scheme 4.22 are isolated.

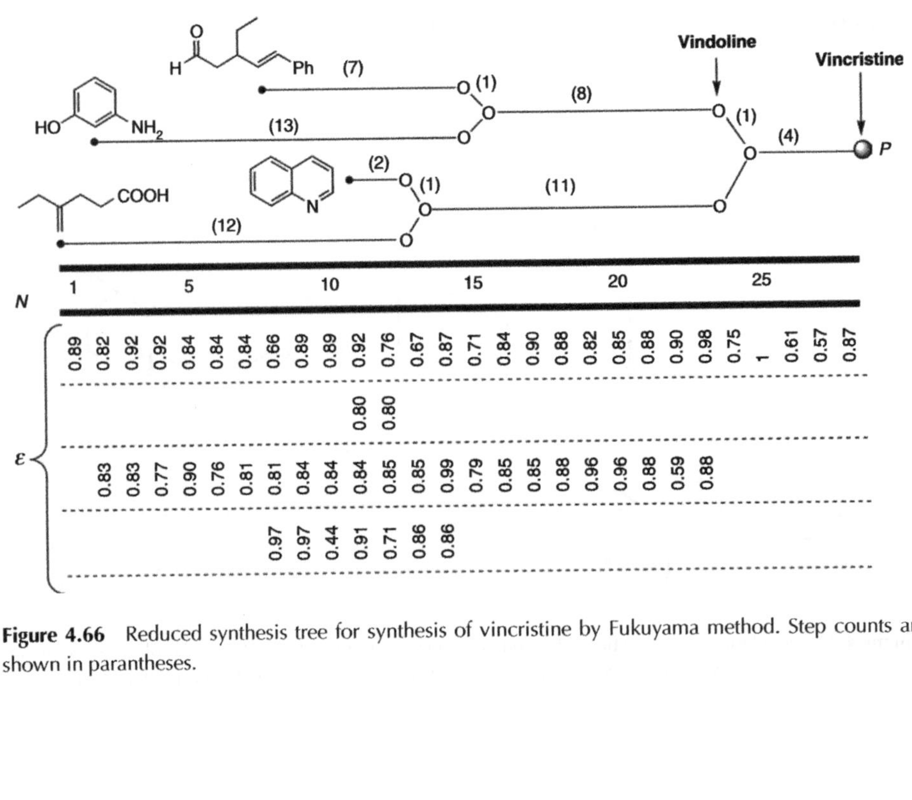

Figure 4.66 Reduced synthesis tree for synthesis of vincristine by Fukuyama method. Step counts are shown in parantheses.

Table 4.32 Summary of reaction metrics and synthesis tree parameters for vinblastine synthesis plans ranked according to overall kernel (maximum) *RME*[a].

Plan	N	M	I	δ	β	μ_1	% *AE*	% $\langle \varepsilon_T \rangle$	% *RME*
Kuehne (1991)	16	19	29	0.369	0.917	−443.14	25.3	13.7	5.0
Fukuyama (2002)	27	56	100	0.468	0.912	−844.13	7.6	1.0	0.3

[a] See Appendix for definitions of terms.

Table 4.33 Summary of reaction metrics and synthesis tree parameters for vincristine synthesis plans ranked according to overall kernel (maximum) *RME*[a].

Plan	N	M	I	δ	β	μ_1	% *AE*	% $\langle \varepsilon_T \rangle$	% *RME*
Fukuyama (2004)	28	57	101	0.440	0.959	−850.27	7.7	0.5	0.1

[a] See Appendix for definitions of terms.

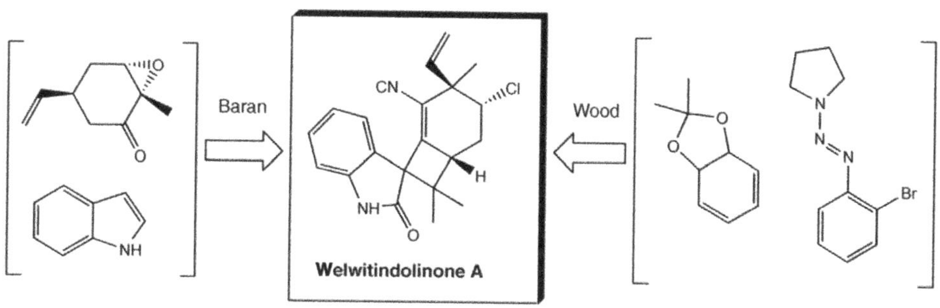

Figure 4.67 Synthesis map showing starting materials used for the synthesis of welwitindolinone A.

overall kernel RME even though it is twice as long and uses twice the number of input materials. The poor performing yields ranging from 30–57% appear in the first five steps in the Baran plan. It is here that the attenuation effect is most felt since the earliest stages are conducted at a larger scale than the later stages. The lowest yield for a reaction reported in the Wood plan is 69%. An hypsicity analysis of both plans, summarized by Figure 4.68, indicates that the *HI* values are −1.91 and +0.63 for the Baran and Wood plans, respectively. The implication here is that though there is only one formal reduction in the Baran plan using $NaBH_3CN$ there is a net overall oxidation required from starting materials to reach the target structure throughout the progress of the synthesis plan.

The Wood plan is closer to the isohypsic condition yet it formally involves two oxidation steps with NaOCl and Dess–Martin reagent, and two reduction steps with NH_4^+ $HB(OAc)_3$ and $NaBH_3CN$. The apparent paradox is explained by noticing from Figure 4.68 that the range of oxidation number changes in the Wood plan is smaller than in the Baran plan. These results imply that to achieve isohypsicity it is not enough to simply reduce the number of redox reactions in a plan. One needs to co-ordinate the goals of reducing the overall step

Table 4.34 Summary of reaction metrics and synthesis tree parameters for welwitindolinone A ranked according to overall kernel (maximum) *RME*[a].

Plan	*N*	*M*	*I*	δ	β	μ_1	% *AE*	% $\langle \varepsilon_T \rangle$	% *RME*
Wood (2004)	23	23	42	0.362	0.936	+27.09	6.8	**2.5**	**0.5**
Baran (2005)	**10**	**10**	**20**	**0.403**	**0.867**	**−99.31**	**12.2**	0.3	0.1

[a] See Appendix for definitions of terms.

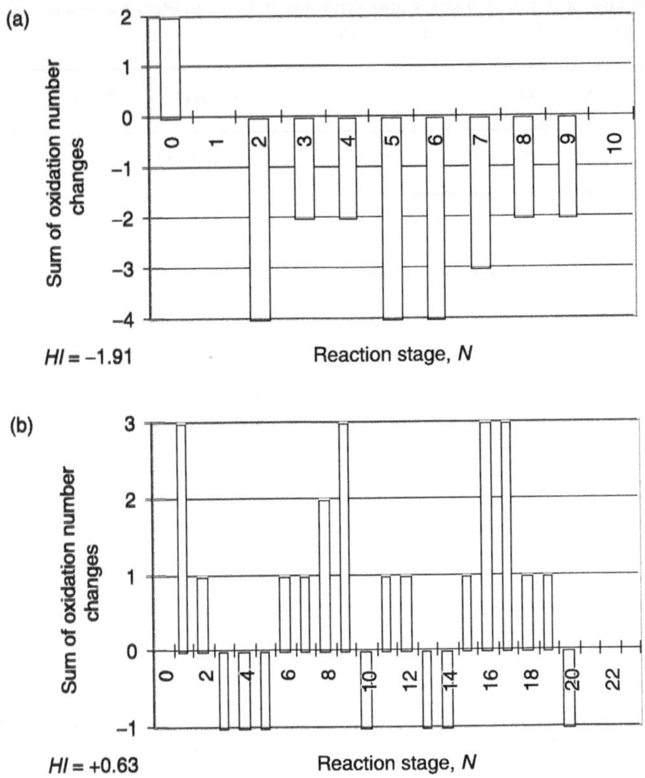

Figure 4.68 Hypsicity profiles for the Baran (a) and Wood (b) synthesis plans for welwitindolinone A.

count, reducing the number of redox reactions, reducing the range of oxidation number changes and reducing the accumulation of gains and losses in oxidation number differences over the course of the synthesis plan.

4.4 Concluding remarks

Although the above examples represent a small snapshot of synthesis analysis some very interesting patterns emerge. Figure 4.69 shows the results of a survey of 138 synthesis plans

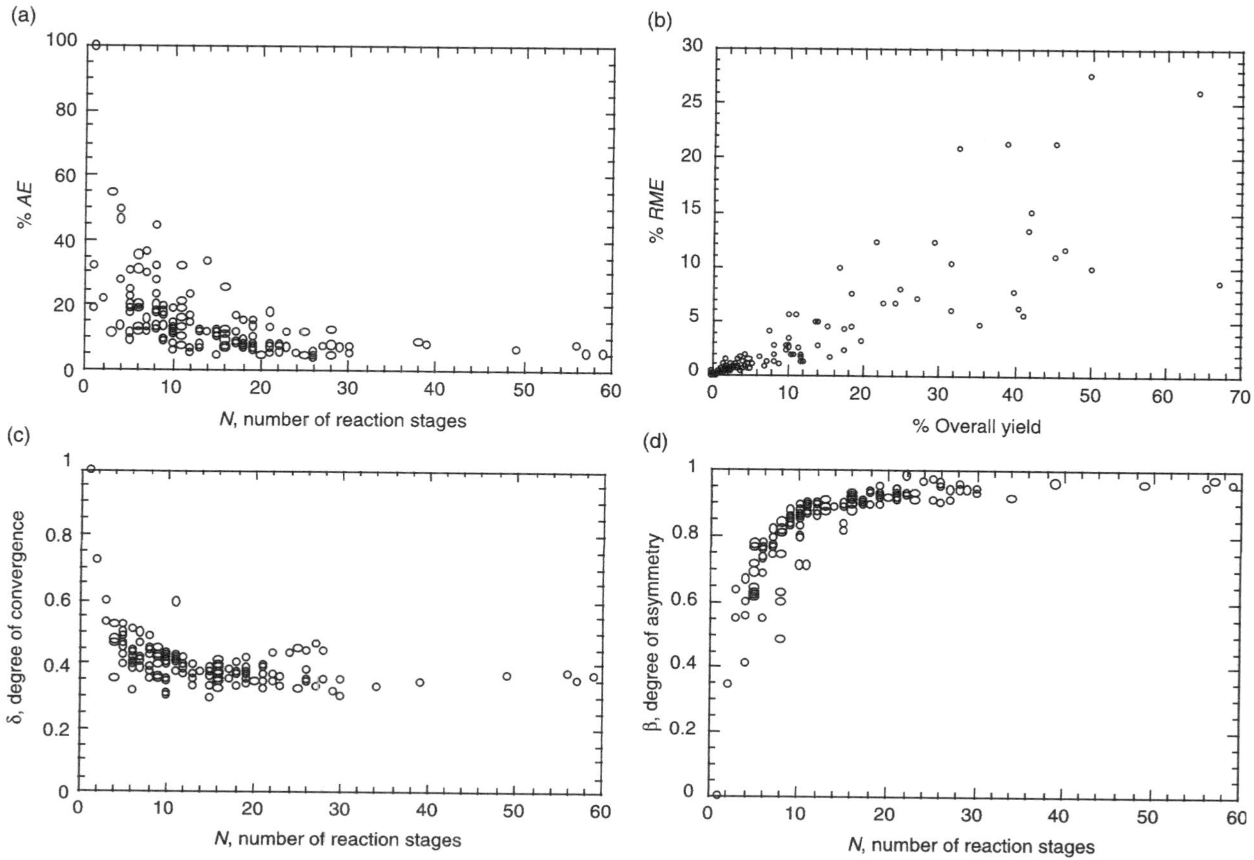

Figure 4.69 Relationships between % *AE* and number of reaction stages (a), % RME and % overall yield (b), degree of convergence and number of reaction stages (c), and degree of asymmetry and number of reaction stages (d) for 138 synthesis plans to 29 target molecules.

to 29 target molecules evaluated so far by the methods presented above. An exponential-like decay is observed when % *AE* is plotted against number of reaction stages *N* (Figure 4.69a). There is a one-to-one correspondence between the increase in % overall yield along the longest branch of a synthesis plan versus that in % overall kernel RME (Figure 4.69b). These results are not unexpected. What is most intriguing are the patterns for the degree of convergence and degree of asymmetry which result from the shapes of synthesis tree diagrams. It is observed that the limiting value for the degree of convergence is around 30% again following an exponential-like decay pattern (Figure 4.69c). The dependence of degree of asymmetry on number of reaction stages is less scattered and shows an exponential-like growth dependence. Long synthesis plans are asymmetric (upper right corner) and short ones are more symmetric (lower left corner) of Figure 4.69d. It is early at this stage to draw any meaningful conclusions about the functional predictability of these trends; however, this may change as more synthesis plans are included in this compilation.

Since the synthesis tree approach essentially chronicles the possible manufacturing origins of a given target structure in a sort of genealogical progression, two important questions arise. One deals with the 'fair' comparison of various synthesis plans to a compound given that this target node is common to all plans yet each plan begins with in principle a different set of source nodes. It is evident that the commonality of starting materials increases the further back one goes since all routes eventually lead to the same set of feedstock materials that constitute the set of genesis compounds from which all others are made. This means that head-to-head comparisons of synthesis plans become less open ended the more the plans begin with the same set of starting materials. In principle, one should go all the way back to the feedstocks from the petroleum and bulk chemicals industries to do a complete analysis as far as possible. A second corollary point deals with how far back one should go to starting purchased progenitor compounds in order to ascertain the 'greenness' of an entire plan. When a chemical company interested in achieving a 'green' synthesis of a target molecule purchases starting materials made by other companies, there is no fool proof way of knowing how 'green' the synthesis plans or processes were for those starting materials. That company can only control the 'greenness' of its forward synthesis plan to the target of interest given a particular set of starting materials. The problem essentially is where to draw the line of responsibility for checking to see if a purchased starting material was also produced in a 'green' way. In addition, environmentally conscientious firms may also consider how companies dispose of their generated wastes before purchasing reagents and intermediates. In effect there is an 'inheritance' of production built into a total metrics analysis for a given target which is difficult to assess unless all parties involved share information, subscribe to the green chemistry philosophy, and carry out detailed metrics calculations as described here for each stretch along the synthesis route beginning with feedstock materials. This is problematic if suppliers opt not to disclose their manufacturing processes or synthetic plans for legitimate proprietary reasons.

The main highlights of this chapter are as follows: (a) reaction metrics that gauge the material, energy and synthetic elegance performances of individual reactions and synthesis plans have been defined and illustrated with several examples; (b) conditions for recycling options for dealing with byproducts of reactions have been determined; (c) probability functions for achieving threshold RME values for any given reaction have been determined given minimum *AE* and minimum reaction yields constraints; and (d) true optimization is achieved when all three classes of metrics work synergistically in the positive direction so that the goals of material efficiency, practicality and synthetic elegance coincide. The reaction mass efficiency (*RME*) metric has been factored into its four contributing parameters: reaction yield (ε), atom economy (AE), excess reagent contribution ($1/SF$), and auxiliary material contribution (MRP). The greatest attenuating factor is MRP since solvents represent the bulk mass of reaction materials used. For synthesis plans, the magnitude of the kernel RME is affected most strongly by step count (N), followed by reaction yields (ε_j) and reaction atom economies ($(AE)_j$), followed by mass of excess reagents. The dependence on N is implicit whereas for the other variables the dependences are explicit. The magnitude of the overall true RME is again most strongly attenuated by auxiliary materials consumption such as reaction solvent and post-reaction materials in the work-up and purification phases. The transition from a new philosophical way of thinking about doing chemistry to the actual use of green chemistry as a standard protocol is made easier and more convincing when there exists a simple method of evaluating performance and depicting the results in a pictorial representation in a standard and straightforward way. Intelligent decision making and ranking is possible when it is based on sound numerical analysis. In the present context these numerical methods are based on basic arithmetic, geometry of triangles and the construction of radial pentagons and tree diagrams to guide the computation. The power of metrics analysis is that it can help to guide the chemist in optimizing a reaction or plan in the right direction.

Acknowledgements

I acknowledge the following people who have encouraged me and offered constructive criticism on several aspects of this ongoing endeavour, particularly, Prof. Paul T. Anastas (Yale University), Dr. Neal G. Anderson (Practical Process R&D, Jacksonville, OR), Dr. Joseph D'Antuono (Row2 Technologies, NJ), Dr. Floyd H. Dean (Ontario Research Foundation, retired), Prof. C.J. Li (McGill University), Dr. Christopher R. Schmid (Eli Lilly and Company, Indianapolis, IN), Dr. Peter L. Spargo (Scientific Update, UK), Dr. Thuy van Pham, and Prof. John C. Warner (University of Massachusetts, Boston). I also thank my undergraduate students who have taken the Industrial and Green Chemistry course at York University for their enthusiasm and pointed questions especially Neeshma Dave-Puzzo, Dragana Djokic, Julia Izhakova, Julia Krainova, Jelena Loncar, Maija Elina Lukkari, Joy McCourt, Nathaniel Nowacki, Eran Obadia, Rosa Park, Matteo Perri, Murtuzaali Sayed, Jordan Schwartz, Ethan Tumarkin, Yodita Zemicael and Barbara Zimmerman.

Appendix

(A) List of symbols and definitions

α threshold reaction mass efficiency (RME) value
a molecular weight of reaction substrate
a_{j+1} ordinate of $(j+1)$th reactant input in synthesis plan
AE atom economy
β asymmetry (skewness) parameter

c_j mass of catalyst for jth reaction
δ degree of convergence relative to single step MCR
$\$_j$ unit cost of the jth input reactant on a per gram basis
ε_j reaction yield with respect to limiting reagent for jth reaction step
ε^* reaction yield for recycling reaction
$\varepsilon_{pseudo\text{-}overall}$ pseudo-overall reaction yield
E_m environmental impact factor based on mass
E_{mw} environmental impact factor based on molecular weight
f_1 ratio of masses of catalysts and all solvents used in product forming reaction to mass of target product
f_2 ratio of masses of catalysts and all solvents used in product forming and recycling reactions to mass of target product
f_j fraction of jth isomeric product in product mixture
FTE fraction of total energy input
$\Phi_{product}$ fraction of total energy input that is directed to forming product
Φ_{waste} fraction of total energy input that is directed to forming waste
ϕ_j mass of excess reagent in jth reaction
ϕ excess moles of reagent used
G number of stages with parallel reactions
HI hypsicity (oxidation level) index
I number of reactant input structures in a synthesis plan
ℓ number of parallel linear sequences
L number of parallel reactions
μ_1 first molecular weight moment
M number of reaction steps in a synthesis plan
MCR multicomponent reaction
m_P mass of target product
MRP materials recovery parameter
$MW(P_j)$ molecular weight of jth intermediate product in synthesis plan
$MW(P_N)$ molecular weight of target product in synthesis plan
n number of points arranged vertically that correspond to reactant input structures from which centroids of intermediate products are calculated
N number of reaction stages in a synthesis plan
$(\mathrm{Ox})_j$ oxidation number of atom j in a given chemical structure
p molecular weight of target product P in synthesis plan
P co-ordinates of target product in synthesis plan
P_{mcr} co-ordinates of target product in a single step MCR
Ψ total actual input energy in kWh consumed in synthesis plan
$\Psi_{product}$ total actual input energy in kWh consumed in synthesis plan that is directed toward forming product
Ψ_{waste} total actual input energy in kWh consumed in synthesis plan that is directed toward forming waste
Ψ_j total actual input energy in kWh for jth reaction in synthesis plan
$\Psi_{product,j}$ actual input energy in kWh for jth reaction that is directed toward forming product

$\Psi_{waste,j}$	actual input energy in kWh for jth reaction that is directed toward forming waste
q	molecular weight of reaction byproduct
ρ_{rel}	relative rate of convergence
r_j	molecular weight of jth reactant in synthesis plan
RMC	raw material cost
RME	reaction mass efficiency
s_j	mass of solvent for jth reaction
$\$_A$	cost per unit mass of reagent A
SF	stoichiometric factor taking into account reagents used in excess: $SF = 1$ for stoichiometric reactions, $SF > 1$ for non-stoichiometric reactions
t_j	reaction time for jth reaction
T_{total}	total optimal reaction time for entire synthesis
θ_p	angle subtended at point P in synthesis tree
θ_{mcr}	angle subtended at point P_{mcr} in synthesis tree
w	mass of unreacted starting materials
$\bar{w}$	mass of total waste produced in a reaction
$\bar{w}_j$	mass of waste in jth reaction step
x	moles of target product in synthesis plan (defines target scale of P), or moles of reagents used (stoichiometric conditions) for a given reaction
X	molecular weight of X group in Merrifield polypeptide synthesis
y	moles of target product after reaction
$\langle y \rangle$	ordinate of point P in synthesis plan
z	moles of reagents used (non-stoichiometric conditions)
ω_j	mass of all post-reaction materials in work-up and purification phases for jth reaction

(B) Derivation of general RME expression (equation (4.1))

A general chemical transformation is given by

$$A + B \rightarrow P + Q$$

where A and B are reagents, P is the target product of interest and Q is the byproduct. If a, b, p and q represent the corresponding molecular weights of all chemical species, c, s and ω represent the masses of reaction solvent, catalyst and all post-reaction materials, respectively, and x, z and y represent the number of moles of A, B, and P where A is the limiting reagent so that $z > x$, and the reaction yield is $\varepsilon = y/x$, then we can write a conservation of mass law as

$$ax + bz + c + s + \omega = py + qy + w + c + s + \omega$$

where w is the mass of waste due to unreacted starting materials and side products.

By definition, the atom economy is given by

$$AE = \frac{p}{p+q} = \frac{p}{a+b}$$

the excess moles of B that are used is $\phi = z - x$, the mass of product collected is $m_p = py$, and the stoichiometric factor is given by

$$SF = 1 + \frac{b\phi}{x(a+b)}$$

The environmental impact factor based on mass, E_m, is then

$$\begin{aligned}
E_m &= \frac{\text{total mass of waste}}{\text{mass of target product}} \\
&= \frac{qy + w + c + s + \omega}{py} \\
&= \frac{ax + bz + c + s + \omega - py}{py} \\
&= \frac{ax + bz + c + s + \omega}{py} - 1 \\
&= \frac{ax + bx + b\phi + c + s + \omega}{py} - 1 \\
&= \frac{a + b + (b\phi + c + s + \omega)/x}{p(y/x)} - 1 \\
&= \frac{a + b + (b\phi + c + s + \omega)/x}{p\varepsilon} - 1 \\
&= \frac{1 + (b\phi + c + s + \omega)/(x(a+b))}{(p\varepsilon)/(a+b)} - 1 \\
&= \frac{[1 + (b\phi)/(x(a+b))] + (c + s + \omega)/(x(a+b))}{(AE)\varepsilon} - 1 \\
&= \frac{(SF) + (c + s + \omega)/(x(a+b))}{(AE)\varepsilon} - 1 \\
&= \frac{(SF)(1 + (c + s + \omega)/(x(a+b)(SF)))}{(AE)\varepsilon} - 1 \\
&= \frac{(SF)(1 + (c + s + \omega)/((y/\varepsilon)(p/(AE))(SF)))}{(AE)\varepsilon} - 1 \\
&= \frac{(SF)(1 + ((AE)\varepsilon(c + s + \omega))/(m_p(SF)))}{(AE)\varepsilon} - 1
\end{aligned}$$

Define the materials recovery parameter, MRP, as

$$MRP = \frac{1}{1 + ((AE)\varepsilon(c + s + \omega))/(m_p(SF))}$$

Then,

$$E_m = \frac{(SF)(1/MRP)}{\varepsilon(AE)} - 1 = \frac{1}{(\varepsilon(AE)(MRP))/(SF)} - 1 = \frac{1}{RME} - 1$$

where $RME = (\varepsilon(AE)(MRP))/(SF)$.

References

1. Constable, D.J.C., Curzons, A.D., Freitas dos Santos, L.M. *et al.* (2001) Green Chemistry Measures for Process Research and Development. *Green Chemistry*, **3**, 7–9.
2. Constable, D.J.C., Curzons, A.D., Cunningham, V.L. (2002) Metrics to 'Green' Chemistry – Which Are the Best? *Green Chemistry*, **4**, 521–527.
3. Curzons, A.D., Constable, D.J.C., Mortimer, D.N., Cunningham, V.L. (2001) So You Think Your Process is Green, How Do You Know? – Using Principles of Sustainability to Determine What is Green – A Corporate Perspective. *Green Chemistry*, **3**, 1–6.
4. Eissen, M., Hungerbühler, K., Dirks, S., Metzger, J. (2003) Mass Efficiency As Metric for the Effectiveness of Catalysts. *Green Chemistry*, **5**, G25–G27.
5. Eissen, M. & Metzger, J.O. (2002) Environmental Performance Metrics for Daily Use in Synthetic Chemistry, *Chemistry: A European Journal*, **8**, 3580–3585.
6. Metzger, J.O. & Eissen, M. (2004) Concepts on the Contribution of Chemistry to a Sustainable Development. Renewable Raw Materials. *Comptes Rendus de l'Académie des Sciences Paris, Série IIc, Chimie*, **7**, 569–581.
7. Metzger, J.O. (1998) Solvent-Free Organic Syntheses. *Angewandte Chemie International Edition*, **37**, 2975–2978.
8. Eissen, M., Mazur, R., Quebbemann, H.G., Pennemann, K.H. (2004) Atom Economy and Yield of Synthesis Sequences. *Helvetica Chimica Acta*, **87**, 524–535.
9. Eissen, M., Metzger, J.O., Schmidt, E., Schneidewind, U. (2002) 10 Years After Rio – Concepts on the Contribution of Chemistry to a Sustainable Development. *Angewandte Chemie International Edition*, **41**, 414–436.
10. Eissen, M. (2001) *Bewertung der Umweltverträglichkeit organisch-chemischer Synthesen.* PhD thesis, Universität Oldenburg.
11. Steinbach, A. & Winkenbach, R. (2000) Choose Processes for Their Productivity. *Chemical Engineering*, **April**, 94–96, 98, 100, 102, 104.
12. Gronnow, M.J., White, R.J., Clark, J.H., Macquarrie, D.J. (2005) Energy Efficiency in Chemical Reactions: A Comparative Study of Different Reaction Techniques. *Organic Process Research and Development*, **9**, 516–518.
13. Anastas, P.T. & Warner, J.C. (1998) *Green Chemistry: Theory and Practice.* Oxford University Press, Oxford.
14. Tucker, J.L. (2006) Green Chemistry, a Pharmaceutical Perspective. *Organic Process Research and Development*, **10**, 315–319.
15. Tundo, P., Anastas, P., StC. Black, D. *et al.* (2000) Synthetic Pathways and Processes in Green Chemistry. Introductory Overview. *Pure and Applied Chemistry*, **72**, 1207–1228.
16. Cornforth, J.W. (1993) The Trouble with Synthesis. *Australian Journal of Chemistry*, **46**, 157–170.
17. Snieckus, V. (1999) Optimization in Organic Synthesis. *Medical Research Reviews*, **19**, 342–347.
18. Snieckus, V. (1983) Whither Organic Synthesis? *Accounts of Chemical Research*, **16**, 186.

19. Seebach, D. (1990) Organic Synthesis – Where Now? *Angewandte Chemie International Edition*, **29**, 1320–1367.
20. Wender, P. & Miller, B.L. (1993) Toward the Ideal Synthesis: Connectivity Analysis and Multibond-Forming Processes. In: *Organic Synthesis: Theory and Applications*, Vol. 2 (ed. T. Hudlicky), pp. 27–65. JAI Press, Greenwich, CT.
21. Nicolaou, K.C. & Snyder, S.A. (2004) The Essence of Total Synthesis. *Proceedings of the National Academy of Sciences of the United States of America*, **101**, 11929–11936.
22. Nicolaou, K.C., Vourloumis, D., Winssinger, N., Baran, P.S. (2000) The Art and Science of Total Synthesis at the Dawn of the Twenty-First Century. *Angewandte Chemie International Edition*, **39**, 44–122.
23. Wender, P.A., Handy, S.T., Wright, D.L. (1997) Towards the Ideal Synthesis. *Chemistry & Industry (London)*, **765**, 767–769.
24. Lavoisier, A. (1778) Mémoire sur la nature du principe qui se combine avec les métaux pendant leur calcination, et qui en augmente le poids. *Mèmoires de l'Académie Royale des Sciences (1775)*, 520–526.
25. Andraos, J. (2005) Unification of Reaction Metrics for Green Chemistry: Applications to Reaction Analysis. *Organic Process Research and Development*, **9**, 149–163.
26. Andraos, J. (2005) Unification of Reaction Metrics for Green Chemistry II: Evaluation of Named Organic Reactions and Application to Reaction Discovery. *Organic Process Research and Development*, **9**, 404–431.
27. Andraos, J. (2005) Erratum to Unification of Reaction Metrics for Green Chemistry: Applications to Reaction Analysis. *Organic Process Research and Development*, **9**, 519.
28. Andraos, J. (2005) On Using Tree Analysis to Quantify the Material, Input Energy, and Cost Throughput Efficiencies of Simple and Complex Synthesis Plans and Networks: Towards a Blueprint for Quantitative Total Synthesis and Green Chemistry. *Organic Process Research and Development*, **10**, 212–240.
29. Andraos, J. & Izhakova, J. (2006) Perspectives on the Application of Green Chemistry Principles to Total Synthesis Design. *Chimica Oggi/The International Journal of Industrial Chemistry*, **24**(6, Suppl.), 31–36.
30. Andraos, J., Sayed, M. (2007) On the Use of 'Green' Metrics in the Undergraduate Organic Chemistry Lecture and Laboratory to Assess the Mass Efficiency of Organic Reactions. *Journal of Chemical Education*, **84**, 1004–1010.
31. Andraos, J. (2007) Gauging Material Efficiency. *Canadian Chemical News*, **59**(4), 14–17.
32. Trost, B.M. (1991) The Atom Economy – A Search for Synthetic Efficiency. *Science*, **254**, 1471–1477.
33. Trost, B.M. (2002) On Inventing Reactions for Atom Economy. *Accounts of Chemical Research*, **35**, 695–705.
34. Trost, B.M. (1995) Atom Economy. A Challenge for Organic Synthesis – Homogeneous Catalysis Leads the Way. *Angewandte Chemie International Edition*, **34**, 259–281.
35. Sheldon, R.A. (2001) Atom Efficiency and Catalysis in Organic Synthesis. *Pure and Applied Chemistry*, **72**, 1233–1246.
36. Trost, B.M., Fredericksen, M.U., Rudd, M.T. (2005) Ruthenium-Catalyzed Reactions – A Treasure Trove of Atom-Economic Transformations. *Angewandte Chemie International Edition*, **44**, 6630–6666.
37. Sheldon, R.A. (1994) Consider the Environmental Quotient. *ChemTech*, **24**(3), 38–47.
38. Sheldon, R.A. (1992) Organic Synthesis – Past, Present and Future. *Chemistry & Industry (London)*, 903–906.
39. Sheldon, R.A. (1997) Catalysis and Pollution Prevention. *Chemistry & Industry (London)*, 12–15.
40. Sheldon, R.A. (2000) Atom Utilisation, E Factors and the Catalytic Solution. *Comptes Rendus de l'Académie des Sciences Paris, Série IIc, Chimie*, **3**, 541–551.

41. Muccioli, G.G., Poupaert, J.H., Wouters, J. *et al.* (2003) A Rapid and Efficient Microwave-Assisted Synthesis of Hydantoins and Thiohydantoins. *Tetrahedron*, **59**, 1301–1307.
42. Miao, W. & Chan, T.H. (2003) Exploration of Ionic Liquids as Soluble Supports for Organic Synthesis. Demonstration with a Suzuki Coupling Reaction. *Organic Letters*, **5**, 5003–5005.
43. Amiard, G. & Heymes, R. (1956) N,N'-Dicyclohexylcarbodiimide. Preparation from Urea and Cyclohexylamine. *Bulletin de la Société Chimique de France*, 1360–1361.
44. Stevens, C.L., Singhal, G.H., Ash, A.B. (1967) Carbodiimides. Dehydration of Ureas. *Journal of Organic Chemistry*, **32**, 2895.
45. Fujibayashi, R. (1995) Preparation of 1,3-Dicyclohexylcarbodiimide (DCC) from 1,3-Dicyclohexylurea (DCU). JP 08231491 (Sumikin Kako KK, Japan).
46. Hussenet, P., Le Goff, P., Sennyey, G. (1996) Two-Step Process for the Preparation of N,N'-Disubstituted Carbodiimides by Phosgenation of Disubstituted Ureas. EP 723955 (Société Nationale des Poudres et Explosifs, France).
47. Smith, M., Moffatt, J.G., Khorana, H.G. (1958) Carbodiimides. VIII. Observations on the Reactions of Carbodiimides with Acids and Some New Applications in the Synthesis of Phosphoric Acid Esters. *Journal of the American Chemical Society*, **80**, 6204–6212.
48. Kumagai, N., Muncipinto, G., Schreiber, S.L. (2006) Short Synthesis of Skeletally and Stereochemically Diverse Small Molecules by Coupling Petasis Condensation Reactions to Cyclization Reactions. *Angewandte Chemie International Edition*, **45**, 3635–3638.
49. Walser, H. (2001) *The Golden Section*. Mathematical Association of American, Washington, DC.
50. Livio, M. (2002) *The Golden Ratio: The Story of Phi, the World's Most Astonishing Number*. Broadway Books, New York.
51. O'Neil, M.J., Smith, A., Heckelman, P.E. (2001) *The Merck Index: An Encyclopedia of Chemicals, Drugs, and Biologicals*, 13th edn. Merck & Co., Inc., New Jersey.
52. Mundy, B.P. & Ellerd, M.G. (1988) *Name Reactions and Reagents in Organic Synthesis*. Wiley, New York.
53. Laue, T. & Plagens, A. (2000) *Named Organic Reactions*. Wiley, New York.
54. Smith, J.G. & Fieser, M. (1990) *Fieser and Fieser's Reagents for Organic Synthesis Collective Index for Volumes 1–12*. Wiley, New York.
55. Li, J.J. (2002) *Name Reactions: A Collection of Detailed Reaction Mechanisms*. Springer-Verlag, Berlin.
56. Andraos, J. (2000) *Named Things in Chemistry and Physics*. http://www.careerchem.com/NAMED/Homepage.html (accessed August 2006).
57. Wöhler, F. (1828) Sur la Formation Artificielle de l'Urée. *Annales de Chimie et de Physique*, **37**(1), 330–334.
58. Wöhler, F. (1828) Ueber künstliche Bildung des Harnstoffs. *Annalen der Physik und Chemie*, **12**(?), 253–256.
59. Posner, G.H. (1986) Multicomponent One-Pot Annulations Forming Three to Six Bonds. *Chemical Reviews*, **86**, 831–844.
60. Ugi, I., Dömling, A., Hörl, W. (1994) Multicomponent Reactions in Organic Chemistry. *Endeavour (New Series)*, **18**(3), 115–122.
61. Armstrong, R.W., Combs, A.P., Tempest, P.A., Brown, S.D., Keating, T.A. (1996) Multiple-Component Condensation Strategies for Combinatorial Library Synthesis. *Accounts of Chemical Research*, **29**, 123–131.
62. Ugi, I. (1998) MCR. XXIII. The Highly Variable Multidisciplinary Preparative and Theoretical Possibilities of the Ugi Multicomponent Reactions in the Past, Now, and in the Future. *Proceedings of the Estonian Academy of Science: Chemistry*, **47**, 107–127.
63. Lombardo, M. & Trombini, C. (1998) One-Pot Multicomponent Reactions. In: *Seminars in Organic Synthesis XXIII Summer School 'A. Corbella'*, pp. 7–32. Società Chimica Italiana, Milan.

64. Dömling, A., Herdtweck, E., Ugi, I. (1998) MCR V. The Seven-Component Reaction. *Acta Chemica Scandinavica*, **52**, 107–113.
65. Ugi, I., Almstetter, M., Bock, M. *et al.* (1998) MCR XVII. Three Types of MCRs and the Libraries – Their Chemistry of Natural Events and Preparative Chemistry. *Croatica Chemica Acta*, **71**, 527–547.
66. Weber, L., Illgen, K., Almstetter, M. (1999) Discovery of New Multi-Component Reactions with Combinatorial Methods. *Synlett*, 366–374.
67. Tietze, L.F. & Modi, A. (2000) Multicomponent Domino Reactions for the Synthesis of Biologically Active Natural Products and Drugs. *Medical Chemical Reviews*, **20**, 304–322.
68. Ugi, I., Dömling, A., Werner, B. (2000) Since 1995 the New Chemistry of Multicomponent Reactions and Their Libraries, Including Their Heterocyclic Chemistry. *Journal of Heterocyclic Chemistry*, **37**, 647–658.
69. Kappe, C.O. (2000) Recent Advances in the Biginelli Dihydropyrimidine Synthesis. New Tricks from an Old Dog. *Accounts of Chemical Research*, **33**, 879–888.
70. Dömling, A. (2000) The Discovery of New Isocyanide-Based Multi-Component Reactions. *Current Opinion in Chemical Biology*, **4**, 318–323.
71. Ugi, I. (2001) Recent Progress in the Chemistry of Multicomponent Reactions. *Pure and Applied Chemistry*, **73**, 187–191.
72. Dömling, A. (2002) Recent Advances in Isocyanide-Based Multicomponent Chemistry. *Current Opinion in Chemical Biology*, **6**, 306–313.
73. Weber, L. (2002) Multi-Component Reactions and Evolutionary Chemistry. *Drug Discovery Today*, **7**, 143–147.
74. Weber, L. (2002) The Application of Multi-Component Reactions in Drug Discovery. *Current Medicinal Chemistry*, **9**, 1241–1253.
75. Orru, R.V.A. & de Greef, M. (2003) Recent Advances in Solution-Phase Multicomponent Methodology for the Synthesis of Heterocyclic Compounds. *Synthesis*, 1471–1499.
76. von Wangelin, A., Neumann, H., Gördes, D., Klaus, S., Strübing, D., Beller, M. (2003) Multicomponent Coupling Reactions for Organic Synthesis: Chemoselective Reactions with Amide-Aldehyde Mixtures. *Chemistry: A European Journal*, **9**, 4286–4294.
77. Ugi, I., Werner, B., Dömling, A. (2003) The Chemistry of Isocyanides, Their Multi-Component Reactions and Their Libraries. *Molecules*, **8**, 53–66.
78. Simon, C., Constantieux, T., Rodriguez, J. (2004) Utilisation of 1,3-Dicarbonyl Derivatives in Multicomponent Reactions. *European Journal of Organic Chemistry*, 4957–4980.
79. Ramachary, D.B. & Barbas, C.F. III (2004) Towards Organo-Click Chemistry: Development of Organocatalytic Multicomponent Reactions Through Combinations of Aldol, Wittig, Knoevenagel, Michael, Diels–Alder and Huisgen Cycloaddition Reactions. *Chemistry: A European Journal*, **10**, 5323–5331.
80. Zhu, J. & Bienyamé, H. (2005) *Multicomponent Reactions*. Wiley-VCH, Weinheim.
81. Dömling, A. (2006) Recent Developments in Isocyanide Based Multicomponent Reactions in Applied Chemistry. *Chemical Reviews*, **106**, 17–89.
82. Ramón, D.J. & Yus, M. (2005) Asymmetric Multicomponent Reactions (AMCRs): The New Frontier. *Angewandte Chemie International Edition*, **44**, 1602–1634.
83. Tejedor, D., González-Cruz, D., Santos-Expósito, A., Marrero-Tellado, J.J., de Armas, P., Garciá-Tellado, F. (2005) Multicomponent Domino Processes Based on the Organocatalytic Generation of Conjugated Acetylides: Efficient Synthetic Manifolds for Diversity-Oriented Molecular Construction. *Chemistry: A European Journal*, **11**, 3502–3510.
84. Dondoni, A. & Massi, A. (2006) Design and Synthesis of New Classes of Heterocyclic C-Glycoconjugates and Carbon-Linked Sugar and Heterocyclic Amino Acids by Asymmetric Multicomponent Reactions (AMCRs). *Accounts of Chemical Research*, **39**, 451–463.

85. Choudary, B.M., Chowdari, N.S., Mahdi, S., Kantam, M.L. (2003) A Trifunctional Catalyst for One-Pot Synthesis of Chiral Diols via Heck Coupling-*N*-Oxidation-Asymmetric Dihydroxylation: Application for the Synthesis of Diltiazem and Taxol Side Chain. *Journal of Organic Chemistry*, **68**, 1736–1746.
86. Kim, S.S. & Rajagopal, G. (2004) Efficient Aerobic Oxidation of Alcohols to Carbonyl Compounds with NHPI/CAN Catalytic System. *Synthetic Communications*, **33**, 2237–2243.
87. Chandrasekhar, S., Narsihmulu, C., Sultana, S.S., Reddy, N.R. (2003) Osmium Tetroxide in Poly(ethylene glycol) (PEG): A Recyclable Reaction Medium for Rapid Asymmetric Dihydroxylation Under Sharpless Conditions. *Chemical Communications*, 1716–1717.
88. Noyori, R., Aoki, M., Sato, K. (2003) Green Oxidation with Aqueous Hydrogen Peroxide. *Chemical Communications*, 1977–1986.
89. Sato, K., Aoki, M., Noyori, R. (1998) A 'Green' Route to Adipic Acid: Direct Oxidation of Cyclohexenes with 30 Percent Hydrogen Peroxide. *Science*, **281**, 1646–1647.
90. Sheldon, R.A., Arends, I.W.C.E., Brink, G.J.T., Dijksman, A. (2002) Green, Catalytic Oxidations of Alcohols. *Accounts of Chemical Research*, **35**, 774–781.
91. Velluz, L., Valls, J., Mathieu, J. (1967) Spatial Arrangement and Preparative Organic Synthesis. *Angewandte Chemie International Edition*, **6**, 778–787.
92. Bertz, S.H. & Sommer, T.J. (1993) Applications of Graph Theory to Synthesis Planning: Complexity, Reflexivity, and Vulnerability. In: *Organic Synthesis: Theory and Applications*, Vol. 2 (ed. T. Hudlicky), pp. 67–92. JAI Press, Greenwich, CT.
93. Hendrickson, J.B. (1977) Systematic Synthesis Design. 6. Yield Analysis and Convergency. *Journal of the American Chemical Society*, **99**, 5439–5450.
94. Bertz, S.H. (2003) Complexity of Synthetic Reactions. The Use of Complexity Indices to Evaluate Reactions, Transforms and Disconnections. *New Journal of Chemistry*, **27**, 860–869.
95. Bertz, S.H. (2003) Complexity of Synthetic Routes: Linear, Convergent and Reflexive Syntheses. *New Journal of Chemistry*, **27**, 870–879.
96. Bertz, S.H. (1981) The First General Index of Molecular Complexity. *Journal of the American Chemical Society*, **103**, 3599–3601.
97. Bertz, S.H. (1983) On the Complexity of Graphs and Molecules. *Bulletin of Mathematical Biology*, **45**, 849–855.
98. Whitlock, H.W. (1998) On the Structure of Total Synthesis of Complex Natural Products. *Journal of Organic Chemistry*, **63**, 7982–7989.
99. Hendrickson, J.B. (1971) A Systematic Characterization of Structures and Reactions for Use in Organic Synthesis. *Journal of the American Chemical Society*, **93**, 6847–6854.
100. Parsons, P.J., Penkett, C.S., Shell, A.J. (1996) Tandem Reactions in Organic Synthesis: Novel Strategies for Natural Product Elaboration and the Development of New Synthetic Methodology. *Chemical Reviews*, **96**, 195–206.
101. Winkler, J.D. (1996) Tandem Diels–Alder Cycloadditions in Organic Synthesis. *Chemical Reviews*, **96**, 167–176.
102. Wasilke, J.C., Obrey, S.J., Baker, R.T., Bazan, G.C. (2005) Concurrent Tandem Catalysis. *Chemical Reviews*, **105**, 1001–1020.
103. Guo, H.C. & Ma, J.A. (2006) Catalytic Asymmetric Tandem Transformations Triggered by Conjugate Additions. *Angewandte Chemie International Edition*, **45**, 354–366.
104. Malacria, M. (1996) Selective Preparation of Complex Polycyclic Molecules from Acyclic Precursors via Radical Mediated or Transition Metal-Catalyzed Cascade Reactions. *Chemical Reviews*, **96**, 289–306.
105. Tietze, L.F. (1995) Domino Reactions in Organic Synthesis. *Chemistry & Industry (London)*, 453–457.
106. Tietze, L.F. & Rackelmann, N. (2004) Domino Reactions in the Synthesis of Heterocyclic Natural Products and Analogs. *Pure and Applied Chemistry*, **76**, 1967–1983.

107. Tietze, L.F. (1996) Domino Reactions in Organic Synthesis. *Chemical Reviews*, **96**, 115–136.
108. Tietze, L.F. & Beifuss, U. (1993) Sequential Transformations in Organic Chemistry: A Synthetic Strategy with a Future. *Angewandte Chemie International Edition*, **32**, 131–163.
109. Tietze, L.F. & Lieb, M.E. (1998) Domino Reactions for Library Synthesis of Small Molecules in Combinatorial Chemistry. *Current Opinion in Chemical Biology*, **2**, 363–371.
110. Grigg, R. & Sridharan, V. (1999) Palladium Catalyzed Cascade Cyclisation – anion Capture, Relay Switches and Molecular Queues. *Journal of Organometallic Chemistry*, **576**, 65–87.
111. Nicolaou, K.C., Edmonds, D.J., Bulger, P.G. (2006) Cascade Reactions in Total Synthesis. *Angewandte Chemie International Edition*, **45**, 7134–7186.
112. Hoffmann, R.W. (2006) Protecting-Group-Free Synthesis. *Synthesis*, 3531–3541.
113. Ireland, R.E. (1969) *Organic Synthesis*. Prentice-Hall, Englewood Cliffs, New Jersey.
114. Serratosa, F. (1990) *Organic Chemistry in Action: the Design of Organic Synthesis*. Elsevier, Amsterdam.
115. Carlson, R. (1992) *Design and Optimization in Organic Synthesis*. Elsevier, Amsterdam.
116. Broadwater, S.J., Roth, S.L., Price, K.E., Kobaslija, M., McQuade, D.T. (2005) One-Pot Multi-Step Synthesis: A Challenge Spawning Innovation. *Organic and Biomolecular Chemistry*, **3**, 2899–2906.
117. Chanon, M., Barone, R., Baralotto, C., Julliard, M., Hendrickson, J.B. (1998) Information Theory Description of Synthetic Strategies in the Polyquinane Series. The Holosynthon Concept. *Synthesis*, 1559–1583.
118. Carey, J.S., Laffan, D., Thomson, C., Williams, M.T. (2006) Analysis of the Reactions Used for the Preparation of Drug Candidate Molecules. *Organic and Biomolecular Chemistry*, **4**, 2337–2347.
119. Sierra, M.A. & de la Torre, M.C. (2000) Dead Ends and Detours En Route to Total Syntheses of the 1990s. *Angewandte Chemie International Edition*, **39**, 1539–1559.
120. Corey, E.J. & Helal, C.J. (1998) Reduction of Carbonyl Compounds with Chiral Oxazaborolidine Catalysts: A New Paradigm for Enantioselective Catalysis and a Powerful New Synthetic Method. *Angewandte Chemie International Edition*, **37**, 1986–2012.
121. Nicolaou, K.C. (2005) Joys of Molecules. 1. Campaigns in Total Synthesis. *Journal of Organic Chemistry*, **70**, 7007–7027.
122. Nicolaou, K.C. (2003) Perspectives in Total Synthesis: A Personal Account. *Tetrahedron*, **59**, 6683–6738.
123. Corey, E.J. (2004) Impossible Dreams. *Journal of Organic Chemistry*, **69**, 2917–2919.
124. Meinwald, J. (2005) Personal Reflections on Receiving the Roger Adams Award in Organic Chemistry. *Journal of Organic Chemistry*, **70**, 4903–4909.
125. Schreiber, S.L. (2000) Target-Oriented and Diversity-Oriented Organic Synthesis in Drug Discovery. *Science*, **287**, 1964–1969.
126. Lee, M.L. & Schneider, G. (2001) Scaffold Architecture and Pharmacophoric Properties of Natural Products and Trade Drugs: Application in the Design of Natural Product-Based Combinatorial Libraries. *Journal of Combinatorial Chemistry*, **3**, 284–289.
127. Hudlicky, T. (1996) Design Constraints in Practical Syntheses of Complex Molecules: Current Status, Case Studies With Carbohydrates and Alkaloids, and Future Perspectives. *Chemical Reviews*, **96**, 3–30.
128. Heathcock, C.H. (1992) Yuzuriha Alkaloids. *Angewandte Chemie International Edition*, **31**, 665–681.
129. Dartt, C.B. & Davis, M.E. (1994) Catalysis for Environmentally Benign Processing. *Industrial Engineering Chemical Research*, **33**, 2887–2899.
130. Noyori, R. (2005) Pursuing Practical Elegance in Chemical Synthesis. *Chemical Communications*, 1807–1811.

131. Federsel, H.J. (2005) Asymmetry on Large Scale: The Roadmap to Stereoselective Processes. *Nature Reviews Drug Discovery*, **4**, 685–697.
132. Abdel-Magid, F.A. (2004) *Chemical Process Research: The Art of Practical Organic Synthesis.* American Chemical Society, Washington.
133. Lee, S. & Robinson, G. (1995) *Process Development: Fine Chemicals From Grams to Kilograms.* Oxford University Press, Oxford.
134. Anderson, N.G. (2000) *Practical Process Research and Development*, Academic Press, San Diego.
135. Fuchs, P.L. (2001) Increase in Intricacy – a Tool for Evaluating Organic Synthesis. *Tetrahedron*, **57**, 6855–6875.
136. Pilipauskas, D.R. (1999) Can the Time from Synthesis Design to Validated Chemistry Be Shortened? *Medical Research Reviews*, **19**, 463–474.
137. Zhang, T.Y. (2006) Process Chemistry: The Science, Business, Logic, and Logistics. *Chemical Reviews*, **106**, 2583–2595.
138. Craig, L.C. (1933) A New Synthesis of Nornicotine and Nicotine. *Journal of the American Chemical Society*, **55**, 2854–2857.
139. Späth, E. & Bretschneider, H. (1928) Eine neue Synthese des Nicotins und einige Bemerkungen zu den Arbeiten Nagais uber Ephedrine. *Chemische Berichte*, **61**, 327–334.
140. Felpin, F.X., Girard, S., Vo-Thanh, G., Robins, R.J., Villieras, J., Lebreton, J. (2001) Efficient Enantiomeric Synthesis of Pyrrolidine and Piperidine Alkaloids from Tobacco. *Journal of Organic Chemistry*, **66**, 6305–6312.
141. Chavdarian, C.G., Sanders, E.B., Bassfield, R.L. (1982) Synthesis of Optically Active Nicotinoids. *Journal of Organic Chemistry*, **47**, 1069–1073.
142. Welter, C., Moreno, R.M., Streiff, S., Helmchen, G. (2005) Enantioselective Synthesis of (+)(R)- and (−)(S)-Nicotine Based on Ir-catalyzed Allylic Amination. *Organic and Biomolecular Chemistry*, **3**, 3266–3268.
143. Loh, T.P., Zhou, J.R., Li, X.R., Sim, K.Y. (1999) A Novel Reductive Aminocyclization for the Synthesis of Chiral Pyrrolidines: Stereoselective Syntheses of (S)-Nornicotine and 2-(2′-Pyrrolidyl)-pyridines. *Tetrahedron Letters*, **40**, 7847–7850.
144. Márquez, F., Llebaria, A., Delgado, A. (2001) Studies on the Diastereoselective Allylation of Aldehydes with Enantiopure 2-Sulfinylallyl Building Blocks. *Tetrahedron Asymmetry*, **12**, 1625–1634.
145. Jacob, P. III (1982) Resolution of (±)-5-Bromonornicotine. Synthesis of (R) and (S)-Nornicotine of High Enantiomeric Purity. *Journal of Organic Chemistry*, **47**, 4165–4167.
146. Baxendale, I.R., Brusotti, G., Matsuoka, M., Ley, S.V. (2002) Synthesis of Nornicotine, Nicotine, and Other Functionalised Derivatives Using Solid-supported Reagents and Scavengers. *Journal of the Chemical Society Perkin Transactions 1*, 143–154.
147. Campos, K.R., Klapers, A., Waldman, J.H., Dormer, P.G., Chen, C.Y. (2006) Enantioselective, Palladium-Catalyzed α-Arylation of N-Boc-pyrrolidine. *Journal of the American Chemical Society*, **128**, 3538–3539.
148. Pictet, A. & Rotschy, A. (1904) Synthese des Nicotins. *Chemische Berichte*, **37**, 1225–1235.
149. Pictet, A. (1900) Ueber die Reduction des Nicotyrins zu inactivem Nicotin. *Chemische Berichte*, **33**, 2355–2356.
150. Pictet, A. & Crépieux, P. (1898) Ueber die Hydrirung des Nicotyrins. *Chemische Berichte*, **31**, 2018–2022.
151. Pictet, A. & Crépieux, P. (1895) Ueber phenyl- und pyridylpyrrole und die constitution des Nicotins. *Chemische Berichte*, **28**, 1904–1912.
152. Ault, R.G., Baird, D.K., Carrington, H.C. *et al.* (1933) Synthesis of D- and of L-Ascorbic Acid and of Analogous Substances. *Journal of the Chemical Society*, 1419–1423.

153. Reichstein, T. & Grüssner, A. (1934) Eine ergiebige Synthese der L-Ascorbinsäure (C-Vitamin). *Helvetica Chimica Acta,* **17**, 311–328.
154. Reichstein, T. & Grüssner, A. (1933) Die Synthese der D-Ascorbinsäure (D-Form des C-Vitamin). *Helvetica Chimica Acta,* **16**, 561–565.
155. Reichstein, T., Grüssner, A., Oppenauer, R. (1933) Synthese der D- und L-Ascorbinsäure (C-Vitamin). *Helvetica Chimica Acta,* **16**, 1019–1033.
156. Haworth, W.N. & Hirst, E.L. (1933) Synthesis of Ascorbic Acid. *Chemistry & Industry (London),* 645–646.
157. Ohle, H. & Berend, G. (1925) Über die Aceton-Verbidungen der Zucker und ihre Derivate, IV. Die Konstitution der Diaceton-galaktose. *Berichte der Deutschen Chemischen Gesellschaft,* **58**, 2585–2589.
158. Weerman, R.A. (1918) L'action de l'hypochlorite de sodium sur les amides d'α-hydroxy-acides et de polyhydroxy-acides, ayant un group hydroxyle à la place α. *Recueil des Travaux Chimiques des Pays-Bas et de la Belgique,* **37**, 16–51.
159. Levene, P.A. & LaForge, F.B. (1914) Note on a Case of Pentosuria. *Journal of Biological Chemistry,* **18**, 319–327.
160. Moffatt, J.G. & Khorana, H.G. (1961) Nucleoside Polyphosphates. XII. The Total Synthesis of Coenzyme A. *Journal of the American Chemical Society,* **83**, 663–675.
161. Davoll, J., Lythgoe, B., Todd, A.R. (1948) Experiments on the Synthesis of Purine Nucleosides. Part XIX. A Synthesis of Adenosine. *Journal of the Chemical Society,* 967–969.
162. Davoll, J., Lythgoe, B., Todd, A.R. (1946) Experiments on the Synthesis of Purine Nucleosides. Part XII. The Configuration of the Glycosidic Centre in Natural and Synthetic Pyrimidine and Purine Nucleosides. *Journal of the Chemical Society,* 833–838.
163. Fischer, E. (1897) Synthese des Hypoxanthins, Xanthins, Adenins und Guanine. *Berichte der Deutschen Chemischen Gesellschaft,* **30**, 2226–2254.
164. Baddiley, J., Kenner, G.W., Lythgoe, B., Todd, A.R. (1944) Experiments on the Synthesis of Purine Nucleosides. Part X. A Synthesis of 9-D-Ribopyranosidoadenine. *Journal of the Chemical Society,* 657–659.
165. Kenner, G.W., Todd, A.R., Weymouth, F.J. (1952) Nucleotides. Part XVII. *N*-Chloroamides as Reagents for the Chlorination of Diesters of Phosphorous Acid. A New Synthesis of Uridine-5′ Pyrophosphate. *Journal of the Chemical Society,* 3675–3681.
166. Johnson, A.W., Miller, G.W., Mills, J.A., Todd, A.R. (1953) Chemistry of the Vitamin B_{12} Group. II. Synthesis of 5,6-Dimethyl-1-α-, -D-Ribofuranosylbenzimidazole. *Journal of the Chemical Society,* 3061–3066.
167. Baddiley, J. & Thain, E.M. (1951) The Synthesis of Pantothenic Acid-2′ and -4′ Phosphates as Possible Degradation Products of Coenzyme A. *Journal of the Chemical Society,* 246–251.
168. Baddiley, J. & Thain, E.M. (1952) Coenzyme A. Part V. A New and Convenient Synthesis of Pantetheine (*Lactobacillus bulgaricus* Factor). *Journal of the Chemical Society,* 800–803.
169. Baddiley, J. & Thain, E.M. (1953) Coenzyme A. Part VIII. The Synthesis of Pantetheine 4′-Phosphate (*Acetobacter* Stimulatory Factor), a Degradation Product of the Coenzyme. *Journal of the Chemical Society,* 1610–1615.
170. Graening, T. & Schmalz, H.G. (2004) Total Syntheses of Colchicine in Comparison: A Journey Through 50 Years of Synthetic Organic Chemistry. *Angewandte Chemie International Edition,* **43**, 3230–3256.
171. Graening, T., Bette, V., Neudörfl, J., Lex, J., Schmalz, H.G. (2005) Total Synthesis of (−)-Colchicine via a Rh-Triggered Cycloaddition Cascade. *Organic Letters,* **7**, 4317–4320.
172. Graening, T., Friedrichsen, W., Lex, J., Schmalz, H.G. (2002) Facile Construction of the Colchicine Skeleton by a Rhodium-Catalyzed Cyclization/Cycloaddition Cascade. *Angewandte Chemie International Edition,* **41**, 1524–1526.
173. Woodward, R.B. (1963) A Total Synthesis of Colchicine. *The Harvey Lectures,* **59**, 31–47.

174. Scott, A.I., McCapra, F., Buchanan, R.L., Day, A.C., Young, D.W. (1965) Total Synthesis of Colchicine Modelled on a Biogenetic Theory. *Tetrahedron*, **21**, 3605–3631.
175. Martel, J., Toromanoff, E., Huynh, C. (1965) A New Synthesis of Deacetamidocolchiceine. *Journal of Organic Chemistry*, **30**, 1752–1759.
176. Schreiber, J., Leimgruber, W., Pesaro, M., Schudel, P., Eschenmoser, A. (1959) Synthesis of Colchicine. *Angewandte Chemie*, **71**, 637–640.
177. Schreiber, J., Leimgruber, W., Pesaro, M., Schudel, P., Eschenmoser, A. (1961) Synthesis of Colchicine. *Helvetica Chimica Acta*, **44**, 540–597.
178. van Tamelen, E.E., Spencer, T.A. Jr., Allen, D.S. Jr., Ovis, R.L. (1959) Total Synthesis of Colchicine. *Journal of the American Chemical Society*, **81**, 6341–6342.
179. van Tamelen, E.E., Spencer, T.A. Jr., Allen, D.S. Jr., Ovis, R.L. (1961) Synthesis of Colchicine. *Tetrahedron*, **17**, 8–34.
180. Evans, D.A., Hart, D.J., Koelsch, P.M. (1978) A New Approach to the Synthesis of Tropolones: Syntheses of Colchicine and β-Dolabrin. *Journal of the American Chemical Society*, **100**, 4593–4594.
181. Evans, D.A., Tanis, S.P., Hart, D.J. (1981) Convergent Total Synthesis of (±)-Colchicine and (±)-Desacetamidoisocolchicine. *Journal of the American Chemical Society*, **103**, 5813–5821.
182. Banwell, M.G. (1996) Cyclopropyl Compounds as Chemical Building Blocks: Total Syntheses of the Alkaloids (−)-Colchicine, Imerubrine and Grandirubrine. *Pure and Applied Chemistry*, **68**, 539–542.
183. Banwell, M.G., Berak, M., Hockless, D.C.R. (1996) Construction of the Colchicine Framework via Two Consecutive Cyclopropane-Mediated Ring-Expansion Reactions. *Journal of the Chemical Society Perkin Transactions 1*, 2217–2219.
184. Lee, J.C., Jin, S.J., Cha, J. (1998) Total Synthesis of Colchicine. a-Methoxy-Substituted Oxyallyl [4 + 3] Cycloaddition Approach. *Journal of Organic Chemistry*, **63**, 2804–2805.
185. Lee, J.C. & Cha, J. (2000) Synthetic Studies on [4 + 3] Cycloadditions of Oxyallyls. 12. Total Synthesis of (−)-Colchicine by an Oxyallyl [4 + 3] Cycloaddition. *Tetrahedron*, **56**, 10175–10184.
186. Cope, A.C. & Overberger, C.G. (1948) Cyclic Polyolefins. I. Synthesis of Cyclooctatetraene from Pseudopelletierine. *Journal of the American Chemical Society*, **70**, 1433–1437.
187. Willstätter, R. & Waser, E. (1911) Über Cyclo-octatetraen. *Berichte der Deutschen Chemischen Gesellschaft*, **44**, 3423–3445.
188. Willstätter, R. & Heidelberger, M. (1913) Zur Kenntnis des Cyclo-octatetraens. *Berichte der Deutschen Chemischen Gesellschaft*, **46**, 517–527.
189. Reppe, W., Schlichting, O., Klager, K., Toepel, T. (1948) Cyclisierende Polymerisation von Acetylen. I. Über Cyclooktatetraen. *Justus Liebigs Annalen der Chemie*, **560**, 1–92.
190. Smith, A.B. III, Freeze, B.S., Xian, M., Hirose, T. (2005) Total Synthesis of (+)-Discodermolide: A Highly Convergent Fourth-Generation Approach. *Organic Letters*, **7**, 1825–1828.
191. Mickel, S.J., Sedelmeier, G.H., Niederer, D. *et al.* (2004) Large-Scale Synthesis of the Anti-Cancer Marine Natural Product (+)-Discodermolide. Part 1: Synthetic Strategy and Preparation of a Common Precursor. *Organic Process Research & Development*, **8**, 92–100.
192. Mickel, S.J., Sedelmeier, G.H., Niederer, D. *et al.* (2004) Large-Scale Synthesis of the Anti-Cancer Marine Natural Product (+)-Discodermolide. Part 2: Synthesis of Fragments C_{1-6} and C_{9-14}. *Organic Process Research & Development*, **8**, 101–106.
193. Mickel, S.J., Sedelmeier, G.H., Niederer, D. *et al.* (2004) Large-Scale Synthesis of the Anti-Cancer Marine Natural Product (+)-Discodermolide. Part 3: Synthesis of Fragment C_{15-21}. *Organic Process Research & Development*, **8**, 107–112.
194. Mickel, S.J., Sedelmeier, G.H., Niederer, D. *et al.* (2004) Large-Scale Synthesis of the Anti-Cancer Marine Natural Product (+)-Discodermolide. Part 4: Preparation of Fragment C_{7-24}. *Organic Process Research & Development*, **8**, 113–121.

195. Smith, A.B. III, Freeze, S., Brouard, I., Hirose, T. (2003) A Practical Improvement, Enhancing the Large-Scale Synthesis of (+)-Discodermolide: A Third-Generation Approach. *Organic Letters*, **5**, 4405–4408.
196. Smith, A.B. III, Kaufman, M.D., Beauchamp, T.J., LaMarche, M.J., Arimoto, H. (1999) Gram-Scale Synthesis of (+)-Discodermolide. *Organic Letters*, **1**, 1823–1826.
197. Smith, A.B. III, Beauchamp, T.J., LaMarche, M.J. *et al.* (2000) Evolution of a Gram-Scale Synthesis of (+)-Discodermolide. *Journal of the American Chemical Society*, **122**, 8654–8664.
198. Paterson, I., Delgado, O., Florence, G.J., Lyothier, I., Scott, J.P., Sereinig, N. (2003) 1,6-Asymmetric Induction in Boron-Mediated Aldol Condensations: Application to a Practical Total Synthesis of (+)-Discodermolide. *Organic Letters*, **5**, 35–38.
199. Paterson, I. & Lyothier, I. (2004) Total Synthesis of (+)-Discodermolide: An Improved Endgame Exploiting a Still–Gennari-Type Olefination with a C1–C8 β-Ketophosphonate Fragment. *Organic Letters*, **6**, 4933–4936.
200. Paterson, I., Florence, G.J., Gerlach, K., Scott, J.P., Sereinig, N. (2001) A Practical Synthesis of (+)-Discodermolide and Analogues: Fragment Union by Complex Aldol Reactions. *Journal of the American Chemical Society*, **123**, 9535–9544.
201. Nerenberg, J.B., Hung, D.T., Somers, P.K., Scheiber, S.L. (1993) Total Synthesis of the Immunosuppressive Agent (−)-Discodermolide. *Journal of the American Chemical Society*, **115**, 12621–12622.
202. Hung, D.T., Nerenberg, J.B., Schreiber, S.L. (1996) Syntheses of Discodermolides Useful for Investigating Microtubule Binding and Stabilization. *Journal of the American Chemical Society*, **118**, 11054–11080.
203. Lee, D., Sello, J.K., Schreiber, S.L. (2000) Pairwise Use of Complexity-Generating Reactions in Diversity-Oriented Organic Synthesis. *Organic Letters*, **2**, 709–712.
204. Xu, R., Chu, G., Bai, D. (1996) Total Synthesis of (±)-Epibatidine. *Tetrahedron Letters*, **37**, 1463–1466.
205. Lee, C.L.K. & Loh, T.P. (2005) Gram-Scale Synthesis of (−)-Epibatidine. *Organic Letters*, **7**, 2965–2967.
206. Kosugi, H., Abe, M., Hatsuda, R., Uda, H., Kato, M. (1997) A Study of Asymmetric Protonation with Chiral β-Hydroxy Sulfoxides. Asymmetric Synthesis of (−)-Epibatidine. *Chemical Communications*, 1857–1858.
207. Fletcher, S.R., Baker, R., Chambers, M.S. *et al.* (1994) Total Synthesis and Determination of the Absolute Configuration of Epibatidine. *Journal of Organic Chemistry*, **59**, 1771–1778.
208. Fletcher, S.R., Baker, R., Chambers, M.S., Hobbs, S.C., Mitchell, P.J. (1993) Synthesis of (+)- and (−)-Epibatidine. *Chemical Communications*, 1216–1218.
209. Zhang, C. & Trudell, M.L. (1996) A Short and Efficient Total Synthesis of (±)-Epibatidine. *Journal of Organic Chemistry*, **61**, 7189–7191.
210. Olivo, H.F. & Hemenway, M.S. (1999) Total Synthesis of (±)-Epibatidine Using a Biocatalytic Approach. *Journal of Organic Chemistry*, **64**, 8968–8969.
211. Aoyagi, S., Tanaka, R., Naruse, M., Kibayashi, C. (1998) Total Synthesis of (−)-Epibatidine Using an Asymmetric Diels–Alder Reaction with a Chiral *N*-Acylnitroso Dienophile. *Journal of Organic Chemistry*, **63**, 8397–8406.
212. Aggarwal, V.K. & Olofsson, B. (2005) Enantioselective a-Arylation of Cyclohexanones with Diaryl Iodonium Salts: Application to the Synthesis of (−)-Epibatidine. *Angewandte Chemie International Edition*, **44**, 5516–5519.
213. Evans, D.A., Scheidt, K.A., Downey, C.W. (2001) Synthesis of (−)-Epibatidine. *Organic Letters*, **3**, 3009–3012.
214. Pandey, G., Bagul, T.D., Sahoo, A.K. (1998) [3 + 2] Cycloaddition of Nonstabilized Azomethine Ylides. 7. Stereoselective Synthesis of Epibatidine and Analogues. *Journal of Organic Chemistry*, **63**, 760–768.

215. Szántay, C., Kardos-Balogh, Z., Moldvai, I., Szántay, C. Jr., Temesvári-Major, E., Blaskó, G. (1996) A Practical Enantioselective Synthesis of Epibatidine. *Tetrahedron*, **52**, 11053–11062.
216. Kotian, P.L. & Carroll, F.I. (1995) Synthesis of (+)- and (−)-Epibatidine. *Synthetic Communications*, **25**, 63–71.
217. Huang, D.F. & Shen, T.Y. (1993) A Versatile Total Synthesis of Epibatidine and Analogs. *Tetrahedron Letters*, **34**, 4477–4480.
218. Clayton, S.C. & Regan, A.C. (1993) A Total Synthesis of (±)-Epibatidine. *Tetrahedron Letters*, **34**, 7493–7496.
219. Corey, E.J., Loh, T.P., Achyutha Rao, S., Daley, D.C., Sarahar, S. (1993) Stereocontrolled Total Synthesis of (+)- and (−)-Epibatidine. *Journal of Organic Chemistry*, **58**, 5600–5602.
220. Giblin, G.M.P., Jones, C.D., Simpkins, N.S. (1997) A Concise Stereoselective Synthesis of Epibatidine Employing Conjugate Addition to an Alkenyl Sulfone Intermediate as the Key Step. *Synlett*, 589–590.
221. Broka, C. (1993) Total Synthesis of Epibatidine. *Tetrahedron Letters*, **34**, 3251–3254.
222. Hoashi, Y., Yabuta, T., Takemoto, Y. (2004) Bifunctional Thiourea-Catalyzed Enantioselective Double Michael Reaction of γ,δ-Unsaturated β-Ketoester to Nitroalkene: Asymmetric Synthesis of (−)-Epibatidine. *Tetrahedron Letters*, **45**, 9185–9188.
223. Holland, J.M., Lewis, M., Nelson, A. (2003) Desymmetrization of a Centrosymmetric Diepoxide: Efficient Synthesis of a Key Intermediate in a Total Synthesis of Hemibrevetoxin B. *Journal of Organic Chemistry*, **68**, 747–753.
224. Nakata, T., Nomura, S., Matsukura, H., Morimoto, M. (1996) Stereoselective Synthesis of the C- and CD-Ring Systems of Hemibrevetoxin B. *Tetrahedron Letters*, **37**, 217–220.
225. Hodgson, D.M. & Galano, J.M. (2005) Enantioselective Access to Isoquinuclidines by Tropenone Desymmetrization and Homoallylic Radical Rearrangement: Synthesis of (+)-Ibogamine. *Organic Letters*, **7**, 2221–2224.
226. Trost, B.M., Godleski, S.A., Genet, J.P. (1978) A Total Synthesis of Racemic and Optically Active Ibogamine. Utilization and Mechanism of a New Silver Ion Assisted Palladium Catalyzed Cyclization. *Journal of the American Chemical Society*, **100**, 3930–3931.
227. Imanishi, T., Yagi, N., Shin, H., Hanaoka, M. (1981) 1.6-Dihydro-3(2H)-Pyridinones as Synthetic Intermediates. A Novel Total Synthesis of (±)-Ibogamine and (±)-Epiibogamine. *Tetrahedron Letters*, **22**, 4001–4004.
228. Imanishi, T., Shin, H., Yagi, N., Hanaoka, M. (1980) 1,6-Dihydro-3(2H)-Pyridinones as Symmetric Intermediates. Formal Synthesis of (±)-Tabersonine and (±)-Catharanthine. *Tetrahedron Letters*, **21**, 3285–3288.
229. Hendrickson, J.B. & Wang, J. (2004) A New Synthesis of Lysergic Acid. *Organic Letters*, **6**, 3–5.
230. Moldvai, I., Temesvári-Major, E., Incze, M., Szentirmay, E., Gács-Baitz, E., Szántay, C. (2004) Enantioefficient Syntheses of A-Ergocryptine: First Direct Synthesis of (+)-Lysergic Acid. *Journal of Organic Chemisry*, **69**, 5993–6000.
231. Ninomiya, I., Hashimoto, C., Kiguchi, T., Naito, T. (1985) Photocyclisation of Enamides. Part 24. Total Synthesis of (±)-Isofumigaclavine B and (±)-Lysergic Acid. *Journal of the Chemical Society Perkin Transactions I*, 941–948.
232. Kiguchi, T., Hashimoto, C., Naito, T., Ninomiya, I. (1982) A New Synthesis of (±)-Lysergic Acid. *Heterocycles*, **19**, 2279–2282.
233. Kornfeld, E.C., Fornefeld, E.J., Kline, G.B. *et al.* (1956) The Total Synthesis of Lysergic Acid. *Journal of the American Chemical Society*, **78**, 3087–3114.
234. Kurihara, T., Terada, T., Harusawa, S., Yoneda, R. (1987) Synthetic Studies of (±)-Lysergic Acid and Related Compounds. *Chemical Pharmaceutical Bulletin*, **35**, 4793–4802.
235. Kurihara, T., Terada, T., Yoneda, S. (1986) A New Synthesis of (±)-Lysergic Acid. *Chemical Pharmaceutical Bulletin*, **34**, 442–3.

236. Ramage, R., Armstrong, V.W., Coulton, S. (1981) A New Synthetic Route to (±)-Lysergic Acid. *Tetrahedron*, **37**, 157–164.
237. Cacchia, S., Ciattini, G., Norera, E., Ortar, G. (1988) A Concise, Palladium-Catalyzed Approach to (±)-Lysergic Acid. *Tetrahedron Letters*, **29**, 3117–3120.
238. Rebek, J., Jr., Tai, D.F., Shue, Y.K. (1984) Synthesis of Ergot Alkaloids from Tryptophan. *Journal of the American Chemical Society*, **106**, 1813–1819.
239. Oppolzer, W., Francotte, E., Battig, K. (1981) Total Synthesis of (±)-Lysergic Acid by an Intramolecular Imino-Diels–Alder Reaction. *Helvetica Chimica Acta*, **64**, 478–481.
240. Rice, K.C. (1980) Synthetic Opium Alkaloids and Derivatives. A Short Total Synthesis of (±)-Dihydrothebainone, (±)-Dihydrocodeinone, and (±)-Nordihydrocodeinone as an Approach to a Practical Synthesis of Morphine, Codeine, and Congeners. *Journal of Organic Chemistry*, **45**, 3135–3137.
241. Iijima, I., Minamikawa, J., Jacobson, A.E., Brossi, A., Rice, K.C. (1978) Studies in the (+)-Morphinan Series. 4. A Markedly Improved Synthesis of (+)-Morphine. *Journal of Organic Chemistry*, **43**, 1462–1463.
242. Rice, K.C. (1977) A Rapid, High-Yield Conversion of Codeine to Morphine. *Journal of Medicinal Chemistry*, **20**, 164–165.
243. Hong, C.Y., Kado, N., Overman, L.E. (1993) Asymmetric Synthesis of Either Enantiomer of Opium Alkaloids and Morphinans. Total Synthesis of (−)- and (+)-Dihydrocodeinone and (−)- and (+)-Morphine. *Journal of the American Chemical Society*, **115**, 11028–11029.
244. Parker, K.A. & Fokas, D. (2006) Enantioselective Synthesis of (−)-Dihydrocodeinone: A Short Formal Synthesis of (−)-Morphine. *Journal of Organic Chemistry*, **71**, 449–455.
245. Trost, B.M., Tang, W., Toste, F.D. (2005) Divergent Enantioselective Synthesis of (−)-Galanthamine and (−)-Morphine. *Journal of the American Chemical Society*, **127**, 14785–14803.
246. Gates, M. (1950) The Synthesis of Ring Systems Related to Morphine. III. 5,6-Dimethoxy-4-cyanomethyl-1,2-naphthoquinone and its Condensation with Dienes. *Journal of the American Chemical Society*, **72**, 228–234.
247. Gates, M. & Tschudi, G. (1956) The Synthesis of Morphine. *Journal of the American Chemical Society*, **78**, 1380–1393.
248. Gates, M. & Tschudi, G. (1952) The Synthesis of Morphine. *Journal of the American Chemical Society*, **74**, 1109–1110.
249. Uchida, K., Yokoshima, S., Kan, T., Fukuyama, T. (2006) Total Synthesis of (±)-Morphine. *Organic Letters*, **8**, 5311–5313.
250. Kaburagi, Y., Tokuyama, H., Fukuyama, T. (2004) Total Synthesis of (−)-Strychnine. *Journal of the American Chemical Society*, **126**, 10246–10247.
251. Nagata, H., Miyazawa, N., Ogasawara, K. (2001) Tandem Single-Step Construction of Chiral Hexahydrophenanthrenes: A Concise Route to (+)-Ferruginol. *Organic Letters*, **3**, 1737–1740.
252. Nagata, H., Miyazawa, N., Ogasawara, K. (2001) A Concise Route to (−)-Morphine. *Chemical Communications*, 1094–1095.
253. White, J.D., Hrnciar, P., Stappendeck, F. (1997) Asymmetric Synthesis of (+)-Morphine. The Phenanthrene Route Revisited. *Journal of Organic Chemistry*, **62**, 5250–5251.
254. Taber, D.F., Neubert, T.D., Rheingold, A.L. (2002) Synthesis of (−)-Morphine. *Journal of the American Chemical Society*, **124**, 12416–12417.
255. Trauner, D., Bats, J.W., Werner, A., Mulzer, J. (1998) Synthesis of Enantiomerically Pure Morphine Alkaloids: The Hydrophenanthrene Route. *Journal of Organic Chemistry*, **63**, 5908–5918.
256. Blakemore, P.R. & White, J.D. (2002) Morphine, the Proteus of Organic Molecules. *Chemical Communications*, 1159–1168.
257. Zezula, J. & Hudlicky, T. (2005) Recent Progress in the Synthesis of Morphine Alkaloids. *Synlett*, 388–405.

258. Weller, D.D. & Rapoport, H. (1976) A Practical Synthesis of Codeine from Dihydrothebainone. *Journal of Medicinal Chemistry*, **19**, 1171–1175.
259. Gates, M. (1953) The Conversion of Codeinone to Codeine. *Journal of the American Chemical Society*, **75**, 4340–4341.
260. Baxendale, I.R., Deeley, J., Griffiths-Jones, C.M., Ley, S.V., Saaby, S., Tranmer, G.K. (2006) A Flow Process for the Multi-Step Synthesis of the Alkaloid Natural Product Oxomaritidine: A New Paradigm for Molecular Assembly. *Chemical Communications*, 2566–2568.
261. Ley, S.V., Schucht, O., Thomas, A.W., Murray, P.J. (1999) Synthesis of the Alkaloids (±)-Oxomaritidine and (±)-Epimaritidine Using an Orchestrated Multi-Step Sequence of Polymer Supported Reagents. *Journal of the Chemical Society Perkin Transactions I*, 1251–1252.
262. Kusama, H., Hara, R., Kawahara, S. *et al.* (2000) Enantioselective Total Synthesis of (−)-Taxol. *Journal of the American Chemical Society*, **122**, 3811–3820.
263. Morihara, K., Hara, R., Kawahara, S. *et al.* (1998) Enantioselective Total Synthesis of Taxol. *Journal of the American Chemical Society*, **120**, 12980–12981.
264. Nicolaou, K.C., Claiborne, C.F., Nantermet, P.G., Couladouros, E.A., Sorensen, E.J. (1994) Synthesis of Novel Taxoids. *Journal of the American Chemical Society*, **116**, 1591–1592.
265. Nicolaou, K.C., Nantermet, P.G., Ueno, H., Guy, R.K., Couladouros, E.A., Sorensen, E. (1995) Total Synthesis of Taxol. 1. Retrosynthesis, Degradation, and Reconstitution. *Journal of the American Chemical Society*, **117**, 624–633.
266. Nicolaou, K.C., Liu, J.J., Yang, Z. *et al.* (1995) Total Synthesis of Taxol. 2. Construction of A and C Ring Intermediates and Initial Attempts to Construct the ABC Ring System. *Journal of the American Chemical Society*, **117**, 634–644.
267. Nicolaou, K.C., Yang, Z., Liu, J.J. *et al.* (1995) Total Synthesis of Taxol. 3. Formation of Taxol's ABC Ring Skeleton. *Journal of the American Chemical Society*, **117**, 645–652.
268. Nicolaou, K.C., Ueno, H., Liu, J.J. *et al.* (1995) Total Synthesis of Taxol. 4. The Final Stages and Completion of the Synthesis. *Journal of the American Chemical Society*, **117**, 653–659.
269. Nicolaou, K.C. & Guy, R.K. (1995) The Conquest of Taxol. *Angewandte Chemie International Edition*, **34**, 2079–2090.
270. Mukaiyama, T., Shiina, I., Iwadare, H. *et al.* (1999) Asymmetric Total Synthesis of Taxol. *Chemistry: A European Journal*, **5**, 121–161.
271. Danishefsky, S.J., Masters, J.J., Young, W.B. *et al.* (1996) Total Synthesis of Baccatin III and Taxol. *Journal of the American Chemical Society*, **118**, 2843–2859.
272. Masters, J.J., Link, J.T., Snyder, L.B., Young, W.B., Danishefsky, S.J. (1995) A Total Synthesis of Taxol. *Angewandte Chemie International Edition*, **34**, 1723–1726.
273. Holton, R.A., Somoza, C., Kim, H.B. *et al.* (1994) First Total Synthesis of Taxol. 1. Functionalization of the B Ring. *Journal of the American Chemical Society*, **116**, 1597–1598.
274. Holton, R.A., Kim, H.B., Somoza, C. *et al.* (1994) First Total Synthesis of Taxol. 2. Completion of the C and D Rings. *Journal of the American Chemical Society*, **116**, 1599–1600.
275. Denis, J.N., Corres, A., Greene, A.E. (1991) Direct, Highly Efficient Synthesis from (S)-(+)-Phenylglycine of the Taxol and Taxotere Side Chains. *Journal of Organic Chemistry*, **56**, 6939–6942.
276. Denis, J.N., Greene, A.E., Guénard, D., Guéritte-Voegelein, F., Mangatal, L., Potier, P. (1988) A Highly Efficient, Practical Approach to Natural Taxol. *Journal of the American Chemical Society*, **110**, 5917–5919.
277. Denis, J.N., Greene, A.E., Serra, A.A., Luche, M.J. (1986) An Efficient, Enantioselective Synthesis of the Taxol Side Chain. *Journal of Organic Chemistry*, **51**, 46–50.
278. Commerçon, A., Bézard, D., Bernard, F., Bourzat, J.D. (1992) Improved Protection and Esterification of a Precursor of the Taxotere and Taxol Side Chains. *Tetrahedron Letters*, **33**, 5185–5188.

279. Guéritte-Voegelein, F., Sénilh, V., David, B., Guénard, D., Potier, P. (1986) Chemical Studies of 10-Deacetyl Baccatin III. Hemisynthesis of Taxol Derivatives. *Tetrahedron*, **42**, 4451–4460.
280. Kingston, D.G.L., Chaudhary, A.G., Gunatilaka, A.A.L., Middleton, M.L. (1994) Synthesis of Taxol from Baccatin III via an Oxazoline Intermediate. *Tetrahedron Letters*, **35**, 4483–4484.
281. Gou, D.M., Liu, Y.C., Chen, C.S. (1993) A Practical Chemoenzymatic Synthesis of the Taxol C-13 Side Chain N-Benzoyl-(2R,3S)-3-phenylisoserine. *Journal of Organic Chemistry*, **58**, 1287–1289.
282. Kanazawa, A.M., Denis, J.N., Greene, A.E. (1994) Direct, Stereoselective Synthesis of the Protected Paclitaxel (Taxol) Side Chain and High-Yield Transformation to Paclitaxel. *Chemical Communications*, 2591–2592.
283. Gennari, C., Carcano, M., Donghi, M., Mongelli, N., Vanotti, E., Vulpetti, A. (1997) Taxol Semisynthesis: A Highly Enantio- and Diastereoselective Synthesis of the Side Chain and a New Method for Ester Formation at C-13 Using Thioesters. *Journal of Organic Chemistry*, **62**, 4746–4755.
284. Ojima, I., Habus, I., Zhao, M. (1991) Efficient and Practical Asymmetric Synthesis of the Taxol C-13 Side Chain, *N*-Benzoyl-(2R,3S)-3-phenylisoserine, and its Analogues via Chiral 3-Hydroxy-4-aryl-b-lactams Through Chiral Ester Enolate-Imine Cyclocondensation. *Journal of Organic Chemistry*, **56**, 1681–1683.
285. Pictet, A. & Gams, A. (1909) Synthèse de la Papavérine. *Comptes Rendus de l'Académie des Sciences Paris*, **149**, 210–212.
286. Pictet, A. & Gams, A. (1909) Synthese des Papaverins. *Berichte der Deutschen Chemischen Gesellschaft*, **42**, 2943–2952.
287. Kropp, W. & Decker, H. (1909) Über Benzyl-phenacetursäure-Derivate. *Berichte der Deutschen Chemischen Gesellschaft*, **42**, 1184–1192.
288. Wahl, H. (1950) Sur la Synthèse Industrielle de la Papavérine. *Bulletin de la Société Chimique de France*, 680–682.
289. Wahl, H. (1951) La Synthèse Industrielle de la Papavérine et des Composés Voisins. *Bulletin de la Société Chimique de France*, D1–D8.
290. Kindler, K. & Peschke, K. (1934) New and Improved Methods for the Synthesis of Pharmacologically Important Amines. IX. Synthesis of Papaverine and Tetrahydropapaverine. *Archiv für Pharmazie*, **272**, 236–241.
291. Pal, B.C. (1958) Industrial Synthesis of Papaverine. *Journal of Scientific and Industrial Research*, **17A**, 270–276.
292. Redel, J. & Bouteville, A. (1949) Sur Les Esters des Papavérine et Norpapavérine Carboxylées Avec le Diéthylaminoéthanol. *Bulletin de la Société Chimique de France*, 443–446.
293. Dean, F.H. (1974) Unpublished results from Ontario Research Foundation.
294. Sheehan, J.C. & Henery-Logan, K.R. (1959) The Total Synthesis of Penicillin V. *Journal of the American Chemical Society*, **81**, 3089–3093.
295. Sheehan, J.C. & Johnson, D.A. (1954) The Synthesis of Substituted Penicillins and Simpler Structural Analogs. VIII. Phthalimidomalonaldehydic Esters: Synthesis and Condensation with Penicillamine. *Journal of the American Chemical Society*, **76**, 158–160.
296. Leach, B.E. & Hunter, J.H. (1953) L- and D-Penicillamine Hydrochlorides. *Biochemical Preparations*, **3**, 111–118.
297. Yu, Q.S. & Lu, B.Y. (1994) Total Synthesis of Racemic Physostigmine, Physovenine and its Sulfur Analog by the Oxindole-5-O-tetrahydropyranyl Ether Route. *Heterocycles*, **39**, 519–525.
298. Julian, P.L. & Pikl, J. (1935) Studies in the Indole Series. IV. The Synthesis of D,L-Eserethole. *Journal of the American Chemical Society*, **57**, 563–566.
299. Julian, P.L., Pikl, J., Boggess, D. (1934) Studies in the Indole Series. II. The Alkylation of 1-Methyl-3-formyloxindole and a Synthesis of the Basic Ring Structure of Physostigmine. *Journal of the American Chemical Society*, **56**, 1797–1801.

300. Julian, P.L. & Pikl, J. (1935) Studies in the Indole Series. III. On the Synthesis of Physostigmine. *Journal of the American Chemical Society*, **57**, 539–544.
301. Julian, P.L. & Pikl, J. (1935) Studies in the Indole Series. V. The Complete Synthesis of Physostigmine (Eserine). *Journal of the American Chemical Society*, **57**, 755–757.
302. Wunberg, J.B.P.A. & Speckamp, W.N. (1978) New Total Synthesis of DL-Physostigmine (DL-Eserine) Via Regioselective $NaBH_4$-reduction of Imides. *Tetrahedron*, **34**, 2399–2404.
303. Trost, B.M. & Zhang, Y. (2006) Molybdenum-Catalyzed Asymmetric Allylation of 3-Alkyloxindoles: Application to the Formal Total Synthesis of (−)-Physostigmine. *Journal of the American Chemical Society*, **128**, 4590–4591.
304. Matsuura, T., Overman, L.E., Poon, D.J. (1998) Catalytic Asymmetric Synthesis of Either Enantiomer of the Calabar Alkaloids Physostigmine and Physovenine. *Journal of the American Chemical Society*, **120**, 6500–6503.
305. Kawahara, M., Nishida, A., Nakagawa, M. (2000) An Efficient Synthesis of Optically Active Physostigmine from Tryptophan Via Alkylative Cyclization. *Organic Letters*, **2**, 675–678.
306. Mukai, C., Yoshida, T., Sorimachi, M., Odani, A. (2006) Co2(CO)8-Catalyzed Intramolecular Hetero-Pauson-Khand Reaction of Alkynecarbodiimide: Synthesis of (±)-Physostigmine. *Organic Letters*, **8**, 83–86.
307. Takano, S., Meriya, M., Ogasawara, K. (1991) Enantiocontrolled Total Synthesis of (−)-Physovenine and (−)-Physostigmine. *Journal of Organic Chemistry*, **56**, 5982–5984.
308. Node, M., Hao, X.J., Fuji, K. (1991) A Chiral Total Synthesis of (−)-Physostigmine. *Chemistry Letters*, **20**, 57–60.
309. Shishido, K., Shitara, E., Komatsu, H., Hiroya, K., Fukumoto, K., Kametani, T. (1986) Total Syntheses of (±)-Physovenine and (±)-Physostigmine. An Application of Tandem Electrocyclic-[3,3]Sigmatropic Reaction of Benzocyclobutenes. *Journal of Organic Chemistry*, **51**, 3007–3011.
310. Rege, P.D. & Johnson, F. (2003) Application of Vicarious Nucleophilic Substitution to the Total Synthesis of dl-Physostigmine. *Journal of Organic Chemistry*, **68**, 6133–6139.
311. Marino, J.P., Bogdan, S., Kimura, K. (1992) The Enantioselective Synthesis of (−)-Physostigmine via Chiral Sulfoxides. *Journal of the American Chemical Society*, **114**, 5566–5572.
312. Woodward, R.B. & Doering, W.E. (1945) The Total Synthesis of Quinine. *Journal of the American Chemical Society*, **67**, 860–874.
313. Rabe, P. (1911) Uber die partielle Synthese des Cinchonins. Zur Kenntnis der Chinaalkaloids XV. *Berichte der Deutschen Chemischen Gesellschaft*, **44**, 2088–2091.
314. Rabe, P., Huntenburg, W., Schultze, A., Volger, G. (1931) Die Totalsynthese der China-Alkaloide Hydro-chinin und Hydro-chinidin Zur Kenntnis der China-Alkaloide, XXV. *Berichte der Deutschen Chemischen Gesellschaft*, **64**, 2487–2500.
315. Stork, G., Niu, D., Fujimoto, A. *et al.* (2001) The First Stereoselective Total Synthesis of Quinine. *Journal of the American Chemical Society*, **123**, 3239–3242.
316. Raheem, I.T., Goodman, S.N., Jacobsen, E.N. (2004) Catalytic Asymmetric Total Syntheses of Quinine and Quinidine. *Journal of the American Chemical Society*, **126**, 706–707.
317. Igarashi, J., Katsukawa, M., Wang, Y.-G., Acharya, H.P., Kobayashi, Y. (2004) Stereocontrolled Synthesis of Quinine and Quinidine. *Tetrahedron Letters*, **45**, 3783–3786.
318. Kaufman, T.S. & Ruveda, E.A. (2005) The Quest for Quinine: Those Who Won the Battles and Those Who Won the War. *Angewandte Chemie International Edition*, **44**, 854–885.
319. Pearlman, B.A. (1979) A Method for Effecting the Equivalent of a de Mayo Reaction with Formyl Acetic Ester. *Journal of the American Chemical Society*, **101**, 6398–6404.
320. Pearlman, B.A. (1979) A Total Synthesis of Reserpine. *Journal of the American Chemical Society*, **101**, 6404–6408.
321. Gómez, A.M., López, J.C., Fraser-Reid, B. (1994) Serial Radical Cyclization of Pyranose-Derived Dienes in the Stereocontrolled Synthesis of Densely Functionalized Cyclohexanes. A Route to Woodward's Reserpine Precursor. *Journal of Organic Chemistry*, **59**, 4048–4050.

322. Chu, C.S., Liao, C.C., Rao, P.D. (1996) A Formal Synthesis of (−)-Reserpine from Methyl Vanillate. *Chemical Communications*, 1537–1538.
323. Hanessian, S., Pan, J., Carnell, A., Bouchard, H., Lesage, L. (1997) Total Synthesis of (−)-Reserpine Using the Chiron Approach. *Journal of Organic Chemistry*, **62**, 465–473.
324. Mehta, G. & Reddy, D.S. (2000) A Formal Synthesis of Reserpine: Hydrindane Approach to the Woodward's Ring-E Precursor. *Journal of the Chemical Society Perkin Transactions I*, 1399–1404.
325. Mehta, G. & Reddy, D.S. (1998) Reserpine Synthesis: A Protocol for the Stereoselective Construction of the Densely Functionalized Ring-E. *Journal of the Chemical Society Perkin Transactions I*, 2125–2126.
326. Martin, S.F., Grzejszczak, S., Rueger, H., Williamson, S.A. (1985) Total Synthesis of (±)-Reserpine. *Journal of the American Chemical Society*, **107**, 4072–4074.
327. Wender, P.A., Schaus, J.M., White, A.M. (1980) General Methodology for *cis*-Hydroisoquinoline Synthesis: Synthesis of Reserpine. *Journal of the American Chemical Society*, **102**, 6157–6159.
328. Fowler, F.W. (1972) Synthesis of 1,2- and 1,4-Dihydropyridines. *Journal of Organic Chemistry*, **37**, 1321–1323.
329. Sparks, S.M., Gutierrez, A.J., Shea, K.J. (2003) Preparation of Perhydroisoquinolines Via the Intramolecular Diels–Alder Reaction of *N*-3,5-Hexadienoyl Ethyl Acrylimidates: A Formal Synthesis of (±)-Reserpine. *Journal of Organic Chemistry*, **68**, 5274–5285.
330. Stork, G., Tang, P.C., Casey, M., Goodman, B., Toyota, M. (2003) Regiospecific and Stereoselective Syntheses of (±)-Reserpine and (−)-Reserpine. *Journal of the American Chemical Society*, **127**, 16255–16262.
331. Woodward, R.B., Bader, F.E., Bickel, H., Frey, A.J., Kierstead, R.W. (1958) The Total Synthesis of Reserpine. *Tetrahedron*, **2**, 1–57.
332. Chen, F.E. & Huang, J. (2005) Reserpine: A Challenge for Total Synthesis of Natural Products. *Chemical Reviews*, **105**, 4671–4706.
333. Wender, P.A., Schaus, J.M., White, A.M. (1987) General Methodology for *cis*-Hydroisoquinoline Synthesis. 3. A Sixteen Step Synthesis of Reserpine. *Heterocycles*, **25**, 263–270.
334. Charest, M.G., Siegel, D.R., Myers, A.G. (2005) Synthesis of (−)-Tetracycline. *Journal of the American Chemical Society*, **127**, 8292–8293.
335. Charest, M.G., Lerner, C.D., Brubaker, J.D., Siegel, D.R., Myers, A.G. (2005) A Convergent Enantioselective Route to Structurally Diverse 6-Acetoxytetracycline Antibiotics. *Science*, **308**, 395–398.
336. Hanessian, S., Desilets, D., Bennani, Y.L. (1990) A Novel Ring-Closure Strategy for the Carbapenems: the Total Synthesis of (+)-Thienamycin. *Journal of Organic Chemistry*, **55**, 3098–3103.
337. Hanessian, S., Bedeschi, A., Battistini, C., Mongelli, N. (1985) A New Synthetic Strategy for the Penems. Total Synthesis of (5R,6S,8R)-6-(a-Hydroxyethyl)-2-(hydroxymethyl)penem-3-carboxylic Acid. *Journal of the American Chemical Society*, **107**, 1438–1439.
338. Reider, P.J. & Grabowski, E.J.J. (1982) Total Synthesis of Thienamycin: A New Approach from Aspartic Acid. *Tetrahedron Letters*, **23**, 2293–2296.
339. Salzmann, T.N., Ratcliffe, R.W., Christensen, B.G., Bouffard, F.A. (1980) A Stereocontrolled Synthesis of (+)-Thienamycin. *Journal of the American Chemical Society*, **102**, 6161–6163.
340. Sletzinger, M., Liu, T., Reamer, R.A., Shinkai, I. (1980) A Facile Transformation of Bicyclic Keto Esters to Bisprotected (±)-8-Epithienamycin Via Enol Activation. *Tetrahedron Letters*, **21**, 4221–4224.
341. Melillo, D.G., Cvetovich, R.J., Ryan, K.M., Sletzinger, M. (1986) An Enantioselective Approach to (+)-Thienamycin from Dimethyl 1,3-Acetondicarboxylate and (+)-α-Methylbenzylamine. *Journal of Organic Chemistry*, **51**, 1498–1504.
342. Holmes, H.L. (1950) Chemistry of the Tropane Alkaloids. In: *The Alkaloids: Chemistry and Physiology*, Vol. 1 (eds. R.H.F. Manske & H.L. Holmes), pp. 288–292. Academic Press, New York.

343. Willstätter, R. (1901) Synthese des Tropidins. *Berichte der Deutschen Chemischen Gesellschaft*, **34**, 129–144.
344. Willstätter, R. (1901) Umwandlung von Tropindin in Tropin. *Berichte der Deutschen Chemischen Gesellschaft*, **34**, 3163–3165.
345. Willstätter, R. & Bommer, M. (1921) Eine vollständige Synthese von r-Ekgonin und von Tropinon. *Justus Liebigs Annalen der Chemie*, **422**, 15–35.
346. Willstätter, R. & Pfannenstiel, A. (1921) Über Succinyldiessigsäureester. *Justus Liebigs Annalen der Chemie*, **422**, 1–15.
347. Willstätter, R. & Racke, F. (1921) Zur Kenntnis des Invertins. *Justus Liebigs Annalen der Chemie*, **425**, 1–135.
348. Robinson, R. (1917) Synthesis of Tropinone. *Journal of the Chemical Society*, **111**, 762–768.
349. Karrer, P. & Alagil, H. (1947) Eine neue Synthese des Tropinons. *Helvetica Chimica Acta*, **30**, 1776–1779.
350. Parker, W., Raphael, R.A., Wilkinson, D.I. (1959) Acetylenic Routes to Tropinone, pseudo-Pelletierine, and Lobelanine. *Journal of the Chemical Society*, 2433–2437.
351. Rejzek, M., Stockman, R.A., van Maarseveen, J.H., Hughes, D.L. (2005) Combining Two-Directional Synthesis and Tandem Reactions: Desymmetrisation by Intramolecular Cycloaddition/Triazoline Fragmentation. *Chemical Communications*, 4661–4662.
352. Yokoshima, S., Ueda, T., Kobayashi, S. *et al.* (2002) Stereocontrolled Total Synthesis of (+)-Vinblastine. *Journal of the American Chemical Society*, **124**, 2137–2139.
353. Kobayashi, S., Ueda, T., Fukuyama, T. (2000) An Efficient Total Synthesis of (−)-Vindoline. *Synlett*, 883–886.
354. Kuehne, M.E., Matson, P.A., Bornmann, W.G. (1991) Enantioselective Syntheses of Vinblastine, Leurosidine, Vincovaline and 20′-Epi-vincovaline. *Journal of Organic Chemistry*, **56**, 513–528.
355. Bandarage, U.K., Kuehne, M.E., Glick, S.D. (2001) Chemical Synthesis and Biological Evaluation of 18-Methoxycoronaridine (18-MC) as a Potential Anti-Addictive Agent. *Current Medicinal Chemistry Central Nervous System Agents*, **1**, 113–123.
356. Kuboyama, T., Yokoshima, S., Tokoyama, H., Fukuyama, T. (2004) Stereocontrolled Total Synthesis of (+)-Vincristine. *Proceedings of the National Academy of Sciences of the United States of America*, **101**, 11966–11970.
357. Ready, J.M., Reisman, S.E., Hirata, M. *et al.* (2004) A Mild and Efficient Synthesis of Oxindoles: Progress Towards the Synthesis of Welwitindolinone A Isonitrile. *Angewandte Chemie International Edition*, **43**, 1270–1272.
358. Reisman, S.E., Ready, J.M., Hasuoka, A., Smith, C.J., Wood, J.L. (2006) Total Synthesis of (±)-Welwitindolinone A Isonitrile. *Journal of the American Chemical Society*, **128**, 1448–1449.
359. Baran, P.S. & Richter, J.M. (2005) Enantioselective Total Syntheses of Welwitindolinone A and Fischerindoles I and G. *Journal of the American Chemical Society*, **127**, 15394–15396.

Chapter 5
Mass Balances and Life Cycle Assessment

Marco Eissen, Georg Geisler, Bruno Bühler, Christian Fischer, Konrad Hungerbühler, Andreas Schmid and Erick M. Carreira

5.1 Introduction

Resource efficient syntheses and production processes of fine chemicals contribute significantly to sustainable development.[1,2] This has also been pointed out in Agenda 21: 'Promote efficient use of materials and resources, taking into account the life cycles of products, in order to realize the economic and environmental benefits of using resources more efficiently and producing fewer wastes' (Agenda 21,[3] Chapter 9.18).

Yield and other mass-related metrics such as atom economy, reaction mass efficiency and mass intensity have been examined by Constable *et al.*[4] with regard to their significance concerning 'greenness' and costs. The importance of using a (product) concentration term, which can be mass intensity or mass index, is additionally emphasized by Laird *et al.*[5] This is in compliance with Winterton,[6] who in his 'twelve more green chemistry principles' demands the establishment of full mass balances.

Mass balancing provides an overview of improvement options for resource efficiency of syntheses and processes. Resource efficiency in existing manufacturing processes is often considered for economic reasons but the latitude for further process optimization is limited and the costs of changes are high.[7] However, degrees of freedom increase for an improvement of resource efficiency in the earlier stages of product development, namely pilot and laboratory stages.[7] Unfortunately, knowledge decreases as degrees of freedom are increasing. Knowledge about future processes which are still in the laboratory stage is rather limited, and in most cases only experimental data from literature or theoretical ideas are available.

The earlier the information on resource efficiency is used in synthesis design, the more efficient process development will be. Therefore, appropriate metrics are initially applied to a choice of literature protocols and preliminary experimental data. As knowledge about the process increases, for example, during scale up, it is successively fed into the mass balance. In addition, environmental impacts associated with the mass balance can be evaluated.

The preparation of such mass balances can become very tedious. For this reason, the Environmental Assessment Tool for Organic Syntheses (EATOS[8]) was developed to provide a convenient software tool for mass balancing of substrates, solvents, auxiliary materials, catalysts, and so on.

$$\text{Mass Index } S^{-1} = \sum \text{Raw materials (kg)/Product (kg)} \tag{5.1}$$

$$\text{Environmental Factor } E = \sum \text{Waste (kg)/Product (kg)} \tag{5.2}$$

$$\text{Cost Index CI} = \sum \text{(Raw materials (kg)} \cdot \text{Costs (Euro/kg))/Product (kg)} \tag{5.3}$$

The research chemist both in academia and in industry profits from the application of metrics such as mass index (equation (5.1)), environmental factor (equation (5.2)) and cost index (equation (5.3)). Therefore, one purpose of this chapter is to demonstrate how to apply such metrics and what kind of information can be obtained from them. Some of their potential application areas are indicated in Box 5.1.

Box 5.1 Application areas and motivations of software-assisted mass balancing.

Research at the university or in an industry

Motivation 1: You have read inspiring literature and have an idea.
→ use metrics to obtain a preview on what you can expect in terms of resource demand, raw material costs and which benchmark is aimed for.

Motivation 2: You have carried out a new synthesis and do not know whether the protocol is an improvement compared to current approaches.
→ use metrics to

(a) document differences regarding resource demands (and their costs).
(b) identify relevant weak points for optimization purposes.
(c) emphasize the challenges you see for future research in the scientific community.
(d) point out the great potential for future improvements to the funding organisations, that is, calculate scenarios.

Motivation 3: You are examining several parameters, many synthesis protocols or even different synthesis sequences.
→ use metrics to maintain an overview of all alternatives in a comparative manner.

Research in an industry

Motivation 4: In case the yield and your instinct are currently the only clues for the quality of your ideas or syntheses.
→ use a software delivering metrics in order to start first optimization efforts. This way, the choice of syntheses with which you would want cost-sensitive experts to calculate, is already a superior one.

Motivation 5: You are using EXCEL or similar non-tailored software for quantification purposes.
→ save time and energy using an appropriate software.

Motivation 6: The overwhelming number of possible ideas in literature forces you to disregard many of them.
→ use an appropriate software to store data, that is, to retain an overview also of those syntheses that do not immediately fit well into your experience background. They might become interesting when an earlier favourite fails.

Box 5.1 Continued.

Motivation 7: You are engaged in the scale-up process.
→ use metrics to

(a) identify the hierarchy of problems.
(b) document the improvements in efficiency.

Education

Motivation 8: You are giving a lecture.
→ use metrics to

(a) give a comprehensive view of all relevant aspects (solvents, catalyst (preparation), etc.).
(b) demonstrate the effect of side reactions on total raw material demand.
(c) compare alternative pathways.

Motivation 9: You are in charge of the educational laboratory.
→ use metrics to sensitize students towards

(a) a responsible way of handling earth resources.
(b) a self-evident integration of alternative routes.
(c) dangers concerning substance-specific properties.

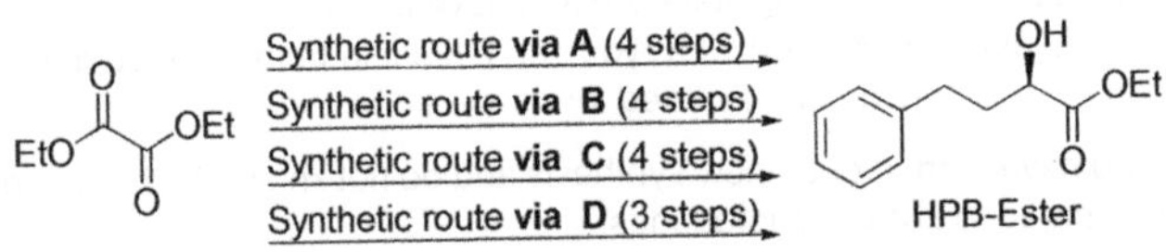

Scheme 5.1 Four synthetic routes via A–via D to ethyl (R)-2-hydroxy-4-phenylbutyrate (HPB-Ester) according to Blaser *et al.* (Figure 3 in reference [10]). In each route, one reaction is a reduction step which is performed using biochemical (A and B) and chemical (C and D) methods, respectively.

A different view on resource efficiency can be obtained if mass balances are converted into comparable environmental impact scores using impact assessment methods. Life Cycle Assessment (LCA, e.g., Guinee, ISO 14001-3) is the most comprehensive of these methods. LCA provides impact categories for most environmental impacts considered relevant today, such as global warming or toxicity to humans. Typically, a 'cradle-to-grave' scope is applied in LCA. Furthermore, mass flows linked to energy demands are included in the assessment (e.g., CO_2 emissions from steam production). Considering LCAs of chemical production, it is necessary to compile mass and energy balances not only for syntheses or downstream processes but for all steps from the extraction of resources ('cradle', e.g., oil production) to waste treatment ('grave') (e.g., Geisler *et al.* [9]).

However, conducting a complete LCA is very time consuming. Fortunately, simple mass balancing may already indicate most relevant requirements for action. This was illustrated by one case study concerning a pharmaceutical intermediate (Scheme 5.1). Compared to LCA, the system boundaries of mass balancing are relatively narrow. For instance, the case study focuses on the chemical reaction and does not take into account the production of substrates. In this respect, a look at two reduction reactions of the synthetic routes via B and via C is worthwhile (Scheme 5.1).

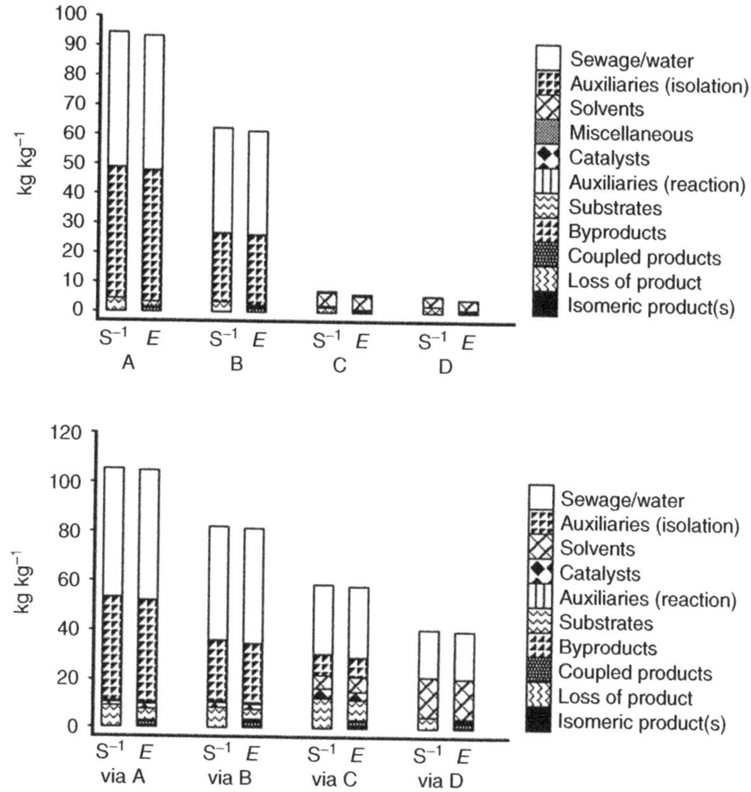

Figure 5.1 Mass indices S^{-1} and environmental factors E of the enantioselective reduction steps A–D (top) and synthesis sequences via A–via D (bottom) using EATOS according to Blaser *et al.* (Figures 10 and 8 in reference [10]). Auxiliaries (isolation) are solvents.

The biocatalytic reduction step B in synthetic route B demands more raw materials (mass index S^{-1}, see equation (5.1)) and generates more waste (environmental factor E, see equation (5.2)) as compared to reduction step C (Figure 5.1). Solvents used to perform the extraction of the product from the aqueous phase in reduction step B are denoted as auxiliaries in Figure 5.1. These solvents and the aqueous phase dominate the mass balances as well as the environmental scores in Figure 5.2 (M4, M8).

In contrast to the mass balance of the reduction step C (Figure 5.1) showing the impact of solvents, product loss, and isomeric product formation, the LCA methodology (Figure 5.2) considers catalyst production (M1) and enantiomeric purification (M7). While product loss and isomeric product formation (Figure 5.1) correspond to the efforts associated with the enantiomeric purification (Figure 5.2), the effort for catalyst synthesis does not become obvious from the mass balance. The reason is that catalyst production was not considered in this examination, although it is possible using the EATOS software, as will be seen in Case study 2. The amount of time required for an LCA is considerably higher than for a mass balance. Nevertheless, similar impact factors have been identified with both methodologies; the mass balancing (Figure 5.1) and the LCA methodology (Figure 5.2).

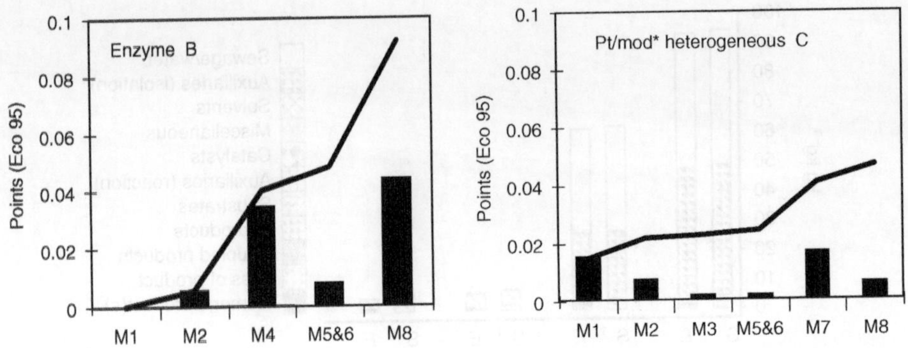

Figure 5.2 Environmental scores of reduction steps in routes B and C (Scheme 5.1) according to the Eco-Indicator 95 evaluation method (Figure 2 in reference [11]). M1 = catalyst, M2 = reduction, M3 = catalyst removal, M4 = extraction, M5 = solvent drain off, M6 = rectification, M7 = enantiomeric purification, M8 = solvent recycling. Reproduced from Jödicke [11], Copyright 1999, with permission from Elsevier.

Therefore, evaluating the potential of mass balancing and LCA in order to improve the efficient utilization of resources is of importance for process development. One needs to determine what the qualitative differences between the two methodologies are. This pertains to all stages of scale-up, that is, to the size increase of the processing volume from the laboratory via the pilot plant to the operation stage (increasing of the batch size[12]). Hence, in this chapter we will juxtapose mass balances from alternative synthesis scenarios as well as from different stages of scale-up in an industrial case study. Concerning the application of both methodologies, implications with regard to process development for fine chemicals are discussed. Thus, one purpose of this chapter is to compare fields of use for mass balancing and LCA in one case study with respect to data demand and information gained.

5.2 Methods

5.2.1 *Mass balancing (MB)*

Inputs and outputs assessed in mass balancing are shown in Figure 5.3. The software EATOS[8] was used to calculate all mass balances of processes. Outputs of EATOS are the mass index[4,13] (equation (5.1), mass of raw material per mass of product output), and the environmental factor E[14] (equation (5.2), mass of waste output per mass of product output). EATOS also allows the calculation of cost indices (e.g., reference [15]) (equation (5.3), cost of raw material per mass of product output).

5.2.2 *Life Cycle Assessment (LCA) considering the impact category global warming*

The boundary for LCA is substantially broader than that for mass balancing (Introduction). Impact categories of LCA are for example global warming, ozone depletion,

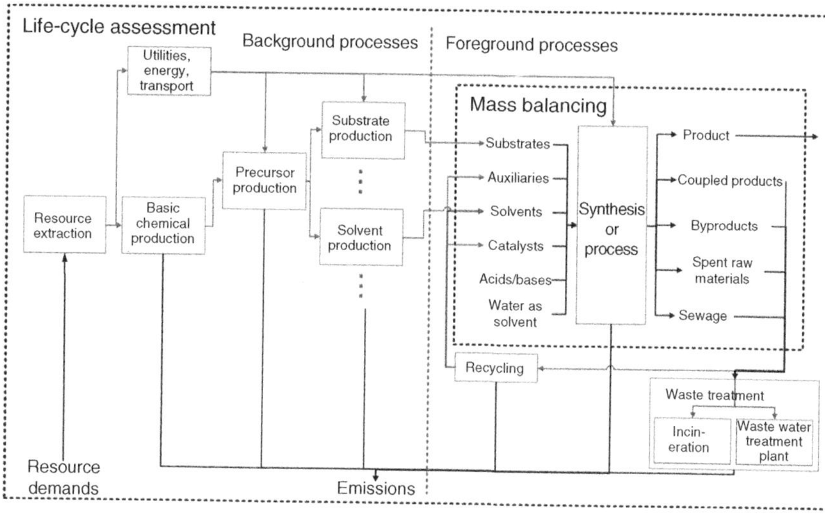

Figure 5.3 Diagram showing boundaries for Mass Balance (encompassed by black dashed lines) and LCA with processes and flows included. Black arrows and flow names show inputs and outputs of the methods, grey arrows and boxes represent processes analysed to set up mass and energy balances. The process networks for the supply of energy, resources and so on are greatly simplified.

acidification, ecotoxicology, eutrophication, and so forth. While the human health impact score was considered in the literature,[9] only global warming was taken into account in this case study. For an LCA, utility and energy demands as well as waste treatment are considered for the foreground process (Figure 5.3). Additionally, all background processes for the supply of raw materials and their precursors down to the resources used as material feedstock or for energy generation are included. Transport operations are also assessed. Using this Life-Cycle Inventory data, a list of cumulated emissions per kg of product can be calculated. Emissions into air from all processes are then evaluated for their potential contribution to global warming, using the global warming potential of the CML-baseline method with a time horizon of 100 years.[16] Emissions of carbon dioxide from the consumption of fossil fuels generally dominate the global warming potential, which is also an indicator of fossil fuel use. Depending on the scope, LCA may also cover the use phase of the chemical produced, for example application as a pesticide (cradle-to-grave scope). However, this work focuses on the production (cradle-to-gate scope).

5.2.3 Case studies

Mass balances of four case studies were evaluated. Additionally, an LCA is considered for the last case study. Case study 1 and Case study 2 are academic and include a comparison between alternative synthesis protocols. In the industrial Case study 3 and Case study 4, a comparison of the steps of a scale up was possible.

5.3 Case studies

Mass indices and environmental factors (equations (5.1) and (5.2)) have been introduced in Section 5.1. For confidentiality reasons, neither chemical names nor exact quantities are specified concerning the industrial case studies. Instead, masses are expressed relatively to input amounts at the laboratory scale. The unit (kg kg^{-1}) expresses how many kilograms of substance are needed to produce one kilogram of product. Abbreviations used in captions of the figures are explained in Box 5.2.

5.3.1 *Case study 1: Acetylene additions to imines*

5.3.1.1 *Mass balancing*

The addition of terminal acetylenes to imines is an important reaction because of the importance of these products as building blocks. Conventionally, the addition reaction shown in Scheme 5.2 is performed with stoichiometric amounts of butyllithium in a step that is, separate from the subsequent nucleophilic addition reaction (see (b)).[17] Carreira has recently developed a procedure that utilizes an iridium catalyst to effect the addition reaction to a wide range of aldimines and ketimines (see (a)).[18]

Box 5.2 Abbreviations.

EATOS	Environmental Assessment Tool for Organic Syntheses
LCA	Life Cycle Assessment
LCI	Life Cycle Inventory
MB	Mass balancing
S^{-1}	Mass index (see Section 5.1)
E	Environmental factor (see Section 5.1)
Lit.	Literature protocol
Prep. Scale	Preparative scale
L	Laboratory scale
P	Pilot scale
O	Operation scale

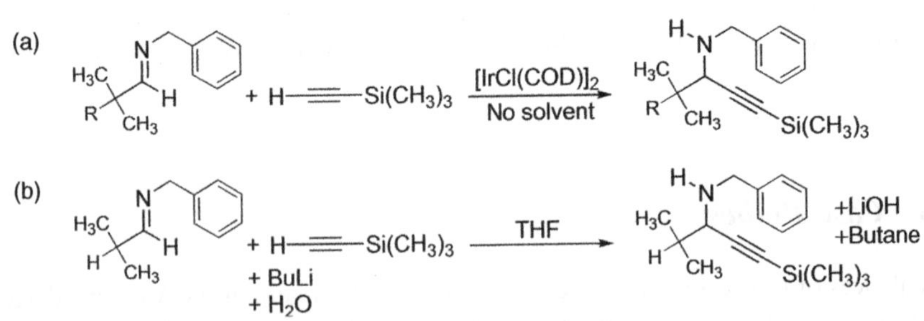

Scheme 5.2 Synthesis of acetylene derivatives (1a R = CH_3; 1b R = H) via a catalytic (a) and a stoichiometric conversion (b).

The mass balance of the processes (Figure 5.4) shows that the catalytic procedure (Scheme 5.2a) is much more resource efficient than the stoichiometric conversion (Scheme 5.2b). As expected, integrating the synthesis of the iridium catalyst results in an increase of the overall waste production (compare (a) II and (a) III, Figure 5.4).

The iridium catalyst is very expensive (98.1 Euro for 0.25 g), therefore, the overall price of synthesis by means of the iridium catalyst (Figure 5.5a) is much higher than for the classical reaction (Figure 5.5b).

In subsequent work aimed at improving the efficiency of the reaction, it has been noted that the addition of sub-stoichiometric quantities of MgI_2 can impart substantial benefits to the process. Thus, in the presence of 2 mol% MgI_2 under otherwise identical conditions the synthesis of a similar amine can proceed with as little as 0.5 mol% Ir(I) catalyst at a faster rate.[21]

Therefore, efforts to reduce costs for raw materials have been most important. The production of the expensive catalyst starting from iridium trichloride is one possibility (a) III, Figure 5.5. A change in the catalytic system is another chemical approach. By addition of a co-catalyst, the quantity of the iridium catalyst could be reduced by a factor of more than five (Figure 5.5(a) I and (a) II). However, measuring up with protocol (b) is still a challenge. Research efforts have to be directed to cost reduction, for example by developing an effective recycling system for the catalyst.

5.3.2 Case study 2: Enantioselective epoxidation

In Case study 2, mass balancing is used to compare a biocatalytic (a) and a chemical catalytic (b) enantioselective epoxidation reaction (Scheme 5.3).

The biocatalytic approach is based on recombinant *Escherichia coli* growing in an aqueous mineral medium (Scheme 5.4). In Scheme 5.4, microbial growth is translated into a stoichiometric equation for biocatalyst synthesis. One needs to consider that biological safety regulations for recombinant class 1 organisms (no danger for humans and the environment) have to be followed with respect to biocatalyst handling.

The recombinant (*E. coli* cells) contains a heterologous styrene monooxygenase catalysing the highly enantioselective epoxidation (e.e. >99%) of styrene and substituted styrenes with molecular oxygen as oxygen donor.[23,24] In the case of styrene epoxidation, 2-phenylethanol is formed as a minor byproduct. During biocatalysis, the addition of glucose is required for growth and intracellular cofactor (NADH) regeneration.[25,26] At the same time glucose serves as the main substrate for biocatalyst production and is mainly converted into biomass and CO_2 whereas acetic acid accumulated as a minor side product. In the chemical approach, a Jacobson catalyst based on a (salen)Mn-complex is used allowing an e.e. of 84–86% for (*S*)-epoxyindane synthesis.[27] Scheme 5.5 shows the synthesis of this Jacobson catalyst.[28] For mass balancing, the procedure for (*S*)-epoxyindane synthesis was adapted and applied to (*S*)-styrene oxide synthesis. The reaction is carried out in methylene chloride with 4-phenylpyridine *N*-oxide as an additive and sodium hypochlorite as the oxygen donor.

In the biocatalytic system, a second organic phase consisting of bis(2-ethylhexyl)phthalate and containing the substrate is added at a phase ratio of 1:1. This procedure enables *in situ* product extraction and protects the microbial cells from toxic effects of the substrate and

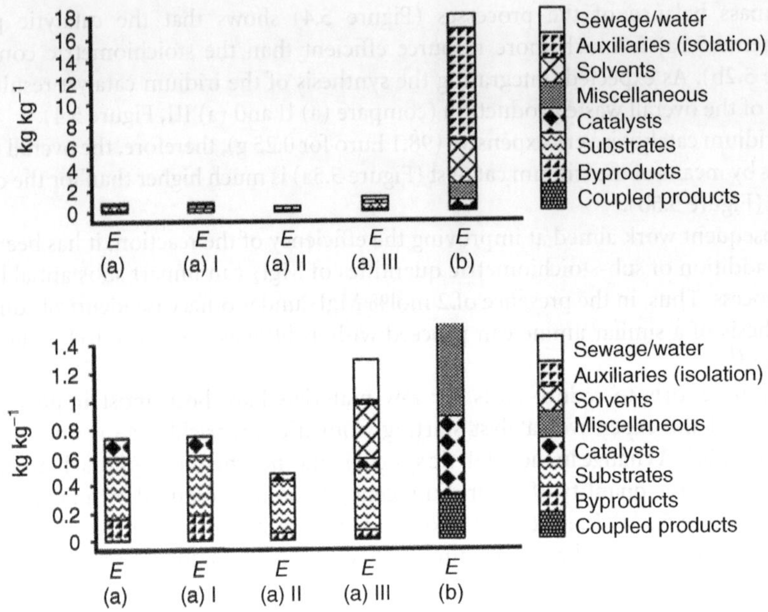

Figure 5.4 Environmental factor *E* for the synthesis of the acetylene derivatives 1 (Scheme 5.2), using the software EATOS.[8] The bottom figure is a zoomed version of the top one. In brief (a) 5 mol% of $[IrCl(COD)]_2$ and 84% yield of 1b; (a) I 5 mol% of $[IrCl(COD)]_2$ and 85% yield of 1a; (a) II 1 mol% of $[IrCl(COD)]_2$; 3 mol% of MgI_2 and 96% yield of 1a; (a) III bases on (a) II and considers the synthesis of the iridium catalyst starting from iridium trichloride according to the literature.[19] (b) idealized conventional procedure to produce 1b using butyllithium. Details: (a) Under argon, 33.6 mg $[IrCl(COD)]_2$ (0.05 mmol) was placed in a 10-mL Schlenk flask, 162 mg benzyl-isobutylidene-amine (1.0 mmol) and 0.28 mL trimethylsilylacetylene (2.0 mmol) were added and the reaction stirred for 24 h at room temperature (RT). Purification by column chromatography on silica gel (hexanes/ethyl acetate 30:1) provided 226 mg of 1b (84% yield) as a colourless oil. (a) I Under argon, 33.6 mg $[IrCl(COD)]_2$ (0.05 mmol) was placed in a 10-mL Schlenk flask, 176 mg benzyl-(2,2-dimethyl-propylidene)-amine (1.0 mmol) and 0.28 mL trimethylsilylacetylene (2.0 mmol) were added and the reaction stirred for 24 h at RT. Purification by column chromatography on silica gel (hexanes/ethyl acetate 50:1) provided 233 mg of 1a (85% yield) as a colourless oil. (a) II Under argon, 6.7 mg $[IrCl(COD)]_2$ (0.01 mmol) and 8.4 mg MgI_2 (0.03 mmol) were placed in a 10-mL Schlenk flask. Subsequently, 176 mg benzyl-(2,2-dimethyl-propylidene)-amine (1.0 mmol) and 0.19 mL trimethylsilylacetylene (1.3 mmol) were added and the reddish brownish suspension stirred for 24 h at RT. Purification by column chromatography on silica gel (hexanes/ethyl acetate 40:1) gave 263 mg of 1a (96% yield) as a colourless oil. (a) III bases on (a) II and considers the synthesis of the iridium catalyst starting from iridium trichloride according to the literature.[19] (b) The synthesis protocol was performed according to the literature[20] yielding 50% of product 1b. Assuming that this procedure could be optimized the calculation here demonstrates a relatively ideal 'butyllithium' protocol: 100% yield; substrates: benzyl-isobutylidene-amine (0.006 mol), trimethylsilylacetylene (0.006 mol), 2.721 g of a 1.5 molar solution of *n*-butyllithium in *n*-hexane (0.006 mol, i.e., a purity of about 14%), water (0.006 mol); solvent: 7 mL tetrahydrofurane; catalyst: boron trifluoride diethyl etherate (0.006); auxiliary materials: sodium hydroxide (10%aq) (2 mL), diethyl ether (15 mL), sodium sulphate (5 g). Commentary: 1a (see (a) I to (a) III) cannot be accessed via the 'butyllithium' protocol. In order to make possible a comparison with protocol (b), protocol (a) is shown here being comparable with protocol (a) I.

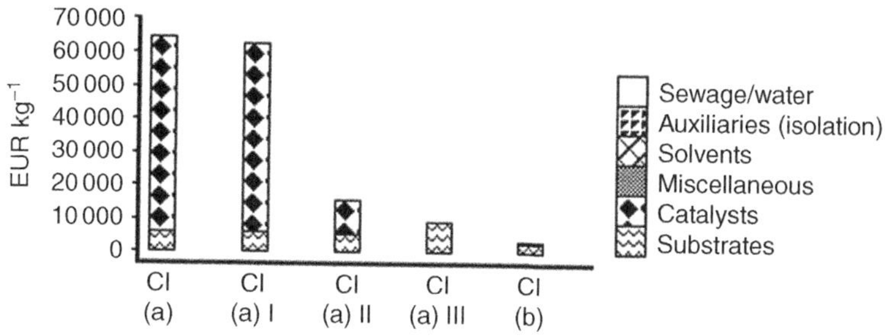

Figure 5.5 Cost Index in the synthesis of the acetylene derivatives **1** (Scheme 5.2), using the software EATOS.[8] Prices according to the Aldrich catalog (2002) were used. For a description of the synthesis a) to b) see Figure 5.4. The price of the imines was not considered.

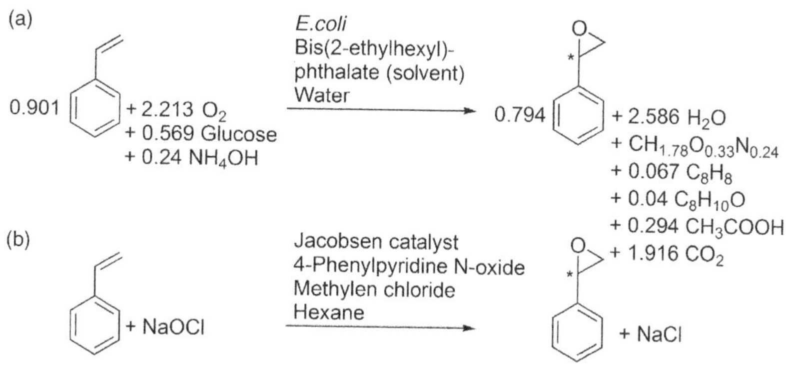

Scheme 5.3 Biocatalytic (a) and chemical catalytic (b) styrene epoxidation.

$$27C_6H_{12}O_6 + 24NH_4OH + 52O_2 \xrightarrow{E.\ coli\ \text{(trace), water, salts, thiamine}} C_{100}H_{178}O_{33}N_{24} + 133H_2O + 62CO_2$$

Scheme 5.4 Stoichiometry of growth of recombinant *E. coli*. The composition of *E. coli* is assumed to be the same as that of *E. aerogenes*. The ratio of atoms is 100 C:178H:33O:24N, that is, the 'molecular formula' is $C_{100}H_{178}O_{33}N_{24}$.[22] An initial small quantity of bacteria is used as an inoculum. Salts contained in the growth medium are KH_2PO_4, K_2HPO_4, Na_2HPO_4, NH_4Cl, NaCl, $MgSO_4 \cdot 7H_2O$. Furthermore, the medium contained low amounts of thiamine and of a trace element solution consisting of $CuCl_2$, $MnCl_2 \cdot H_2O$, $ZnSO_4$, H_3BO_3, $Na_2EDTA \cdot 2H_2O$, $CaCl_2 \cdot H_2O$, $FeSO_4 \cdot H_2O$, $Na_2MoO_4 \cdot 2H_2O$, and HCl(aq).

the product. The product can then be isolated by phase separation and vacuum distillation of the organic phase. In the chemical catalytic system, product isolation includes the input of hexane, dichloromethane, aqueous saturated sodium chloride, sodium sulphate, and small amounts of calcium hydride and also concludes with a vacuum distillation step.

Cost evaluation based on mass balancing results in clearly lower cost indices for the biocatalytic process as compared to the chemical catalytic process (Figure 5.6). For three of the substances of the chemical procedure (Jacobsen catalyst (catalyst), 4-phenylpyridine

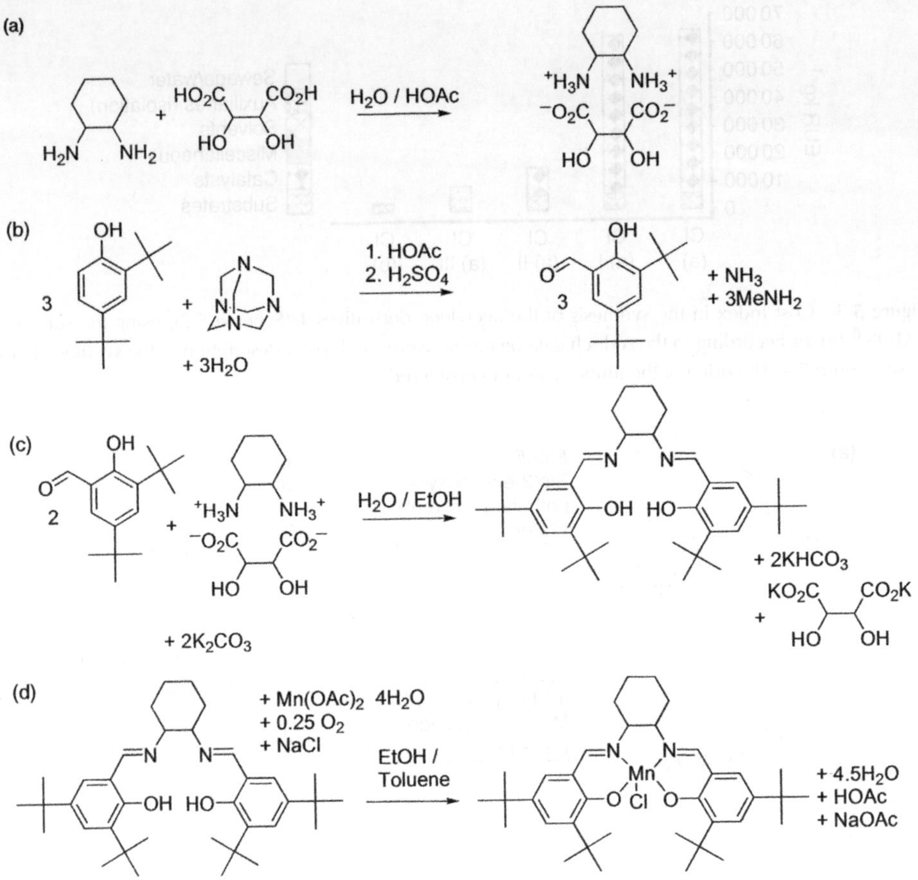

Scheme 5.5 Synthesis of the Jacobsen catalyst.

N-oxide (catalyst) and aqueous solution of sodium hypochlorite), the price an industrial enterprise would have to pay was not available. Hence, prices were based on the Aldrich catalogue. Because the procurement quantities in the Aldrich catalogue are small and, therefore, listed substances are relatively expensive, their prices were multiplied by 0.1. The costs for the chemical catalytic process are dominated by catalyst costs. The high water costs in the chemical process are due to the cost of the substrate sodium hypochlorite. Because a commercially available solution is considered here, its price is split up into hypochlorite ('Substrates', Figure 5.6) and water ('Sewage/Water', Figure 5.6). Furthermore, costs for the biocatalytic process can be considerably reduced by solvent recycling, which is considered to be efficient based on the phase separation data[26] (bottom graph in Figure 5.6). From this cost evaluation, it becomes clear that the biocatalytic approach is a superior alternative from an economic point of view.

The picture does look different when environmental factors are considered: a slightly higher *E*-factor has been obtained for the biocatalytic as compared to the chemical catalytic procedure (Figure 5.7). Both approaches show a high contribution of solvent use, sewage,

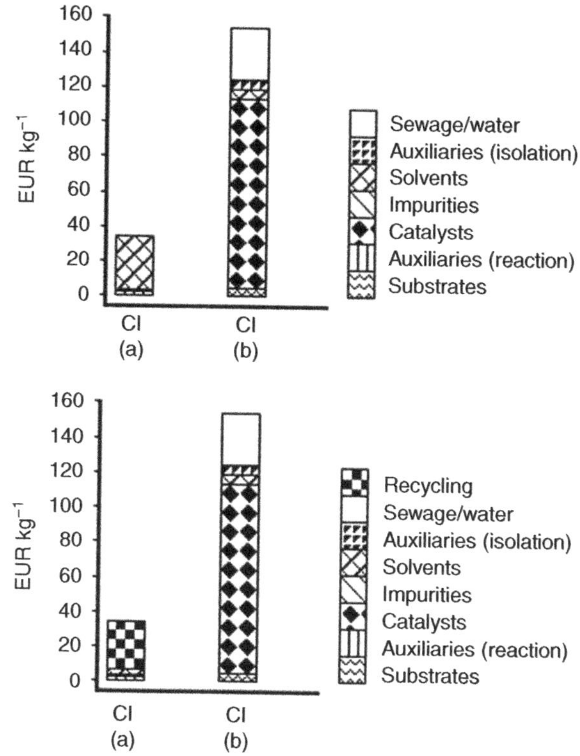

Figure 5.6 Cost Indices CI for the biocatalytic (a) and chemical catalytic (b) synthesis of (*S*)-styrene oxide (Scheme 5.3), using the software EATOS.[8] The bottom figure shows the cost savings achieved by solvent recycling.

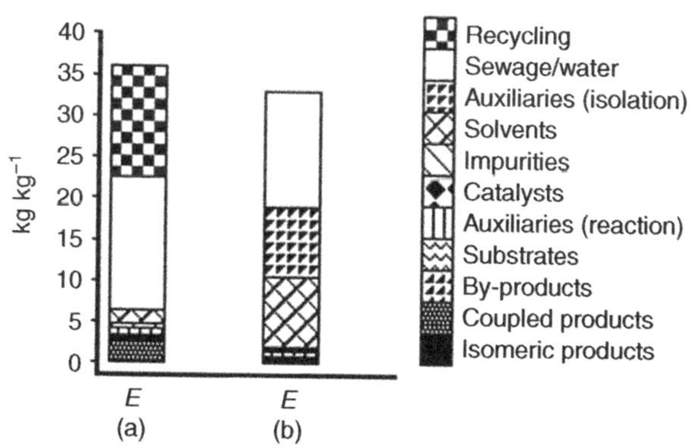

Figure 5.7 Environmental factors *E* for the biocatalytic (a) and chemical catalytic (b) synthesis of (*S*)-styrene oxide (Scheme 5.3) using the software EATOS.[8] The reduction of the environmental factor achieved by solvent recycling is indicated.

and, in the chemical procedure, auxiliaries used for product isolation. As indicated for the biocatalytic process, the contribution of solvent use can be considerably reduced by solvent recycling. This may also be true for the chemical process but has not been investigated so far. Besides the mass index (equation (5.1)), the environmental factor (equation (5.2)) and the cost index (equation (5.3)), the software EATOS provides two metrics, which evaluate qualitative aspects linked with the substances used. This is the environmental index for the input (EI_{in}) and the environmental index for the output (EI_{out}). They shall not be discussed in detail here. However, they have been examined in this case study. The results show that the choice of solvents, aside from their being recyclable, and the achieved product concentration are important improvement options. The solvent chosen so far for the biocatalytic approach is of fossil origin and problematic with respect to human toxicity and bioaccumulation potential. It may be replaced by a solvent or solvent mixture of a renewable origin with a similarly low log P_{ow} (log octanol–water partition coefficient). Increasing the product concentration would not only reduce the amount of solvent wastes and sewage per kg of product, but also the amount of biomass produced as a coupled product in the biocatalytic approach.

A major difference in the evaluation of the two approaches concerns catalyst synthesis. Whereas catalyst production is integrated in the biocatalytic procedure (Scheme 5.4) and thus also contained in the cost index and the environmental factor, it is not considered in the chemical catalytic approach. A more realistic approach is to include the synthesis of the Jacobsen catalyst (Scheme 5.5) in the mass balance. In Figure 5.8, resources used for catalyst production are separately indicated ('Further Syntheses'). For the biocatalytic procedure, water dominates the environmental factor. The environmental factor increases for the chemical procedure, whereas the cost index, when representing only the raw material costs, declines if the (salen)Mn-catalyst is assumed to be synthesized and not bought.

5.3.3 Case study 3: Dieckmann condensation

This case study deals with an industrial Dieckmann condensation (intramolecular ester condensation; Scheme 5.6) considering the laboratory and the operation scale.

The industrial reaction in Scheme 5.6a reflects a typical example, which is often performed in a similar way for training purposes in the organic laboratory with HCl instead of H_2SO_4 as a co-substrate (Scheme 5.6b). The substrates in both syntheses are different, but molecular weights of substrate **2** and adipic acid ethyl ester in (b) are of a similar magnitude. Interestingly, the yield (75%) of the Dieckmann condensation according to the literature protocol (Scheme 5.5b[29]) is more than 10 percentage points lower than the yield achieved for the industrial reaction in operation scale, although it is significantly higher (>20%) than in the laboratory scale of the industry case.

Figure 5.9 shows a comparison of the laboratory and operation scale of the production of product **3** (Scheme 5.6).

It is noticeable that the use of water is not indicated in the laboratory scale, although salts have to be washed just as in the operational process. Presumably, the laboratory recipe was not meticulously kept. Reduced substrate demand (reduction to about 60%, Table 5.1) caused by a higher yield lowers the demand of base and acid (93% and 95%, respectively, related to the input of substrate **2**) and, thus, salt formation and water demand.

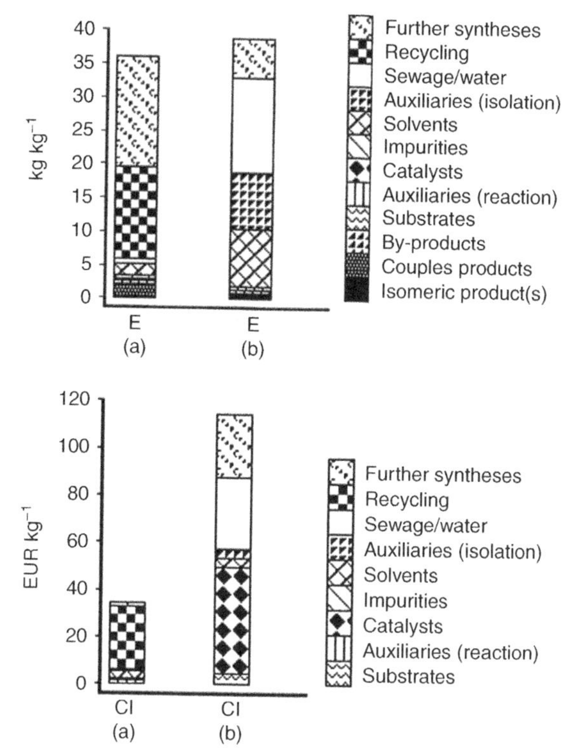

Figure 5.8 Environmental factors *E* (top figure) and cost indices *CI* (bottom figure) for the biocatalytic (a) and chemical catalytic (b) syntheses of (*S*)-styrene oxide (Scheme 5.3) including the synthesis of the Jacobsen catalyst and of the bacteria (Scheme 5.4) as further syntheses. Waste produced during biocatalyst synthesis is indicated. However, it has to be considered that biocatalyst and product synthesis cannot be separated.

(a) Substrate **2** + NaOEt + $0.5H_2SO_4$ ⟶ Product **3** + EtOH + $0.5Na_2SO_4$

(b)

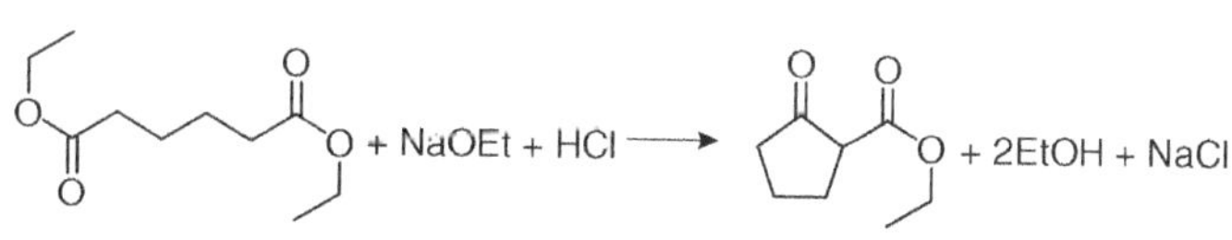

Scheme 5.6 Industrial Dieckmann condensation reaction (a), and a literature protocol (b) used in an educational laboratory.[29] (A similar protocol is given in reference [30] (experiment 4023).) Of course, base and acid are not entered simultaneously.

Water use is identical in the operation scale and in the literature procedure. If the yield was increased in the literature protocol simply by applying an extraction procedure, which is not intended in the current recipe, the amount of water needed (always related to one kilogram of product) would be less than in the operational process. Therefore, the currently generated sewage quantity in operation scale appears to be higher than necessary. Since the reduction of waste-treatment costs and the improvement of the environmental performance are

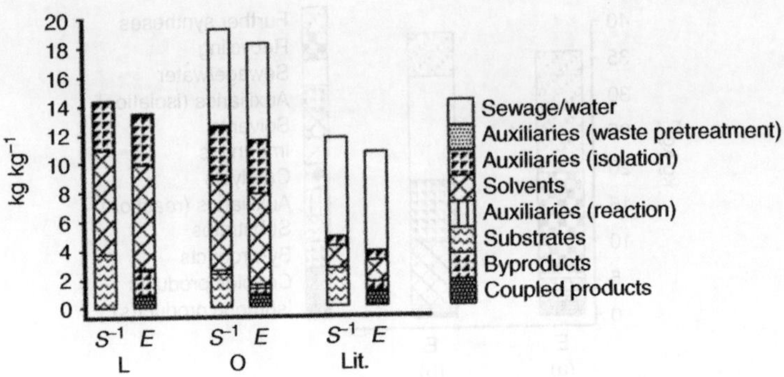

Figure 5.9 Mass indices S^{-1} and environmental factors E for the Dieckmann condensation reactions depicted in Scheme 5.6, using the software EATOS[8]: Results for laboratory (L) (≈0.2 kg) and operation (O) (≈1500 kg) scale, as well as for a literature protocol (Lit.) (scale: 0.012 kg[29]) are shown.

Table 5.1 Demand of selected substances related to the quantity used in the laboratory scale.

		Reference value: 1 kg of product	Reference value: 1 kg of substrate **2**
Substance	L (%)	O (%)	O (%)
Substrate **2**	100	61	100
NaOEt	100	57	93
H_2SO_4	100	58	95
Solvent 1	100	59	96
Solvent 1[a]	100	53	87
Auxiliary (Solvent 2)	100	103	169
Auxiliary (Solvent 2)[a]	100	28	46

L = laboratory, O = operation.
[a] Recycling considered.

always important objectives, this is potentially interesting information that might lead to an adaptation of the process. As can be seen in this example, the mass-related metrics such as mass index and environmental factor provide clues for improvement options, especially when alternative syntheses are taken into account.

Only a little effort is necessary to reduce solvent 1 demand used during reaction scale-up. The quantity used in the laboratory stage was reduced to 59% in the operation stage (Table 5.1). However, related to substrate 2, 96% of solvent 1 is still used. Thus, 87% of the original quantity of solvent will be fed to the incinerator for disposal, while the recycle rate is only 9.1% (from 96% to 87%, Table 5.1). Considering that there is a factor of five difference in solvent 1 demand between the operation scale and the literature procedure (see the segments 'Solvent' of the mass index, S^{-1}, in Figure 5.10), the potential for optimization becomes obvious. Again, mass index and environmental factor clearly show a potential need for action. Of course, in case it is necessary to maintain a certain dilution to keep the

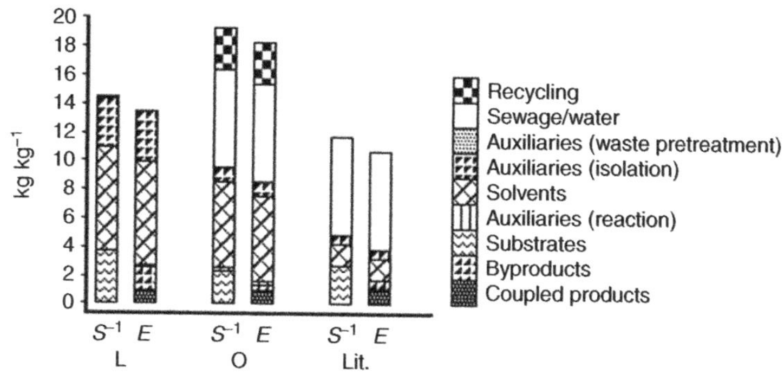

Figure 5.10 Mass indices S^{-1} and environmental factors E for the Dieckmann condensation reactions depicted in Scheme 5.6, using software EATOS[8]: Laboratory (L) and operation scale (O) in which recycled material is presented separately and a literature protocol (Lit.).[29]

$$\underset{\mathbf{4}}{\text{R–CO–CH}_2\text{–CO–R}'} + \underset{\mathbf{5}}{\text{R}''\text{–COCl}} + 2\text{NaOH} + \text{HCL} \xrightarrow[\text{Catalyst} \; \longrightarrow \; \text{Auxiliary material}]{\text{Auxiliary material}} \underset{\mathbf{6}}{\text{Product}} + 2\text{NaCl} + 2\text{H}_2\text{O}$$

Scheme 5.7 Overall stoichiometry of a three-step route to product **6**. Base and acid are used in stoichiometric amounts, but, of course, in different reaction steps.

yield as high as it is in the current operation, the amount of solvent cannot be reduced. Nevertheless, these metrics at least highlight the potential for solvent reduction.

In contrast to the quantity of solvent 1 used during the reaction, the quantity of extraction solvent 2 (work up) increases during scale up (Laboratory 100% → Operation 103%), especially when it is related to substrate **2** (Laboratory 100% → Operation 169%). Compared to the yield obtained from the literature protocol in which an extraction procedure is missing, an efficient extraction seems to be important in order to achieve sufficient product accumulation. However, as the mass index and the environmental factor demonstrate with respect to the possibility for reducing the volume of water used (see above), solvent 2 demand should be able to be reduced as well, since less water use means less solvent is required for extraction. Still, at least the recycle rate of solvent 2 is as high as 72.8% (from 169% to 46%, Table 5.1), regarding the current data of the technical operation scale.

5.3.4 Case study 4: Process scale up and resource efficiency of an industrial product

The production of the agrochemical **6** (Scheme 5.7) is carried out batchwise via a three-step protocol. Mass balancing has been conducted for three stages of development: Laboratory-, pilot- and operation scale. An LCA was available for the operation stage only. A description of this LCA including data sources and data acquisition methods was published by Geisler *et al.* [9] (product A in reference [9] corresponds to product **6** here). Many parameters in the Life-Cycle Inventory (LCI) are estimated, especially utility demands and yields of processes for the production of precursors.[9] Uncertainty in these estimations was illustrated in a

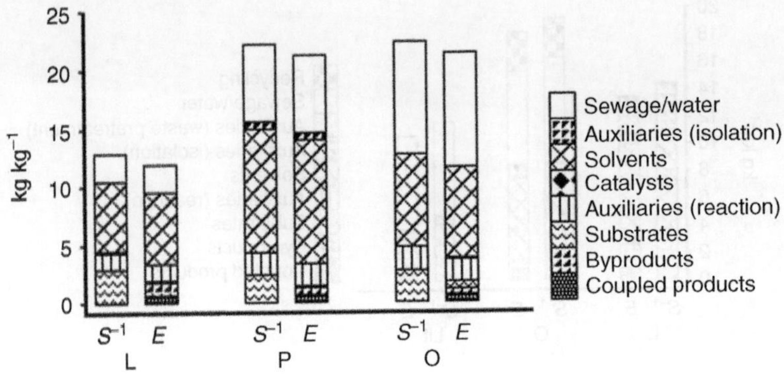

Figure 5.11 Mass indices S^{-1} and environmental factors E at different process development stages for the synthesis depicted in Scheme 5.7. The parameters have been determined using the software EATOS[8]: (L) laboratory (≈0.1 kg), (P) pilot (≈400 kg), and (O) operation (≈1000 kg) scale.

best- and a worst-case scenario of production efficiency. In this contribution, only the best-case scenario is shown, because this was found to represent real process efficiencies better than the worst-case scenario in a verification using data from the company. In contrast to Case study 3, data concerning the pilot scale are available.

5.3.4.1 Mass balancing

Substrate **4** reacts with substrate **5** in the presence of an auxiliary material to yield product **6** (Scheme 5.7). In subsequent rearrangement reactions, stoichiometric amounts of mineral base and acid (both quoted as substrates) are needed to recycle the auxiliary material. In Figure 5.11, all materials given in Scheme 5.7 are illustrated. On operation scale, sodium hydroxide is used in even higher excess than in the pilot plant in order to increase recycling efficiency. Substrate excess (essentially sodium hydroxide) is indicated in the corresponding segment of the E-column (Figure 5.11), in spite of the fact that sodium hydroxide will finally be neutralized in the downstream process.

The yield increases by 15 percentage points during scale up to the pilot scale and by about 2 more percentage points in the final process. This improvement is reflected in the demand of **4** and **5**, which declines to 80% in the pilot scale and 78% in the operation scale (Table 5.2). The excess of substrates (0.137 kg kg^{-1} (L), 0.176 kg kg^{-1} (P), 0.589 kg kg^{-1} (O), Figure 5.12) can mainly be attributed to sodium hydroxide and hydrochloric acid inputs.

Consumed solvent quantities are shown in Figure 5.13. In the course of scale-up, the demand for solvent 1 during reaction is reduced almost by a factor of two (Figure 5.13). Solvent recycling diminishes the loss of solvent 1 to 0.17 kg kg^{-1}. Solvent 2 was introduced in the pilot stage. Its quantity differs only slightly comparing pilot and operation scale. Its recyclability is lower than that of solvent 1.

Table 5.2 and Figure 5.14 not only document changes in solvent use but also in the demand for other selected substances. The changes are given relative to the amount used in the laboratory stage. Reference values are one kilogram of product (columns 3

Table 5.2 Substance demand related to the substrate and product quantities in the laboratory scale for the reaction shown in Scheme 5.7.

		Reference value: 1 kg of product		Reference Value: 1 kg of substrates **4,5**	
Substance[a]	L (%)	P (%)	O (%)	P (%)	O (%)
Diketone **4**	100	80	78		
Acid chloride **5**	100	80	79		
Base	100	86	152	108	193
Acid	100	93	87	116	111
Auxiliary (R)	100	129	88	162	112
Auxiliary (R)[b]	100		10		13
Catalyst	100	80	73	100	92
Catalyst[b]	100		24		30
Solvent 1	100	93	54	c	c
Solvent 1[b]	100	73	3	c	c
Solvent 2		100	94	c	c
Solvent 2[b]			12		
Water/Sewage	100	282	411	354	522

[a] The list of substances is an excerpt from the recipe. Therefore, a segment represented in Figure 5.10 such as 'Catalysts' may comprise further materials; L = laboratory, P = pilot, O = operation stage.
[b] Recycling considered.
[c] No data are indicated here for confidentiality reasons.

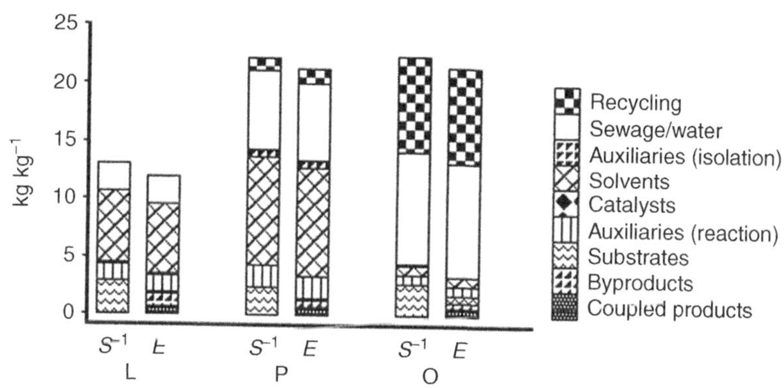

Figure 5.12 Mass indices S^{-1} and environmental factors E at different process development stages for the synthesis depicted in Scheme 5.7. Separate representation of recycled material for the synthesis shown in Scheme 5.7 on (L) laboratory-, (P) pilot-, and (O) operation scale.

and 4, similarly as in Figure 5.11), and one kilogram of substrates **4** and **5** (columns 5 and 6).

When related to the consumption of substrates **4** and **5**, acid and base are used in higher amounts in the operation scale than in the laboratory scale. Still, raw material demand was significantly reduced in absolute figures (not shown for confidentiality reasons).

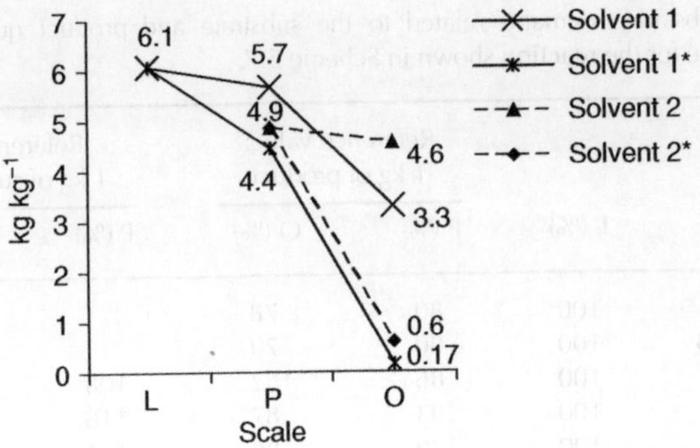

Figure 5.13 Solvent demands during the development of a process for the synthesis route in Scheme 5.7. For percentage changes see Table 5.2 (footnote 'b' recycling considered). Due to the introduction of a second solvent in the pilot and operation scale solvent demand is higher than in the laboratory scale (laboratory scale 6.1 kg kg^{-1} versus pilot scale 10.6 kg kg^{-1} and operation scale 7.9 kg kg^{-1}, see also Figure 5.10).

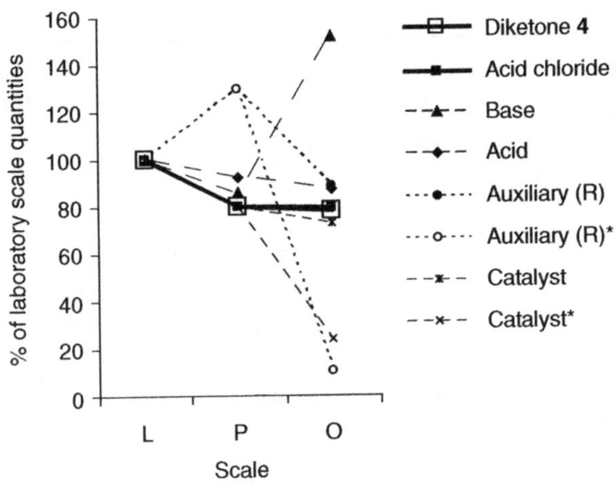

Figure 5.14 Graphic representation of parts of columns 2–4 in Table 5.2 (footnote 'b' recycling considered).

5.3.4.2 Life Cycle Assessment

The assessment covered all processes leading to the final product **6** (cradle-to-gate; Figure 5.3). Substance- and energy-specific contributions to the total global-warming score are given in Figure 5.15 for the operation scale of the process.

The relative contribution of any flow in this life cycle to the total global-warming score of the production of a unit mass of product **6** is shown. The highest shares in the impact score are exhibited by the two substrates, because their production is taken into account, comprising multi-step synthesis routes starting from crude oil. All other input mass flows show relatively small contributions (maximum of 4% for solvent 2). This is due to relatively high

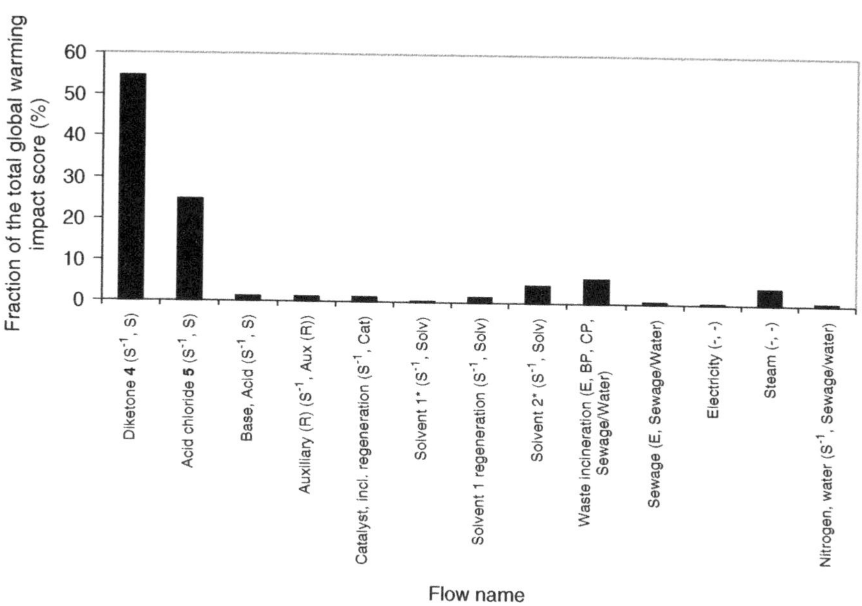

Figure 5.15 Fractions of global-warming scores (%) for all substance flows (see Table 5.2) concerning the operation scale. Corresponding mass metrics of the flows in Figure 5.10 are indicated in parentheses.

recycle rates (Table 5.2), especially for the solvents: relatively low emissions of greenhouse gases occur during solvent regeneration, which is achieved by distillation. In contrast to mass balancing, additional information has been used to split up waste flows into organic wastes (incineration) and aqueous wastes (wastewater treatment), as well as to calculate utility demands.[9] Therefore, additional contributions to the impact-score become apparent due to CO_2 emissions during waste incineration and from the combustion of fossil fuels for steam generation (Figure 5.3).

The dominant contribution of substrates to the global warming potential (Figure 5.15) is in striking contrast to the visualization in mass balancing (Figure 5.11). This is due to the differences in balance regions (Figure 5.3) of the two methods. Regarding solvent utilization, the picture generated by LCA and mass balancing is comparable: solvent use amounts to only 6.1% altogether of the global warming potential (Figure 5.15) as a consequence of efficient solvent recycling. This optimization effort is also apparent in the mass balance presented in Figure 5.13 and demonstrates the reduction of solvent utilization during scale-up.

The mass-related metrics shown in Figure 5.11 indicate that the amount of a substrate (see also byproduct formation), an auxiliary material for reaction, and of a solvent have to be reduced. The detailed view of the mass indices S^{-1} of the pilot scale, for example, the segments 'Substrates' and 'Aux (R)' and the size of segments 'Substrates' (excess) and 'Aux (R)' of the environmental factor E, deliver the information listed in Table 5.2; 108% base and 162% auxiliary (R) are used. The measure to increase base addition for recycling purposes was successful: at the expense of 193% base, much auxiliary material 'Aux (R)' was saved in operation scale (reduction from 162% to only 13%). This leads to an overall

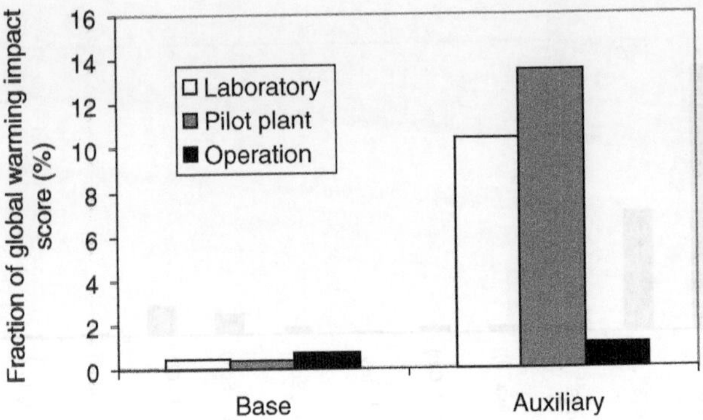

Figure 5.16 Fractions of global-warming scores (% of total impact) for Auxiliary (R) and base used to recycle Auxiliary (R) at different stages of process development. Corresponding mass metrics can be found in Figure 5.10 (see column S^{-1}): The base is used as substrate (S); and the auxiliary material (Aux (R)) is used during the reaction.

reduction in raw material consumption. The increased use of base for increased recycling of Auxiliary (R) is specifically assessed for its global warming potential in Figure 5.16, comparing all three stages of scale-up. The CO_2-equivalent mass emitted by using greater than stoichiometric amounts of base and acid in the operation stage is small compared to the amount of CO_2-equivalent mass economized by recycling more Auxiliary (R). Hence, this optimization was shown to be desirable with the LCA methodology as well as by using the mass balance. In general, LCA includes the quality of substances (e.g., the energy needed for their production), which is not achieved through mass balancing. However, mass indices require less data and effort to calculate and are often sufficient to indicate relevant optimization potentials.

5.4 Discussion

5.4.1 Mass balancing results

Mass indices and environmental factors allow a detailed comparison of different synthetic strategies to potential alternatives. This comparison may reveal significant advantages related to resource utilization or serious problems concerning raw material costs (Case study 1). A comparison may also, at the very least, raise the question of whether or not the optimization potential is already exploited in the process (Case study 3). Mass indices and environmental factors facilitate greater process understanding and document the changes made to a recipe in the laboratory or to a process during scale up. For instance, the increase of sodium hydroxide in Case study 4 enabled a more efficient use of the auxiliary material.

In some cases, the changes in the recipe from pilot to operation scale are substantial. In Case study 4, the introduction of a second solvent is accompanied by a better yield, at the expense of a five-fold higher sewage production. Consequently, raw material demand to produce one kilogram of product is diminished, and the introduction of a second solvent can be considered resource efficient. The optimization of auxiliary material utilization is also important. Auxiliary material use increased significantly in the pilot stage of Case study 4, but was brought under control in the operation stage by using more sodium hydroxide.

Of course, the more resource-intensive a synthesis or process is, that is, the further away it is from the ideal synthesis,[31–33] the higher its optimization potential. For instance, as can be seen in the comparison of four routes (chemical and biochemical) to ethyl (*R*)-2-hydroxy-4-phenylbutyrate[10] (Scheme 5.1), biochemical transformations often have a high water and solvent demand in early stages of development. This is also true for the epoxidation reaction in Case study 2, in which the importance of solvent choice and final product concentration became obvious. Case study 2 also shows that biocatalytic approaches can be very interesting from an economic point of view, but are not necessarily ecologically sound without target-oriented process development and optimization. In addition, the considerable potential for optimization of biochemical transformations can be assessed in the example of the enantioselective enzymatic hydrolysis of (*S*)-2-ethoxy-3-(4-hydroxyphenyl)propanoic acid by H.-J. Deussen *et al.*[34] The authors juxtaposed different stages of development from laboratory to the pilot plant process. Using the production of one kilogram of enantiomer as a basis, they demonstrated that water demand in a 44 kg as compared to a 70 g scale could be reduced from 200 L to 5 L (factor 40), solvent (MTBE) demand from 154 L to 6–8 L (factor $\approx$20), and enzyme demand from 143–743 g to 17 g (minimum factor $\approx$8).

As a first approach, based on experiences with other projects, a person using mass balance information can anticipate appropriate estimates for the efficiency of solvent demand, sewage production and so on. Therefore, a comparison of several synthetic approaches becomes easier and, at the same time, more complete. Evaluations of alternatives were carried out in the literature (e.g., references [8,10,15,35–42] marginally in references [1,43,44]) and here in Case study 1–3. However, comparing syntheses in the limited balance regions (Figure 5.3) of mass balancing does not show the complete picture when complicated substances such as catalysts are employed. The more preceding reaction chains are considered, the better. For example, in Case study 2 the balance region of the chemical catalytic procedure was extended to include the synthesis sequence of the catalyst. While it is common practice to neglect the production of the catalyst, the turnover frequency (TOF) and turn over number (TON) are considered important metrics in evaluating the efficiency of a catalyst, the overall synthetic scheme and the overall process. However, these metrics, like others such as atom economy,[45] can only be considered as single metrics without any linkages between them. They have a limited scope with partly overlapping information content – they do not complement one another to provide a consistent assessment. Weighting the relevance of these metrics as a means of making comparisons amongst them also cannot be easily accomplished. Therefore, a reliable comparison of alternative synthesis protocols or sequences with such metrics is also difficult. Still, if one is only interested in the effectiveness of one or more catalytic systems for academic reasons, regardless of the reaction conditions (temperature, pressure, time, etc.) or production of the catalyst, then the TOF and TON may be of interest.

Mass balancing unifies all metrics such as yield, atom economy, TON, TOF, and so on. Therefore, the application of mass indices and environmental factors presents itself as an integrating solution with moderate data demands. A fair comparison of alternative syntheses can be performed by considering the quantity and the production of the catalyst. The higher the TON or the TOF, the lower the catalyst input, leading to lower expenses for catalyst production, to lower resource demands and less waste production. Such an approach would result in balance regions larger than those shown in Figure 5.3 for mass balancing, but still considerably smaller than those needed for LCA. Figures and tables such as Figure 5.11 prove to be helpful visualization means.

While valuable insights can be gained from mass balancing, the method may not be straightforward enough in some situations to decide whether a change in the mass balance really leads to increased resource efficiency. In Case study 4, for instance, one kilogram of base has different resource demands than the same mass of auxiliary material (Figure 5.15). In instances such as this, such trade-offs are commonly resolved by expert judgement.

5.4.2 Life Cycle Assessment (LCA) results (global warming)

LCA offers valuable additional information because mass flows are converted into comparable units with respect to resource use or environmental impacts. For instance, in Case study 4 it was shown that the increase in acid and base input from pilot to operation stage is worthwhile. This is because the fresh auxiliary material input that would be avoided has considerably higher greenhouse gas emissions associated with its production than is the case for the additional acid and base inputs (Figure 5.16). A full LCA allows a ranking of all inputs with respect to a specific environmental problem, which in turn allows one to set priorities for optimization without additional expert judgement. In Case study 4, such a ranking shows the dominant contribution to global warming impacts from substrate production, followed by solvent and utility supply (Figure 5.15). It is expected that the dominant influence of substrates generally applies to global-warming impacts of fine and speciality chemical production because substrates are themselves products of relatively complex, multi-step synthesis routes. Likewise, solvent recycling is of general importance, because solvents are often used in considerable amounts. In Case study 4, impacts from solvent recycling by distillation were small compared to impacts of fresh solvent production (Figure 5.15), bearing in mind that around 3.1 kg of solvent are supplied by recycling, and only around 0.2 kg is virgin solvent (Figure 5.13).

These findings support the general usefulness of efforts to increase yield and solvent recycling rates, as documented in Case study 4 (see also Geisler *et al.* [9]). However, producing an LCA requires a substantially higher effort for data acquisition as compared to mass balancing.

In addition, the LCA results presented here include considerable uncertainties. First, utility demands are rough estimates,[9] but these are considered to be appropriate in an overview analysis as given here. In contrast to this, a process-specific assessment of energy demands (e.g., reference [46]) may be appropriate where energy-intensive process steps are involved (e.g., reaction with extreme temperatures) or where the initial estimates of energy demands used here dominate the results. Second, we modelled mass and energy balances of solvent recycling, waste incineration and waste water treatment based on generic scenarios. It should be noted, however, that the resource efficiency of solvent recycling may depend strongly on the solvent type and composition of spent solvent, as well as on site-specific waste management options.[9] For example, Capello *et al.*[47] demonstrated how site-specific data could be generalized in an LCA on treatment options for spent solvents. In addition, the greenness of solvents was examined from the literature[48] considering LCA and an EHS-methodology (environmental, health and safety). Third, data on background processes and assessment factors for global warming potentials carries specific uncertainty, which may however, be quantified.[49]

5.4.3 Assessment scopes of MB and LCA

In addition to the comparison of different stages of process development during scale up, the evaluation of alternative syntheses or processes using mass indices and environmental factors or LCA may reveal the lack of resource efficiency and thus call into question the current mode of operation. Decisions that were made in earlier stages of scale-up may then be revised.

With regard to the high leverage of optimization in early stages of product development (Introduction), it appears to be most efficient to perform mass balancing after the first experimental results are obtained in the laboratory or even while planning experiments. This can easily be done using the kind of electronic laboratory notebook as provided by the software EATOS.[8] This facilitates quantifying raw material costs for alternative routes and/or consecutive development stages. In principle, a switch from a paper version to an electronic version of the laboratory notebook appears satisfactory or very satisfactory to 85% of interviewed users of such a tool.[50]

However, the significance of results from such analyses depends on the quality of the input data. For example, laboratory recipes often do not meticulously document solvent and auxiliary input masses. In many cases, water inputs and waste management are not determined before the pilot stage is reached. Estimates similar to those applied in LCA[9] may be used in order to complete a preliminary mass balance. While such estimations cause considerable uncertainty, it seems more appropriate to evaluate alternatives based on preliminary information, that is, experience-based assumptions concerning the production of substrate or catalyst, than to simply ignore potentially important contributions to the mass balance.

Mass balances are also an appropriate basis for LCAs. However, the application of LCA makes sense only if the synthesis routes of precursors are known (Figure 5.3). Full Life-Cycle Inventories (LCIs) can only be estimated then, for example, according to the procedure of Geisler *et al.*[9] However, estimating energy demands and waste treatment for laboratory protocols may not be meaningful, since these protocols will always be modified during scale-up. Therefore, LCA is only considered to be useful for assessments from the pilot stage onwards. A proposal for incorporation of LCA in the product development of pesticides may serve as an example.[51]

If it is already known at the pilot stage in which plant the final operation will be situated, solvent recycling, waste treatment, and utility supply processes should be modelled according to that specific plant. This reduces uncertainty in the scale-up from pilot to operation.

Once mass balances or LCIs are being compiled, further indicators may be applied to these data. In LCA, relatively comprehensive sets of indicators are established, comprising for instance human toxicity and ecotoxicity potentials.[16] Concerning mass balancing, indicators for a first overview of process safety were proposed.[8] Also, environmental indices based on mass balances exist.[8] However, in the current stage of development these assessment methods are merely rough approximations and should only be understood as a means to attract attention to specific problems as shown in Case study 2.

Conclusions

Our case studies prove that optimization objectives generally followed in synthesis design and during scale up show a high potential for increasing resource efficiency. These objectives are, for example, increases of yield and the recycling efficiency of solvents and auxiliary materials.

To determine the best synthetic route and/or the best synthetic protocol (in any stage of scale up), alternative comparison scenarios need to be set up. These are iteratively approximated to reality by means of experimental examination. Mass balancing and/or LCA of those experiments will show whether performance levels of alternatives are retained or surpassed. If serious problems emerge, formerly less efficient alternatives will need to be considered again. It is thus important to keep a quantitative and objective overview of all alternatives in mind. Maintaining an efficient overview of alternatives is a prime strength of mass balancing. Therefore, assisted by means of an appropriate software (e.g., references [8,50]) the determination of mass indices and environmental factors appears to be promising not only for the purpose of accompanying scale-up but also for planning and conducting syntheses.

The considerably higher effort of carrying out LCAs would be worthwhile in two situations. First, individual LCAs may be valuable in cases where complex and potentially relevant trade-offs occur in mass balancing which cannot be resolved by expert judgement. Second, the routine use of LCA could be aimed for. This would necessitate a dedicated effort to compile inventory data for frequently used processes, such as waste treatment. Once this is achieved, calculating LCAs for existing or new processes is relatively easy.

A combined use of mass balancing and LCA may however be the most promising option. Mass balancing may then be used to screen possibly large numbers of alternative synthesis routes. LCA would be valuable for a more detailed evaluation of process alternatives in later stages of scale up, where more data are available.

The scope to increase resource efficiency during scale up was shown to be large in the case studies assessed here. Therefore, efforts to implement methods such as mass balancing or LCA would pay off soon, since this would help to fully tap optimization potentials concerning ecology and economy.

Acknowledgements

We are very grateful to Dr. Hässig, Dr. Schneider and Dr. Kulka from Syngenta Crop Protection for providing data on production processes and valuable discussions. Further, we gratefully acknowledge the reviewing of the manuscript by Jürgen O. Metzger.

References

1. Eissen, M., Metzger, J.O., Schmidt, E., Schneidewind, U. (2002) 10 Years after Rio – Concepts on the Contribution of Chemistry to a Sustainable Development. *Angewandte Chemie International Edition*, **41**, 414–436.
2. Poliakoff, M., Fitzpatrick, J.M., Farren, T.R., Anastas, P.T. (2002) Green Chemistry: Science and Politics of Change. *Science*, **297**, 807–810.
3. Note Report of the United Nations Conference on Environment and Development, Rio de Janeiro, 3–14 June 1992, (a) Rio Declaration on Environment and Development; (b) Agenda 21; to be found at http://www.un.org/esa/sustdev.
4. Constable, D.J.C., Curzons, A.D., Cunningham, V.L. (2002) Metrics to Green Chemistry – Which are the Best? *Green Chemistry*, **4**, 521–527.
5. Laird, T., Hermitage, S.A., Tilstam, U. (2003) Some Items of Interest to Process R&D Chemists and Engineers. *Organic Process Research & Development*, **7**, 124–134.
6. Winterton, N. (2001) Twelve More Green Chemistry Principles. *Green Chemistry*, **3**, G73–G75.
7. Heinzle, E. & Hungerbühler, K. (1997) Integrated Process Development: The Key to Future Production of Chemicals. *Chimia*, **51**, 176–183.
8. Eissen, M. & Metzger, J.O. (2002) Environmental Performance Metrics for Daily Use in Synthetic Chemistry. *Chemistry – A European Journal*, **8**, 3580–3585.
9. Geisler, G., Hofstetter, T., Hungerbühler, K. (2003) Production of Fine and Speciality Chemicals: Procedure for the Estimation of LCIs. *International Journal of Life Cycle Assessment*, **9**, 101–112.
10. Blaser, H.-U., Eissen, M., Fauquex, P.F. *et al.* (2003) Comparison of Four Technical Syntheses of Ethyl (R)-2-Hydroxy-4-Phenylbutyrate. In: *Large-Scale Asymmetric Catalysis* (eds H.-U. Blaser & E. Schmidt), pp. 91–104. Wiley-VCH, Weinheim.
11. Jödicke, G., Zenklusen, O., Weidenhaupt, A., Hungerbühler, K. (1999) Developing Environmentally-Sound Processes in the Chemical Industry: A Case Study on Pharmaceutical Intermediates. *Journal of Cleaner Production*, **7**, 159–166.
12. Zlokarnik, M. (2003) Scale-Up in Chemical Engineering. *Ullmann's Encyclopedia of Industrial Chemistry*. Wiley-VCH Verlag GmbH & Co.
13. Curzons, A.D., Constable, D.J.C., Mortimer, D.N., Cunningham, V.L. (2001) So You Think Your Process Is Green, How Do You Know? – Using Principles of Sustainability to Determine What Is Green – a Corporate Perspective. *Green Chemistry*, **3**, 1–6.
14. Sheldon, R.A. (1994) Consider the Environmental Quotient. *CHEMTECH*, **3**, 38–47.
15. Eissen, M., Hungerbühler, K., Dirks, S., Metzger, J.O. (2003) Mass Efficiency as Metric for the Effectiveness of Catalysts. *Green Chemistry*, **5**, G25–G27.
16. Guinée, J.B., Gorrée, M., Heijungs, R. *et al.* (2001) *Life Cycle Assessment: An Operational Guide to the ISO Standards.* Centre of Environmental Science (CML), Leiden University, The Netherlands.
17. Wada, M., Sakurai, Y., Akiba, K.-y. (1984) Addition of Alkynyl Anions to Aldimines Containing A-Hydrogens: A Novel Synthesis of ß-Aminoacetylenes. *Tetrahedron Letters*, **25**, 1083–1084.
18. Fischer, C. & Carreira, E.M. (2001) Direct Addition of TMS-Acetylene to Aldimines Catalyzed by a Simple, Commercially Available Ir(I) Complex. *Organic Letters*, **3**, 4319–4321.
19. Komiya, S. (1997) *Synthesis of Organometallic Compounds: A Practical Guide (Inorganic Chemistry)*. Wiley & Sons, New York.
20. Wada, M., Sakurai, Y., Akiba, K.-y. (1984) Addition of Alkynyl Anions to Aldimines Containing Alpha-Hydrogens: A Novel Synthesis of Beta-Aminoacetylenes. *Tetrahedron Letters*, **25**, 1083–1084.
21. Fischer, C. & Carreira, E.M. (2004) MgI_2 an Additive in Ir(I)-Catalyzed Addition of Silylacetylenes to Imines: Expeditious Synthesis of Propargylic Amines. *Synthesis*, **9**, 1497–1503.
22. Blanch, H.W. & Clark, D.S. (1996) *Biochemical Engineering*. Marcel Dekker, New York.

23. Schmid, A., Hofstetter, K., Feiten, H.J., Hollmann, F., Witholt, B. (2001) Integrated Biocatalytic Synthesis on Gram Scale: The Highly Enantio Selective Preparation of Chiral Oxiranes with Styrene Monooxygenase. *Advanced Synthesis & Catalysis*, **343**(6–7), 732–737.
24. Otto, K., Hofstetter, K., Rothlisberger, M., Witholt, B., Schmid, A. (2004) Biochemical Characterization of StyAB from *Pseudomonas* sp strain VLB120 as a Two-Component Flavin-Diffusible Monooxygenase. *Journal of Bacteriology*, **186**(16), 5292–5302.
25. Park, J.B., Buehler, B., Habicher, T. *et al.* (2006) The Efficiency of Recombinant *Escherichia coli* as Biocatalyst for Stereospecific Epoxidation. *Biotechnology and Bioengineering*, **95**(3), 501–512.
26. Panke, S., Held, M., Wubbolts, M.G., Witholt, B., Schmid, A. (2002) Pilot-Scale Production of (*S*)-Styrene Oxide from Styrene by Recombinant *Escherichia coli* Synthesizing Styrene Monooxygenase. *Biotechnology and Bioengineering*, **80**, 33–41.
27. Larrow, J.F., Roberts, E., Verhoeven, T.R. *et al.* (1999) (1S,2R)-1-Aminoindan-2-ol. *Organic Syntheses*, **76**, 46–56.
28. Larrow, J.F. & Jacobsen, E.N. (1998) (R,R)-N,N'-bis(3,5-di-tert-butylsalicylidene)-1,2-Cyclohexanediaminomanganese(III) Chloride, a Highly Enantioselective Epoxidation Catalyst. *Organic Syntheses*, **75**, 1–11.
29. Tietze, L.F. & Eicher, T. (1991) *Reaktionen und Synthesen*, 2nd Edition. Georg Thieme Verlag, Stuttgart, New York.
30. Bahadir, M., Hopf, H., Nijakowski, A. *et al.* (2006) Sustainability in the Organic Chemistry Lab Course (NOP, Nachhaltigkeit (Sustainability), Organische Chemie (Organic Chemistry), Praktikum (Lab Course)). *An Internet Laboratory Book*, http://www.oc-praktikum.de/index.php?page=entry&lang=en (Last accessed on 04.01.2007).
31. Wender, P.A. (1996) Frontiers in Organic Synthesis: Introduction. *Chemical Reviews*, **96**, 1–2.
32. Gladysz, J.A. (2001) Recoverable Catalysts. Ultimate Goals, Criteria of Evaluation, and the Green Chemistry Interface. *Pure and Applied Chemistry*, **73**, 1319–1324.
33. Hartwig, J.F. (2002) Chemical Synthesis: Raising the Bar for the "Perfect Reaction". *Science*, **297**, 1653–1654.
34. Deussen, H.-J., Zundel, M., Valdois, M. *et al.* (2003) Process Development on the Enantioselective Enzymatic Hydrolysis of *S*-Ethyl 2-Ethoxy-3-(4-hydroxyphenyl)Propanoate. *Organic Process Research & Development*, **7**, 82–87.
35. Lenoir, D. (2006) Selective Oxidation of Organic Compounds – Sustainable Catalytic Reactions with Oxygen and without Transition Metals? *Angewandte Chemie International Edition*, **45**, 3206–3210.
36. Protti, S., Dondi, D., Fagnonia, M., Albini, A. (2007) Photochemistry in Synthesis: Where, When, Why. *Pure and Applied Chemistry*, **79**(11), 1929–1938.
37. Selva, M. & Perosa, A. (2008) Green Chemistry Metrics: A Comparative Evaluation of Dimethyl Carbonate, Methyl Iodide, Dimethyl Sulfate and Methanol as Methylating Agents. *Green Chemistry*, **10**, 457–464.
38. Andraos, J. (2005) Unification of Reaction Metrics for Green Chemistry: Applications to Reaction Analysis. *Organic Process Research & Development*, **9**, 149–163.
39. Andraos, J. (2005) Unification of Reaction Metrics for Green Chemistry II: Evaluation of Named Organic Reactions and Application to Reaction Discovery. *Organic Process Research & Development*, **9**, 404–431.
40. Andraos, J. (2006) On Using Tree Analysis to Quantify the Material, Input Energy, and Cost Throughput Efficiencies of Simple and Complex Synthesis Plans and Networks: Towards a Blueprint for Quantitative Total Synthesis and Green Chemistry. *Organic Process Research & Development*, **10**, 212–240.
41. Andraos, J. & Sayed, M. (2007) On the use of 'Green' Metrics in the Undergraduate Organic Chemistry Lecture and Lab to Assess the Mass Efficiency of Organic Reactions. *Journal of Chemical Education*, **84**, 1004–1010.

42. Kuzemko, M.A., Van Arnum, S.D., Niemczyk, H.J. (2007) A Green Chemistry Comparative Analysis of the Syntheses of (*E*)-4-Cyclobutyl-2-2-(3-Nitrophenyl)Ethenyl Thiazole, Ro 24-5904. *Organic Process Research & Development*, **11**, 470–476.
43. Metzger, J.O. & Eissen, M. (2004) Concepts on the Contribution of Chemistry to a Sustainable Development – Renewable Raw Materials. *Les Comptes Rendus De L Académie Des Sciences – Chimie*, **7**, 569–581.
44. Ranke, J., Bahadir, M., Eissen, M., Konig, B. (2008) Developing and Disseminating NOP: An Online, Open-Access, Organic Chemistry Teaching Resource to Integrate Sustainability Concepts in the Laboratory. *Journal of Chemical Education*, **85**, 1000–1005.
45. Eissen, M., Mazur, R., Quebbemann, H.-G., Pennemann, K.-H. (2004) Atom Economy and Yield of Synthesis Sequences. *Helvetica Chimica Acta*, **87**, 524–535.
46. Schneider, M., Zimmermann, K., Aquino, F., Bonrath, W. (2001) Industrial Application of Nafion-Systems in Rearrangement – Aromatisation, Transesterification, Alkylation, and Ring-Closure Reactions. *Applied Catalysis A-General*, **220**, 51–58.
47. Capello, C., Hellweg, S., Badertscher, B., Hungerbühler, K. (2005) Life-Cycle Inventory of Waste Solvent Distillation: Statistical Analysis of Empirical Data. *Environmental Science and Technology*, **39**, 5885–5892.
48. Capello, C., Fischer, U., Hungerbühler, K. (2007) What is a Green Solvent? A Comprehensive Framework for the Environmental Assessment of Solvents. *Green Chemistry*, **9**(9), 927–934.
49. Geisler, G., Hellweg, S., Hungerbühler, K. (2005) Uncertainty Analysis in Life Cycle Assessment (LCA): Case Study on Plant-Protection Products and Implications for Decision Making. *International Journal of Life Cycle Assessment*, **10**, 184–192.
50. Achour, Z., Laidboeur, T., Gien, O., Musolino, A., Bon, X., Grimaud, B. (2005) Sanofi–Synthelabo Chemical Development and the Development of an Electronic Laboratory Notebook. *Organic Process Research & Development*, **8**, 983–997.
51. Geisler, G., Hellweg, S., Hofstetter, T.B., Hungerbühler, K. (2005) Life-Cycle Assessment in Pesticide Product Development: Methods and Case Study on Two Plant-Growth Regulators from Different Product Generations. *Environmental Science and Technology*, **39**, 2406–2413.

Chapter 6
Process Metrics

David J.C. Constable, Conchita Jimenez-Gonzalez and Alexei Lapkin

6.1 General considerations for process metrics

There has been a considerable amount written about assessing the economic or commercial viability of chemical processes.[1,2] There has also been a considerable amount written about how to comparatively assess processes from a green chemistry or green engineering perspectives,[3–12] and reviews of various metrics systems.[13–16] Each of these has value in helping to answer the general question of which metrics are best. However, one of the immense difficulties with most metrics discussions and the application of metrics is finding the few key metrics that have the greatest impact or potential to 'green' any given process.

In general, this difficulty may be seen as a boundary problem, that is, where does one draw the boundary for a consideration of how 'green' is a process? At this point in time, it is our opinion that any proposed suite of metrics or metrics systems that are not based upon or use life-cycle considerations is fundamentally flawed. Stated slightly differently, a guiding principle of process metrics must be the comprehensive application of a life-cycle approach if one is to appropriately evaluate the 'greenness' of processes. The reader is referred elsewhere in the literature to develop a greater understanding of Life Cycle Inventory and Assessment methodologies (see also Chapter 7 in this book).[17–20]

Another practical consideration for process metrics that is related to a life-cycle view is the tendency of not taking a systems, systems-wide, or holistic view of a process. When reviewing a process flow and instrumentation diagram, it is very easy to forget, or perhaps in the case of students and/or someone early in their career, to not know what a plant might look like. Processes are generally, but not always, a closely linked set of unit operations carried out across a finite time and discrete space. Changes in one particular unit operation, say a reactor, would most likely bring cascade effects in other unit operation such as the separation train. It would be a fallacy to have a set of metrics that do not take into consideration the closely knit network of cause-and-effect relationships that comprise chemical processes.

On very large scales such as in modern petrochemical and chemical plants, there are considerable opportunities to look at the collection of unit operations and processes and develop metrics that will enable one to work towards significant mass and energy integration amongst those processes.[21] This is generally not achievable on smaller, batch multi-purpose chemical plants given the campaign approach most of these facilities employ and the widely varying nature of the processes over time. Metrics in the batch chemical context will therefore have to be very different.

A systems-wide view will also require the collection of more than a single metric and implies a multivariate view of the system. There is an underlying complexity in most chemical operations and it is hard work to find just a few key metrics to describe a given system or operation. When evaluating the 'greenness' of a process, it is not uncommon to encounter trade-offs amongst different metrics, for example, a change in a critical process parameter that brings a better environmental profile can have an adverse effect in the overall safety of the process, or a change that reduces one environmental impact could increase another. A systems approach with the right set of carefully chosen metrics can allow for the evaluation of trade-offs to make the appropriate judgement on the 'greenness' of a process, accounting for the overall lowest impact.

This leads nicely into another general principle for process metrics – one size does not fit all. Stated differently, metrics must be adapted for their context and continuously evaluated as to their utility, applicability, appropriateness, and so on. Metrics should also be tested and validated to ensure that they are successfully driving the desired goals set by the organization. They must also be easily understood and accepted by key stakeholders. Using existing data collection methods or business systems for collecting information for process metrics, for example, should ensure generally broad acceptance of metrics.

Another general principle for good process metrics is that these should promote strategic analysis and continuous improvement. If the metrics are being collected but not evaluated on a regular basis, and decisions based upon the metrics results are not made, there is no point in collecting them. This may seem to be an obvious point, but metrics are not always routinely questioned, assessed, evaluated and evolved to help make strategic decisions or make them more useful to a business.

For the sake of simplicity, Table 6.1 contains a small collection of general categories of interest or concern that should be considered for metrics application in any route or process development. As will be immediately obvious from the table, the list mostly contains categories that would generally be found in many texts on process metrics. It should be

Table 6.1 General areas of interest for process metrics.

Materials	Equipment	Operability	EHS risk	Quality
Physical form and properties (i.e., gas, liquid, solid)	Unit operation type	Throughput/cycle time	Occupational exposure	Purity/impurity profile
Mass (i.e., total, solvent, reactant, process, etc.)	Number of unit operations	Robustness	Environmental – air, water, land	
Inherent hazard (e.g., toxicity, stability, reactivity)	Size (volume)	Energy (i.e., total, heating, cooling, recovery, treatment, etc.)	Safety/process safety	
Cost	Scalability	Ease of cleaning and maintenance		
Renewability Recyclability	Controllability			

EHS = environment, health and safety.

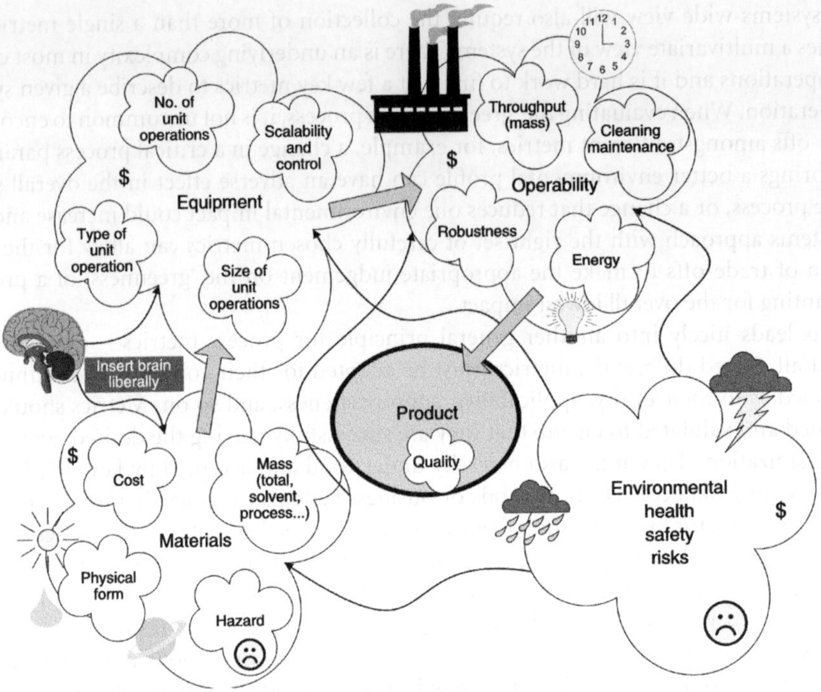

Figure 6.1 Interrelationships between process metrics categories.

noted that each category and each example could be further dissected for additional attributes or areas of concern in ever-increasing levels of detail. This table was not intended to be comprehensive but illustrative and high level. It should also be noted that Life Cycle considerations impact on each category and aspect contained in the table.

Figure 6.1 is presented as an attempt to show the interrelationships between the process metrics categories shown in Table 6.1. Materials (chemicals, solvents, reagents, etc.) choices will directly affect the choice of unit operations and the combination of materials and equipment will directly impact the operability of the process. Materials, equipment and operability all have EHS impacts and opportunities associated with them and ultimately impact product and product quality.

The remainder of this chapter will be an extended discourse on the categories in Table 6.1 and Figure 6.1 and suggested approaches to 'green' metrics development.

6.2 Materials

6.2.1 Physical form and properties

Physical form and properties are directly linked to and impact other aspects of a process; that is, reactor type, type of mixing, throughput; that is, how quickly a chemical will dissolve in a solvent or precipitate out, the ease of separation of two liquids, and so on.

It will also impact on energy for heating or cooling, recovery if applicable, cleaning and wastes.

In terms of metrics for this aspect, one could profitably divide this into three different approaches, depending on the question that metrics are trying to answer:

- summing the number of materials possessing a given attribute,
- summing the mass of materials possessing a given attribute,
- or a combination of these to find a general score for material complexity.

In the case of summing the numbers of materials, one might consider the number of materials of different phase, number of different solvents containing azeotropes or close boiling points, and so on. One could also sum on the basis of mass. Both number and mass could be proxies for the degree of complexity associated with the process. Larger numbers and masses of different materials are likely to necessitate more equipment or energy use, slow throughput and increase general materials management. Another option is to devise a combined metric to provide an index of complexity. Complexity generally drives up cost and increases waste in one part of the life cycle or another.

6.2.2 Mass

This has been a very useful area of metrics exploration. In general terms, one can simply divide mass into key types, such as mass of solvent, water, process chemicals (i.e., not reactants) and reactants, and sum the amount of each. This may seem to be a bit simplistic, but simplicity does have the advantage of allowing one to discover what the key contributions to energy or waste are from a materials perspective, and this can in turn drive one to consider less impactful alternatives.

Some specific metrics for mass have been found to be useful and are described elsewhere.[22] In addition to reaction mass efficiency described elsewhere and in Chapter 4, a useful metrics is Mass Intensity, defined as:

$$\text{Mass Intensity (MI)} = \frac{\text{Total mass into the process (kg)}}{\text{Mass of product (kg)}}$$

Mass Intensity takes into account yield, stoichiometry, solvent and reagents in a reaction mixture and expresses this on a mass/mass basis. In the ideal situation, MI would approach 1. Total Mass is defined as including everything that is put into a reaction vessel, that is, reactants, reagents, solvents, catalysts, and so forth. Total mass also includes all mass used in acid, base, salt and organic solvent washes, and organic solvents used for extractions, crystallizations, or for solvent switching. Many would also include process water in the total mass calculation and others would add water used for cleaning. However, including water can skew mass data in many processes, depending on the process details.

Including water in mass metrics can be a somewhat contentious issue at times. Water by itself does not, in many instances, constitute a significant environmental impact. However, in the case of highly purified water there are generally significant life-cycle impacts related to the chemicals and equipment used to purify the water. This is especially true for such industries as the semi-conductor industry, pharmaceuticals and some parts of the fine chemicals industry. There is also the problem of mixed aqueous–organic reaction mixtures, where separations and subsequent disposal of wastes can lead to increased use of solvent, additional unit operations, waste treatment operations and most certainly much more energy will be consumed. Finally, in many parts of the world, competition for potable water is becoming more of an issue and will continue to be a greater issue in the future. Consequently, metrics for water use are being included more frequently.

Mass Intensity may also be expressed as a percentage by taking the inverse and multiplying by 100. This metric is also known as mass productivity or mass efficiency and provides an indication of how much of the input mass, expressed as a percentage, is integrated into a final, saleable product (how much of what we buy ultimately produces a profit). This is perhaps valuable for its shock value and ease of interpretation, especially for non-technical business people.

Some other useful mass indicators include:

$$\text{Solvent Intensity} = \frac{\text{Total solvent input excluding water}}{\text{Total mass input}}$$

$$\text{Waste Intensity} = \frac{\text{Total waste produced}}{\text{Total mass input}}$$

6.2.3 Inherent hazard

This aspect of materials is a major driver in assessing the overall Environmental, Safety and Health Risk associated with a process. It is also a key driver in equipment design and material of construction, chemical storage and handling, throughput and process robustness. For example, highly reactive materials (e.g., vinyl chloride monomer, aziridines, acetylenes, hydrogen fluoride, etc.) are routinely and safely used, but they do require considerable care during manufacture, storage, transport and use to maintain thermodynamic and kinetic control without causing any adverse events.

As with the case of mass, there are several approaches to metrics for this aspect. One can simply sum numbers and/or mass of chemicals possessing hazards in different areas; for example, process safety, occupational exposure, or environmental hazard. Typically, most companies will use a banding approach for materials that allows a quick identification of the hazard category, and usually marries hazard with a suggested control approach; for example, layers of protection, pressure relief valves, and so on. One is then able to rapidly identify issues and potential opportunities for elimination, substitution, or control.

An example of how this might look for a group of materials is shown in Figure 6.2. Materials are listed according to type and then a hazard ranking applied. This hazard ranking is generally based on an assessment of a variety of potential hazards associated with a given material. It may be a composite ranking based on multiple types of hazards or the ranking can be done for each hazard category individually. In the example shown in Figure 6.2, two different potential means of scoring are illustrated; either a simple weighted average (hazard ranking times mass of material used), and a score based on weighted averages for each category of materials (hazard ranking times mass of [solvent, reagent or process chemical] used) and taking a geometric mean of the scores for each category.

Compsite score as weighted average

7

Score by type of material

Solvent	7
Reactant	3
Process chemical	8
Composite score (geometric mean)	6

Material	Ranking	Mass - kg/kg API	Material class
Acetone	8	15.7	Solvent
Acetonitrile	6	37.9	Solvent
DMF	2	0.5	Solvent
Heptane	9	25.8	Solvent
Hexane	4	3.1	Solvent
N-propanol	5	7.1	Solvent
TBME	6	3.2	Solvent
Reactant 1	4	0.9	Reactant
Reactant 2	1	0.8	Reactant
Reactant 3	4	0.8	Reactant
Reactant 4	4	0.9	Reactant
Reactant 5	1	0.9	Reactant
Reactant 6	4	1.9	Reactant
5% Pd/C	4	0	Process chemical
Acetic anhydride	7	0.7	Process chemical
Activated charcoal	10	0.1	Process chemical
Hexyl lithium	7	1.3	Process chemical
Potassium carbonate	10	2.3	Process chemical
Potassium hydrogen sulphate	4	1.1	Process chemical
Sodium hydroxide	10	0.6	Process chemical

High hazard material – selection of lower hazard material recommended; if substitution is not feasible perform health risk assessment and adopt exposure control strategy to reduce health risks

Hazard material; perform health risk assessment and adopt exposure control strategy to reduce health risks

Relatively low hazard material – perform health risk assessment and adopt exposure control strategy to manage health risks

Figure 6.2 Generic example of hazard scoring for process materials.

As can be seen, a category score may help to focus efforts on areas of the process most in need of attention and accounting for the perceived priorities at a given time.

A note of caution is appropriate at this juncture as many confuse the inherent hazard of materials with risk. This will be covered in slightly greater detail later in this chapter, but it is important to distinguish between hazard and risk. Risk is defined as being a function of the inherent hazard of a material and the potential, or likelihood, for exposure:

$$\text{Risk} = f(\text{the inherent hazard of a material, the potential or likelihood for exposure})$$

It is often expanded to include a severity rating at given probabilities. This is useful to bear in mind from a green perspective because one can use highly hazardous materials but design a process that has less impact and lower risk. For example, one might consider *in situ* generation of phosgene in small volumes that are immediately consumed.[23]

6.2.4 Cost

Cost is probably the one aspect of materials that almost every business would argue is the most important consideration for evaluating a process. If one looks at a bill for materials, it is often very easy to see that process costs are related to just a few chemicals, and this is very helpful in prompting additional research into alternative chemicals, routes or chemical processing technologies to increase the efficiency of use or develop a less costly alternative process.

Apart from the actual purchase cost a business pays a supplier for a chemical, there are costs that are usually not considered. Happily there has been considerable development of approaches to Total Cost Assessment[24] or environmental accounting,[25] and the reader is strongly encouraged to seek out more information about how this is done. Application of Total Cost Assessment will provide a strategic view of trends that can directly affect the long-term viability of a process and is well worth the effort.

A couple of observations are important here to bring a green perspective to cost metrics. First, the current cost of materials is generally not an accurate reflection of the true or overall cost required to deliver that material. This is where a life cycle and systems-based perspective will very readily reveal what costs are not being borne by the user of any given chemical (see Table 6.2). These include liability (type 3 cost), relationship or image costs (type 4 costs), and costs borne by society that we generally do not have a means of pricing, for example, air, water, CO_2, (type 5 costs). It is understood that some costing mechanisms are being implemented for emissions, but these are in their early days and are not broadly accepted or standardized across regions, states or continents.

Second, for many high value or chemically useful materials in commercial use, there are limited supplies, or the supply of these materials is at significant cost to the environment. An example would be in the supply of some transition metals and platinum group metals that are of great interest for their catalytic potential. In addition to the impacts from extracting, refining and employing these materials in catalyst systems, our use of these materials generally results in a portion of these metals being widely dispersed again since complete recovery from process streams is not always possible. In some cases, catalysts are homogeneous and difficult to extract and recover from the reaction mixtures. In those cases where the catalysts are heterogenized, there is likely to be a 5–10% loss of the catalyst with the filter or filter aid. It would be very difficulty and likely costly indeed to collect these materials or to find alternatives that perform the same function.

6.2.5 Renewability

It is fair to say that very few chemicals currently available in routine commercial operations are derived from renewable sources or if from renewable sources, they often

Table 6.2 Total cost accounting – cost types.

Type I Costs	Direct costs. Traditional capital investment and operational costs, including: • Equipment • Labour • Raw materials • Waste treatment
Type II Costs	Indirect and hidden costs: • Reporting • Monitoring • Regulatory (e.g., operating permits and fees)
Type III Costs	Contingent costs: • Liabilities (e.g., clean up) • Lawsuits • Damage to resources • Accidents
Type IV Costs	Internal intangible costs: • Company or brand image • Consumer loyalty • Worker morale • Worker relations • Community relations
Type V Costs	External intangible costs: • Increase of housing costs • Degradation of resources

are associated with considerable life-cycle impacts or exhibit trade-offs of environmental impact, such as reduction of global warming potential at the same time that the eutrophication potentials increase. A contemporary example of this debate is found in the production of corn-based ethanol. Considerable controversy surrounds the sustainability of this particular use for corn, and this controversy is not without good reason.

The take away message from the debate is two-fold. First, assessments of renewability have to be done from a life-cycle perspective. Second, there is considerable difficulty in comparing highly developed chemical processing routes based on non-renewable sources with processes that use potentially renewable materials with processing approaches that are not nearly as developed as a process that has been optimized over the course of many years. This is the current dilemma facing many biologically based fermentation or biotransformation processes when compared to chemical synthetic processes, which is not significantly different from the dilemma facing new versus established technologies. The state-of-the-art is still developing and the comparisons are not always entirely able to be an 'apples-to-apples' comparison. It is also not uncommon for the materials and energy use in a biological or potentially renewable process to be several times as large as a chemical synthesis given the differences in process intensity, separability of the product from the biological matrix and the conversion efficiency of the organisms.

6.2.6 Recyclability

Process recycle includes in-process and post-processing recycle. In general, in-process recycle is the preferred approach if at all possible, but is clearly more common in the petrochemical and commodity chemicals context. For batch chemical operations commonly found in the fine chemical, pharmaceutical and agrochemicals businesses, in-process recycle is generally very difficult to do. The premise in those industries, however, is to reduce the scale of batch operations, make them continuous, and number up to achieve the required volumes. If this can be done successfully, it is more likely to be able to incorporate in-process recycle.

Out-of-process or post-process recycle is commonly employed when possible and is used extensively on and off site. It may be important to distinguish between the two options as each has a different set of issues. In general, off-site recycle is likely to have a greater number of potential impacts related to the transport and management of the solvent on and off the site.

Approaches to metrics for recyclability are similar to approaches for other aspects. One can sum the masses of potentially recyclable chemicals, solvents, water, and so on. Recycling can, however, be potentially problematic depending on the type of reaction mixture, the scale of the process and throughput. For example, aqueous–organic reaction mixtures containing solvents with similar boiling points or those that form azeotropic mixtures may be from a balanced perspective, less desirable to recycle than simply obtaining virgin solvents. Clearly one needs to balance costs and impacts across the life cycle to appropriately evaluate recycling opportunities. In such instances, however, recyclability metrics should drive chemists and engineers to utilize different solvent systems that enable the chemistry while ensuring facile recovery and reuse.

Whether the recycling occurs in-process or out-of-process, amounts of recycled materials will necessarily impact mass metrics, as the net amount of materials used is bound to be reduced by the amount recycled. From a life-cycle viewpoint, it is important to account not only for the reduced impacts, but also for the environmental and resource usage impacts that result from recycling materials (e.g., distillation energy for solvent recovery). Nonetheless, in most cases the benefits from avoiding manufacturing impacts tend to dwarf the energy and resources utilized for recycling the materials.

6.3 Equipment and operability intertwined

At the beginning of this chapter, several points were made about general metrics principles that are particularly applicable within the context of equipment and operability. It is worth to revisit them for just a moment to say that good process metrics for these categories are especially dependent on an understanding of the overall process, and optimization of a process should be done from a multivariate perspective. Metrics in these categories should be seen as having considerable dependencies on each other and on the materials and chemicals used in the process.

It is also generally not the case that Industry is building new plants to accommodate each new process or product. This means that within any industrial sector of the chemical enterprise, the type of unit operations in use is generally fairly small and fixed. Within a multi-purpose chemical plant commonly found in the batch chemical industry, it is common practice for process designers to 'make do' with what is available on a given site to avoid capital expenditure, plant shut down for modifications, and so forth. In terms of metrics, the recommended approach is generally to try and understand at a high level what the complexity is that arises from a given process. There is therefore great opportunity to take relatively standardized and simple approaches to the evaluation of processes.

6.3.1 Type and number of unit operation

These two metrics provide an indication of the complexity of a process. There are a number of different unit operations required in any but the simplest processes and there are a number of options for achieving a desired end. For example, in isolating a solid from a liquid reaction mixture, there are different strategies one might employ including gravity settling and decanting, solvent switching, filtration, centrifugation and variations on or combinations of all of the above. In the batch processing context, apart from the complexity of timing campaigns and overall throughput, each of the above types of unit operations comes with a different set of impacts. These impacts primarily affect the overall mass and energy efficiency of a process and the potential for various EHS impacts; for example, occupational exposure risk, process safety risk and/or environmental risk. Simplicity of a process is one of the principles of inherent safety, and the more complex a process is, the farther it is from being inherently safer.

6.3.2 Size of unit operations

In general, the size of a unit operation can make a difference in the efficiency of mixing, the reaction rate, by-product formation, control of exo- or endotherms, separations, and so on. From a process engineering standpoint, one is interested in obtaining and maintaining a high degree of control over mass and energy transfer in any unit operation, but this may be difficult to achieve unless a chemical plant is designed to be fit-for-purpose. As this is often not the case, there is an inherent inefficiency associated with many processes.

6.3.3 Scalability

Process scalability is an indication of how well a process would handle the rigors of moving to different production sizes as the process develops. Process scalability is widely used in the software industry to describe the ability to produce mass customization of products. Scalability in chemical processes implies sufficient process understanding and control to ensure that when a process moves to different and/or larger or smaller equipment, there is some confidence that product quality and yield will not be adversely affected. In continuous processing context of a large chemical plant, scalability is routinely achieved or else a product would never make it to the market, although scaling up a new technology is by no means a trivial endeavour.

In a multi-purpose chemical plant context, facile transitions from R&D laboratory or bench scale to Pilot scale to Commercial size are not frequently easy to achieve. While the reasons for this are many, the consequences of this lack of scalability can be a reduction in product quality and yield, increased by-product formation, longer cycle times and in some cases an inability to reproduce key product properties such as colour, size or crystal structure.

From a green perspective, decreased yields, by-product formation and inability to reproduce key product properties will invariably increase waste and require greater materials and energy use. Longer cycle times invariably will lead to increased energy use and in some cases, increased materials use.

To measure the scalability of a process, one needs to define the critical quality attributes and to understand the chemistry and processes involved in order to find the limits of acceptability of these critical attributes, and thus, the limits of the scalability of a process. For a chemical process to be functional at large scale it should also be operationally simple, safe and straightforward.

6.3.4 Controllability

In petrochemical and bulk commodity chemical manufacture, real-time process control has been a fact of life for many years. There is considerable understanding of processes and control of process parameters is usually maintained within tight specifications to ensure statistical process control to within six sigma, or the occurrence of one defect in a million. This has been enabled through the use of real-time analytical capability that works with programmable logic circuits to make small changes to various process inputs and physical parameters as required.

In batch chemical operations, this level of real-time process control is rarely achieved, although there are increasing attempts in recent years to achieve greater statistical process control, the industry is generally only able to operate at about three or occasionally four sigma, or one defect in 1000–10 000.

From a green perspective, processes that are not in tight control are obviously going to produce a greater quantity of waste, consume more materials and energy per unit of finished product, and lead to reduced throughput and cycle time. In some cases, not holding the process in control will lead to a failure to meet product specifications with the follow-on need to either reprocess the off-spec product or discarding of the product entirely. Either way, through the additional waste production or the materials consumed, an out-of-control process is a problem.

Metrics for this might include number of excursions from statistical process control, but one very useful metric for controllability is process capability, or more accurately, process capability indices. Process capability compares the output of an in-control process to the specification limits by using capability indices. The comparison is made by forming the ratio of the spread between the process specifications (the specification 'width') to the spread of the process values. In a six-sigma environment, this is measured by six standard deviation units for the process (the process 'width'). A process under control is one where almost all the measurements fall inside the specification limits. The general formula for process capability index is:

$$\text{Process Capability} = C_p = \frac{USL - LSL}{6\sigma}$$

where *USL* is the upper specification limit, *LSL* is the lower specification limit and σ is the standard deviation of the process. Clearly, if $C_p > 1.0$, then the process specifications cover almost all the process measurements. Indirect proxies of controllability could be the amount of waste produced per kg of product, or the amount of materials and/or energy consumed per kg of product caused by excursions outside the control zone, for example, a rejected batch will become waste, and additional mass and energy will be required to replace (or rework) the rejected batch.

6.3.5 Robustness

Simply speaking, process robustness is characterized by the extent to which process excursions adversely affect product quality and yield. A process that is not greatly affected by variations in process temperatures, mixing, minor variations in rates of addition, and so on would be considered robust. Good process understanding through appropriate statistical design of experiment to test process inputs and parameters is the key to understanding process robustness.

The main difference between robustness and controllability is that a capable or controlled process will stay within the desired parameters. A robust process can exhibit excursions outside the control parameters, but the critical attributes will not be significantly affected by the excursion beyond specifications.

When assessing the robustness of a process, several factors that can adversely affect it include: non-selective or side reactions that might produce adverse effects and impurities; physical and chemical stability of the materials involved; and complexity of the separation train of the processes.

6.3.6 Throughput/cycle time

Throughput is in simple terms the average saleable production output per a given time unit. Cycle time is the average time between the release and completion of a job, in other words, the rate at which products are manufactured. Key parameters that affect throughput in a chemical plant include the chemical conversion, yield, capacity and availability of existing equipment, process time, cycle time, number of chemical steps, number of unit operations, plant layout, warehouse processes, raw material availability, process bottlenecks and labour availability, amongst others.

In general, the goal of any well-run manufacturing operation is to ensure that all unit operations on-plant are utilized at capacity for as much time as possible before cleaning and/or maintenance is required. The design goal for any process is to ensure that reactions proceed to completion as rapidly as possible without losing control of the reaction and ensuring yields as close to theoretical as possible. Once again, for continuous processing, high throughputs must be maintained to achieve profitability and economic viability, so throughput is generally at capacity or nearly so.

In the multi-purpose batch chemical plant context, continuous utilization of multiple unit operations on-plant is extremely difficult and rarely achieved. In addition, for many batch chemical operations, it is not uncommon for there to be long reaction times, extended periods of time at reflux, extended filtration times, drying times, and so on. Each of these process parts, steps or stages lengthens the cycle time and is likely to lead to increased consumption of materials, energy and labour. For example, a survey of processes in AstraZeneca established a clear relationship between the number of steps in a synthesis and the throughput; when the number of steps decreased linearly, the throughput increased exponentially.

There can be different approaches to tracking throughput; either volume of product or material manufactured per unit of time, number of products per unit time, mass per unit time, and so on.

In the campaign context of a batch plant, it is important to note that throughput and cycle time should include time required for cleaning, transportation and maintenance as these can dramatically affect the overall times and the utilization of any part of a plant. For example, a simple model developed from the AstraZeneca survey calculates the time for a manufacturing campaign as:

$$\text{Time for Manufacturing} = T_m = \frac{N_B}{P} + misc$$

where N_B is the number of batches, P is the productivity and *misc* accounts for any time needed for additional activity, such as plant cleaning and maintenance.

6.3.7 Energy

In a large petrochemical and chemical manufacturing complex, energy reduction through process integration and modification are constantly pursued given the enormous amounts of energy consumed by many processes. Energy efficiency is comparatively much greater than what is found in the batch chemical environment.

For many batch chemical operations it is not unusual for both heating and cooling to be required for any given step or stage of the synthesis. While this can be avoided through closer attention to the combination of chemistry with reactor type and configuration, it is generally not routinely achieved for a variety of reasons. However, it is generally true that the existence of a large installed base of reactors with their supporting unit operations prevents the implementation of newer technologies. Existing in-ground capital that has been paid for many times over is difficult to stop using unless the gains in efficiency or the reduction in costs are overwhelming.

Energy metrics are usually very similar to those for mass; that is, one can slice the total energy used per kg of product to highlight key materials use, or forms of energy such as that used to heat or that to cool. As mentioned before, it is also important to account for the life-cycle energy requirements, which would include not only processing energy, but also the energy required to produce raw materials, recycle of materials, and waste treatment. Some potential metrics might include:

$$\text{Energy Intensity} = \frac{\text{Total process energy [MJ]}}{\text{kg of final product}}$$

$$\text{Life Cycle Energy} = \frac{\text{Life cycle energy requirements [MJ]}}{\text{kg of final product}}$$

$$= \frac{\sum \text{(process, material manufacturing, recoverying, treatment)}}{\text{kg of final product}}$$

$$\text{Waste Treatment Energy} = \frac{\text{Waste treatment energy requirements [MJ]}}{\text{kg of final product}}$$

$$\text{Solvent Recovery Energy} = \frac{\text{Solvent recovery energy requirements [MJ]}}{\text{kg of final product}}$$

Or can be expressed as a fraction of total energy input, such as:

$$\text{Solvent Energy Ratio} = \frac{\text{Total energy for solvent use and recovery}}{\text{Total energy input}}$$

$$\text{Waste Energy Ratio} = \frac{\text{Total waste produced}}{\text{Total energy input}}$$

6.3.8 Cleaning and maintenance

In a batch chemical plant, because individual unit operations are utilized for multiple products, many pieces of kit may be subjected to long clean-out periods using large solvent volumes and/or aqueous detergents, or both. If possible, clean-in-place protocols as opposed to break down and rebuild, and so forth are preferred. These cleaning materials are often not considered as part of a process and so their use is not optimized in the same manner as are other process-related materials and solvents. Frequency of cleaning, length of cleaning, volumes of solvent, water, detergent, energy use, and so on are all important parameters that affect the real mass intensity of a process, as well as cycle time and throughput, amongst others.

In general, a combination of volume per unit of time and/or energy would be most useful. However, one should also consider these materials in terms of their intrinsic hazards, just as for any process reagent, solvent or reactant.

6.4 EHS hazards and risk

Volumes have been written on the environmental, health and safety hazards and risks that are associated with chemicals and chemical processing. In general, there is a tendency on the part of many to focus on one portion of the EHS risk universe to the exclusion of other areas. For example, as extremely potent or toxic compounds are handled, there may be greater concern about Occupational Hygiene exposures while there is less consideration given to potential process safety risks such as dust explosions, or perhaps there is not enough appreciation of potential environmental risks that might arise through wastes produced through cleaning. Consequently, it is extremely important to look at EHS risks together as an integrated problem and not as isolated disciplines.

A second pitfall is to try and list all of the regulatory constraints or lists on which a particular chemical has the misfortune to be found. In this day and age, most chemicals are on one list or another, or if not, one is lured into a false sense of security because the particular chemical of interest appears to be relatively free of potential issues. However, this usually has more to do with the fact that there has not been sufficient testing done on the material to determine its hazard, or it is not made in sufficient volumes to exceed a regulatory bright line. Recent legislation such as REACH should in time overcome this lack of EHS information, but this will take some time, and for chemicals not manufactured in large quantities, there will still be a lack of information.

A third pitfall is to confuse hazard and risk. There are those that suggest that a very hazardous material cannot or should not be used under any circumstances. While application of the precautionary approach is perhaps desirable, there are situations where materials of higher hazard may be used at a lower level of risk than a larger quantity of material of lesser hazard. Once again, it is a question of taking a more holistic view and carefully weighing alternatives. The classic example of this is in the Boots Ibuprofen process, where hydrogen fluoride and carbon monoxide were used to reduce the number of synthetic steps and increase the mass efficiency while dramatically decreasing waste.[26]

In many countries, it is now common practice in industry to group or band chemicals into certain ranges of hazard, and to apply a certain control technology or set of technologies with that band. Following completion of an appropriate risk assessment, the task of determining appropriate control technology is therefore much more straightforward. Banding is generally done in each of the EHS risk areas, for example, Occupational Hygiene, Process Safety and environmental. This generally simplifies the job of the bench scientist or engineer who is not an expert in assessing EHS hazards and risks but must still ensure that the process they are developing has a minimal amount of EHS risk associated with it.

A final note of caution for developing process metrics and their relationship to toxicology concerns the idea of acute and chronic effects. In general, most attention is directed towards substances that possess acute hazards. For both human and environmental hazards, chronic exposures to chemicals that cause ill health in humans or other species are often difficult to predict and in many cases, conclusively determine cause and effect. Apart from avoiding use of chemicals known to cause ill health from chronic exposures, there are several strategies one may employ to avoid potential problems. The obvious principle and most often expressed by environmentalists is to uniformly apply the precautionary principle; that is, until hazard potential is known, avoid using materials in question. If the precautionary principle is not able to be followed, one may assess chemicals by analogy; that is, if the chemical is structurally similar to a chemical of known chronic hazards, you may want to avoid it. A more sophisticated approach, as was described in Chapter 2, is to use quantitative structure activity relationships (QSAR) to screen out chemicals, but this not only requires an understanding of the compound of interest but also requires a higher degree of understanding of human or ecotoxicity to make good use of the predictive capability of any type of modelling.

6.4.1 Occupational exposure hazard and risk

The easiest means for assessing occupational exposure hazards associated with materials used in a process is through the use of Permissible or Occupational Exposure Limits (OEL or PEL) which go by a variety of names; for example, TLV (U.S. – American Conference of Government Industrial Hygienists), MAK (Germany), or individual company established values. Occupational exposure limits are usually set based on a combination of the inherent toxicological hazard of a chemical and a series of safety factors such as intra-species variability in test results, nature and severity of the effect, adequacy and quality of the information, and so on. These exposure limits generally assume an 8-hour exposure and a healthy workforce population being exposed to the chemical for five days/week.

One can assess the potential for Occupational Exposure risk through close attention to the materials and the unit operations employed in a process, by performing an exposure assessment to be coupled to a given hazard to estimate the occupational exposure risk (e.g., Dow's Exposure Index[27]). A variety of approaches are possible here, from simply summing the number of materials in a given hazard band through more sophisticated approaches that take into account additional toxicological concerns such as the potential for carcinogenicity, mutagenicity, reproductive effects, and so forth. It is also possible to simply sum the mass of materials in a given band or to do a high-level assessment of potential risk based on the mass used, its physical form, the type of unit operations in the process or the potential for accidental release into the workspace.

6.4.2 Process safety hazards risk

There are a variety of process safety risks one needs to assess with chemical processes. In general, these risks will lead to an evaluation of the potential for the process to have precipitous changes in temperature and or pressure that lead to secondary events such as detonations, explosions, over pressurizations, fires, and so forth. The most cost-effective way of avoiding these sorts of risks is through the adoption of inherent safety principles. Inherent safety principles are very similar to and complementary to pollution prevention principles, where one attempts to use a hierarchy of approaches to avoid and/or reduce the risk of an adverse event. The reader is referred elsewhere to a more complete treatment of this important area of process design.[28]

For processes under development, the most cost-effective means of avoiding potential risk is to eliminate those materials that are inherently unsafe; that is, those materials whose physical or physico-chemical properties lead to them being highly reactive or unstable. This is somewhat difficult to achieve for several reasons. First, without a full battery of tests to determine, for example, flammability, upper/lower explosivity limits and their variation with scale, minimum ignition temperatures, and so on, it is almost impossible to tell how a particular chemical will behave in a given process. Second, chemical instability may make a compound attractive to use because its inherent reactivity ensures a reaction proceeds to completion at a rapid enough rate to be useful; that is, the reaction is kinetically and thermodynamically favoured.

The approach to developing metrics for process safety is analogous to those that might be used to assess Occupational Exposure risk. One can cite as well several indices that have been developed as metrics for estimating and ranking the safety of a given process or chemical reaction, such as the DOW fire and explosion index,[29] the Stoessel index[30] for hazard assessment and classification of chemical reactions, the Inherent Safety Index, the Prototype Index for Inherent Safety, amongst others.[31,32]

6.4.3 Environmental hazards and risk

There are a number of strategies one might employ to assess the collective environmental risk associated with a process. Whatever strategy that is employed, however, there are a number of general areas that should be considered as part of the overall assessment. First, one should consider the inherent hazard, fate and effects of materials used in the process. Second, one needs to consider the potential for release from the process and its unit operations. Third, one should consider issues related to the transportation, storage and disposal options related to the materials used in the process. Finally, one should consider the life-cycle impacts of producing those materials.

For the purposes of inherent hazard, fate and effects and a consideration of appropriate metrics, it is helpful to have a good understanding of a compound's chemical properties in mind as this will fundamentally drive overall environmental risk. Fate and effects are very important components of environmental risk that are worth a moments' further consideration. The environmental fate of a chemical is concerned with where a compound will go (air, water, soil) once it is released. Once in the environment, one needs to know what are the potential effects a chemical may have on organisms, including man. In general, insufficient attention is paid to the chemical mechanisms in the environment that impact chemical fate or distribution and most attention is given to the potential effects a chemical will have on a variety of organisms. Just because a compound is inherently hazardous to plants or animals does not mean that once released it will necessarily present a great risk to the environment. It is important to carefully evaluate distribution and degradation mechanisms that will directly affect the potential for exposure from any given chemical. The reader is referred elsewhere for a more complete treatment of this important topic.[33,34]

From an environmental hazard assessment perspective, most chemicals in recent years have been categorized according to their potential for persistence, bioaccumulation and toxicity. Persistence is associated with whether or not a chemical will be resistant to chemical (e.g., hydrolysis, photolysis, etc.) or biological (e.g., biodegradation, metabolism, etc.) degradation or breakdown. Various tests are used to determine a chemical's environmental depletion mechanism.[35] Bioaccumulation is the tendency for chemicals to become increasingly concentrated, usually in fat, as one moves up the food chain from microorganisms to large fish, birds or mammals. The water:octanol partition coefficient, log K_{OW} or D_{OW} (corrected for pH and ionizability of a compound in water) is generally used to estimate the tendency for bioaccumulation. Toxicity is generally the most contentious area of concern and there are a wide variety of toxicity tests that might be used. In general, one conducts multiple acute toxicity (endpoint is lethality) tests at three levels of the food chain; for example, algae, aquatic flea (e.g., *Daphnia magna*), and fish (e.g., *Pimephelas* [fathead minnow], *Onchorynchus* [trout], etc.). There are variations on the type of test depending on the fate of the compound (will it be released to freshwater or marine environment, or applied to land as for a pesticide, etc.) and the type of application for the chemical. Recently, there has been increasing concern about chronic exposures to chemicals, so there has been a movement towards requiring toxicity testing that assesses chronic exposures and different endpoints (e.g., fecundity, endocrine disruption, etc.) The reader is referred elsewhere for an extensive treatment of ecotoxicity testing.[35]

In terms of environmental metrics to assess processes, it is hopefully clear that a considerable testing burden exists to assess potential environmental hazards that lead to a credible risk assessment. At a first pass, one would typically screen compounds from an environmental hazard perspective to assess their tendency for persistence, bioaccumulation and toxicity. Depending on the final application of the compound, one might avoid commercial production of a particular compound, or one might devise processes that would use the compound but control the environmental risk to acceptable levels. For this, it is important to perform a process-specific risk assessment, as the impact of a given chemical, or set of chemicals, will be affected by the inherent hazard, environmental fate, and the specific characteristics of the process, available treatment, volumes, amongst others.

6.5 Quality

6.5.1 *Purity*

Purity is certainly a key driver for most chemical synthetic processes. The synthetic chemist is obviously looking to use reactions that result in very high yields with as few impurities in the isolated intermediate or product as possible. In the ideal situation, a purity of as close to 100% is sought, but this is often a very difficult target, especially for complex target molecules and the processes used to make them. In general, as the complexity of a chemical synthesis increases, the greater the occurrence of difficult separations and the need for one or more recrystallizations of the final product and/or intermediates to achieve the desired purity. Recrystallizations are essential in some cases to achieve the desired purity, crystal structure (including polymorphs) or particle size characteristics, but they obviously lead to an increase in spent solvent. While spent solvent can hopefully be recycled, it usually comes at a cost of energy and the additional capital invested in storage and distillation infrastructure.

With the above in mind, one might track purity in the usual manner of stating the purity or impurity profile of the final product or output of any given process. From a green perspective, however, one might be more interested in the number of isolations and or recrystallizations required for an overall process to achieve the desired purity. The other point to consider is that 'quality' has been defined as the capacity of satisfying the customer's requirements. In some instances, a higher purity will go beyond the customer's needs, so additional purification steps to increase purity can in some cases add to the mass and energy intensity without necessarily adding value added to a customer, and therefore, being wasteful steps. Alternatively, one might look at combinations such as net waste per isolation, or the quantity of energy per isolation as a means of distinguishing between processes. However, while these are more insightful metrics to derive and use, they are more difficult and time consuming metrics to develop.

Conclusions

As can be seen from the preceding discussion, developing green metrics for chemical processes requires a holistic, systems point of view across a range of disciplines. Metrics are also generally context dependent; that is, one kind or one set of metrics does not fit all situations. Instead, different organizations or companies will have to undertake some very hard work to identify, assess and implement metrics that are most applicable to their needs. The good news is that there are a large number of metrics that have already been identified, and many of these will meet the needs of most organizations or companies.

Any one individual is unlikely to possess sufficient knowledge in all areas of interest to identify key metrics so it should be common practice for green metrics to be developed drawing on the resources of cross-disciplinary teams. In addition, to truly drive the right direction towards the design of greener, safer processes, there is the need to resist the temptation of addressing metrics in a compartmentalized manner, as many of these metrics are interrelated. Finally, one should apply the 80/20 rule liberally; that is do not strive for the perfect set of metrics that covers all situations if a few metrics meet your needs most of the time.

References

1. Douglas, J. (1988) *Conceptual Design of Chemical Processes.* McGraw-Hill, New York.
2. Peters, M.S., Timmerhaus, K.D., West, R.E. (2002) *Plant Design and Economics for Chemical Engineers,* 5th Edition. McGraw Hill, New York.
3. Constable, D.J.C., Curzons, A.D., dos Santos, L.M.F. *et al.* (2001) Green Chemistry Measures for Process Research and Development. *Green Chemistry,* **3**, 7–9.
4. CWRT (2005) Baseline metric and other documents posted on the website of the Centre for Waste Reduction Technologies. www.aiche.org/cwrt/. Accessed 2005.
5. IChemE (2003) The Institution of Chemical Engineers. Rugby, UK.
6. Curzons, A.D., Constable, D.J.C., Mortimer, D.N., Cunningham, V.L. (2001) So You Think Your Process is Green, How Do You Know? Using Principles of Sustainability to Determine What is Green – A Corporate Perspective. *Green Chemistry,* **3**, 1–6.
7. Constable, D.J.C., Curzons, A.D., Cunningham, V.L. (2002) Metrics to 'Green' Chemistry – Which are the Best? *Green Chemistry,* **4**, 521.
8. Jiménez-González, C., Curzons, A.D., Constable, D.J.C., Overcash, M.R., Cunningham, V.L. (2001) How Do You Select the 'Greenest' Technology? Development of Guidance for the Pharmaceutical Industry. *Clean Products and Processes,* **3**, 35.
9. Jiménez-González, C., Constable, D.J.C., Curzons, A.D., Cunningham, V.L. (2002) Developing GSK's Green Technology Guidance: Methodology for Case-Scenario Comparison of Technologies. *Clean Technology and Environmental Policy,* **4**, 44.
10. Jimenez-Gonzalez, C., Curzons, A.D., Constable, D.J.C., Cunningham, V.L. (2004) Expanding GSK's Solvent Selection Guide – Application of Life Cycle Assessment to Enhance Solvent Selections. *Clean Technology and Environmental Policy,* **7**(1), 42–50.
11. Jimenez-Gonzalez, C., Curzons, A.D., Constable, D.J.C., Cunningham, V.L. (2004) Cradle-to-Gate Life Cycle Inventory and Assessment of Pharmaceutical Compounds. *International Journal for Life Cycle Assessment,* **9**(2), 114.
12. Butters, M., Catterick, D., Craig, A. *et al.* (2006) Critical Assessment of Pharmaceutical Processes – A Rationale for Changing the Synthetic Route. *Chemical Reviews,* **106**(7), 3002–3027.
13. Marteel, A.E., Davies, J.A., Olson, W.W., Abraham, M.A. (2003) Green Chemistry and Engineering: Drivers, Metrics, and Reduction to Practice. *Annual Review of Environmental Resources,* **28**, 401–428.
14. Wrisberg, N., Haes, H.A.U.d., Bilitewski, B. *et al.* (2002) In: *Analytical Tools for Environmental Design and Management in a Systems Perspective* (eds. N. Wrisberg, H.A.U.d. Haes, U. Triebswetter, P. Eder, R. Clift). Kluwer Academic Publishers, Dordrecht, pp. 45–73.
15. Lapkin, A. (2006) In: *Renewables-Based Technology: Sustainability Assessment* (eds. J. Dewulf & I.H.v. Langenhove). John Wiley & Sons, New York, pp. 39–53.
16. Brunner, N. & Starkl, M. (2004) Decision Aid Systems for Evaluating Sustainability: A Critical Survey. *Environmental Impact Assessment Review,* **24**, 441–469.

17. Udo de Haes, H., Finnveden, G., Goedkoop, M. *et al.* (eds.) (2002) *Life Cycle Impact Assessment: Striving Towards Best Practice.* SETAC Press, Pensacola, FL.
18. Barnthouse, L., Fava, J., Humphreys, K. *et al.* (eds.) (1998) *Life-Cycle Impact Assessment: The State-of-the-Art.* SETAC Press, Pensacola, FL. http://www.setac.org/htdocs/what_pub.html
19. Jiménez-González, C., Kim, S., Overcash, M.R. (2000) Methodology for Developing Gate-to-Gate Life Cycle Inventory Information. *International Journal of Life Cycle Assessment*, **5**(3), 153–159.
20. Jiménez-González, C. & Overcash, M.R. (2000) Energy Sub-Modules Applied in Life-Cycle Inventory of Processes. *Clean Products and Processes*, **2**(1), 57–66.
21. El-Halwagi, M.M. (1997) *Pollution Prevention through Process Integration.* Academic Press, San Diego, CA.
22. Schmidt-Bleek, F. (1993) MIPS Re-visited. *Fresenius Environmental Bulletin*, **2**, 407–412.
23. Ajmera, S.K., Losey, M.W., Jensen, K.F., Schmidt, M.A. (2001) Microfabricated Packed-Bed Reactor for Phosgene Synthesis. *AIChE Journal*, **47**, 1639–1647.
24. Kennedy, M.L. (1997) *Total Cost Assessment for Environmental Engineers and Managers.* John Wiley & Sons, New York.
25. Schaltegger, S. & Burritt, R. (2000) *Contemporary Environmental Accounting: Issues, Concepts and Practice.* Greenleaf Publishing, Sheffield.
26. Cann, M.C. & Connelly, M.E. (2000) *Real World Cases in Green Chemistry.* American Chemical Society, Washington, DC. http://www.chemistry.org/portal/resources/ACS/ACSContent/education/greenchem/case.pdf.
27. AIChE (1998) *Dow's Chemical Exposure Index Guide*, 1st Edition, 38pp. AICheE, New York.
28. See for example Hurme, M. & Rahman, M. (2005) Implementing inherent safety throughout process lifecycle. *Journal of Loss Prevention in the Process Industries*, **18** (4–6), 238–244.
29. AIChE (1994) *Dow's Fire & Explosion Index Hazard Classification Guide*, 7th Edition, 88 pp. AICheE, New York.
30. Stoessel, F. (1993) What is Your Thermal Risk? *Chemical Engineering Progress*, **89**(10), 68–75.
31. Edwards, D.W. & Lawrence, D. (1993) Assessing the Inherent Safety of Chemical Process Routes: Is There a Relation between Plant Costs and Inherent Safety? *Transactions of the Institute of Chemical Engineers*, **71**(B), 252–258.
32. Heikkilä, A.-M., Hurme, M., Järveläinen, M. (1996) Safety Considerations in Process Synthesis. *Computers and Chemical Engineering*, **20**, S115–S120.
33. MacKay, D., Shiu, W.Y., Ma, K., Lee, S.C. (2006) *Handbook of Physical-Chemical Properties and Environmental Fate for Organic Chemicals*, 2nd Edition. CRC Press, Taylor & Francis Group, Boca Raton, FL.
34. Cronin, M. & Livingstone, D. (eds.) (2004) *Predicting Chemical Toxicity and Fate.* CRC Press, Danvers, MA.
35. OECD (2001) *OECD Guidelines for the Testing of Chemicals*, 13th Addendum. Organization for Economic Co-operation and Development (OECD) Publishing.

Chapter 7
Application of LCA in Process Development

Dana Kralisch

7.1 Introduction

Traditionally, chemical process development focused on economic criteria. However, over the last two decades, additional criteria of sustainability have become increasingly important and were integrated in the decision-making process during process development.[1–3] Ecological sustainability, one of the three pillars of sustainability besides economic and social criteria, can be examined using life cycle assessment (LCA). The LCA methodology had been standardised by the International Standard Organisation (ISO) (ISO 14040:2006[4] and ISO 14044:2006[5]) and was defined as the 'compilation and evaluation of the inputs, outputs and potential environmental impacts of a product system throughout its life cycle'.[4] The environmental aspects and potential impacts associated with a process, product or service system are assessed by compiling an inventory of relevant inputs and outputs followed by an evaluation of environmental impacts associated with those inputs and outputs. At this stage, defined impact categories are used covering a wide range of impacts on the environment, including the depletion of exhaustible raw materials, the emission of hazardous substances as well as land use. In the last step, the results of the inventory and impact phases have to be interpreted in relation to the goal and scope of the LCA study.[4]

The holistic cradle-to-grave approach of LCA studies avoids the problems otherwise connected to balancing single life cycle stages, which disregard the impact of the respective upstream or downstream processes. Consequently, a problem shift from one life cycle stage to another is not occurring. In addition, the use of a common functional unit allows for the comparison of the environmental impacts of alternative products, processes or services. The main application of LCA according to the CML in Leiden (Netherlands), is the analysis of the origin of problems related to a particular product, the comparison of parameter improvement of a given product, the design of new products and the possibility to choose between a number of products.[6] This applies to the evaluation of processes as well. Owing to its broad applicability and its validity, the LCA methodology has gained worldwide acceptance as a useful tool for strategic planning, process development as well as policy-making and a wide range of other decision-making situations. For more information about the LCA methodology and its history/development see Section 7.2.

Certainly, the time frame and costs to perform a detailed LCA study often exceed the demands of process development, especially in industry. Owing to this, the method may not be attractive for decision-makers. In addition, a detailed LCA requires an extensive database which is often not available during the development of new processes with for example unknown disposal strategies, especially in the early design phases. The Simplified LCA (SLCA) methodology, as introduced in Section 7.2.4 is a suitable alternative to perform iterative screening and hot-spot detection already at this stage.

Although the methodologies for life cycle product- and process-related assessment are similar, the questions and requirements concerning a process-related assessment may differ. This allows for a quantitative comparison of different technological routes for example the production of the same material or its disposal. Thus, the conventional system boundary is extended to include the life cycles of different technologies. The application of LCA during process development is explained by using a case study dealing with the substitution of conventional batch technology by micro-reaction technology (see Section 7.3).

In addition to environmental criteria, a range of other relevant criteria are used to guide the decision-making process during the phase of process development. Therefore, LCA is often employed as part of an evaluation toolbox. In this context, examples have been published, where the LCA methodology (and other evaluation methods) is coupled with multi-objective optimisation techniques to aid the optimisation process concerning several objectives. In most cases, economic and ecological objectives are taken into account within this approach in order to find for example a trade-off between the Best Practicable Environmental Option (BPEO) and the Best Available Technique not Entailing Excessive Costs (BATNEEC)).[7] This approach offers the potential for enhanced innovation in process development through a selection of best process and material alternatives over the whole life cycle by simultaneously optimising ecological, economic and technical or safety criteria. Thus, the coupling of LCA and other life cycle-based assessment methods with multi-objective optimisation and decision-support provides a potentially powerful decision-making tool for sustainable process development (see Section 7.4 for more detail).

Finally, the coupling of SLCA, further evaluation criteria and multi-objective optimisation are demonstrated with the help of a second case study dealing with the improvement of synthesis strategies for ionic liquids (see Section 7.5).

7.2 The LCA methodology

7.2.1 The life cycle approach and the LCA methodology

Life cycle thinking is an approach covering all life cycle stages of a process, product or service system from extracting and processing raw materials, manufacturing, transportation and distribution, use/reuse, recycling and waste management. This is often described as 'cradle to grave' approach as well. Life cycle thinking is rather a qualitative discussion which only allows for the identification of hot-spots in a life cycle. In contrast, the LCA methodology offers the possibility to quantify and assess the environmental aspects and potential impacts associated with these processes, product or service systems in order to choose the least burdensome.[4] The concept can also be used to optimise the environmental performance of a single product (eco-design) or a company. In this context, the assessment is based on a compilation of an inventory of relevant inputs and outputs of the system, an evaluation of the potential environmental impacts associated with those inputs and outputs and an interpretation of the results of the inventory analysis and impact assessment phases in relation to the goal and scope of the study.[4] It can be applied to guide decision-making on environmental design and improvement and to rank the environmental burdens of alternative products, processes and services. The level of detail as well as the boundaries of an LCA study depends on subject and intention of the study, but the ISO standards have to be followed. At this, ISO 14040:2006[4] provides a summary of practice, applications and limitations of LCA for a broad range of potential users and stakeholders. It describes the principles and frameworks of an LCA study, the stages of inventory (LCI) and impact assessment (LCIA) and the interpretation of results as well as the relationship between the LCA phases in comparison to ISO 14044:2006,[5] which specifies requirements and provides guidelines for LCAs. It is designed for preparing, conducting as well as critical reviewing of life cycle analyses. For this purpose, it provides guidance on the impact assessment phase, the interpretation of results, and the nature and quality of the data collected. In addition, guidelines regarding critical review and report of LCA results as well as conditions for use of value choices and optional elements are given. In the following sections a short overview of the different phases of LCA and its requirements and issues is given.

The LCA methodology comprises four phases:

Goal and Scope Definition. This phase deals with the selection of system boundaries and the setting of the functional unit which describes the primary function(s) fulfilled by a (product) system and can be used as a basis for the comparison of alternative systems.[8]

Life Cycle Inventory (LCI). During LCI, mass and energy balances are performed to quantify the material and energy inputs into the system as well as wastes and emissions from the system.

Life Cycle Impact Assessment (LCIA). In the phase of LCIA, the mass and energy balances quantified in the LCI are aggregated into a limited set of recognised environmental effects. As an example, the impact-oriented CML approach comprises three steps of evaluation: (i) classification, (ii) characterisation and (iii) normalisation (optional) followed by a weighting and valuation of the results. In the classification step, the inputs and outputs quantified during LCI are related to different impact categories according to their anticipated effect on the environment. Impact categories according to CML[8] usually included into LCIA are potentials of Abiotic Resource Depletion (ADP), Global Warming (GWP), Ozone Depletion (ODP), Photochemical Ozone Creation (POCP), Acidification (AP), Eutrophication (EP), Human Toxicity (HTP) and Eco-toxicity (ETP) as well as Land Use (LU). Within the characterisation step, the magnitude as well as the potency of the impacts is taken into account using equivalence factors in relation to a reference compound. The optional normalisation step can be carried out to relate the data to an actual magnitude of impacts in a given area. Further life cycle impact assessment methods are for example the Eco-Indicator 99,[9] the method of Critical Volumes,[10] the Eco-Point method[11] or the EPS method.[12]

Life Cycle Interpretation. The results obtained within the Life Cycle Inventory and/or the Life Cycle Impact Assessment are interpreted in the light of the Goal and Scope Definition (e.g., by means of sensitivity or uncertainty analyses) in order to draw conclusions and make recommendations.[8]

A range of software tools is commercially available to aid the implementation of the LCA methodology (e.g., UMBERTO,[13] GABI[14] and SimaPro[15]). Among others, they contain comprehensive inventory databases, which facilitate the evaluation of up- and downstream stages of the process under investigation (e.g., the Ecoinvent database by Frischknecht *et al.*[16]) Some inventory data is also freely accessible (e.g., from the Association of Plastics Manufacturers[17] or from the German Federal Environmental Agency[18]).

One problem of process and product evaluations with the LCA methodology and their 'cradle to grave' approach constitutes closed-loop systems. The idea of the so-called 'cradle to cradle' approaches is to favour systems in which valuable, high-tech synthetics and mineral resources circulate in cycles of production, use, recovery and remanufacture.[19] Such systems integrate three key design principles reflecting the intelligence of natural systems, which can inform human design as well: (i) waste equals food, (ii) use current solar income, and (iii) celebrate diversity.[20] Although a study of these systems by means of LCA is in principle possible, the evaluation is extremely extensive and the results strongly depend on the allocation rules, determined during the Goal and Scope Definition.

Another problem of LCAs is that they are non-site-specific. The reasons for this lie in the fact that they include the whole life cycle of systems with resources which may originate in different countries and waste products and emissions which may distribute globally. They deal with factual inputs, outputs and the environmental impact potentials of the system under investigation on a global, and, in some cases, regional scale. Yet, they do not address the intrinsic risks resulting from the system itself. However, a combination with risk assessment methods can be used to close this gap.

7.2.2 Short overview of the history of LCA

The roots of the LCA methodology lie in the 1960s, when environmental issues received increasing attention from both government and public. The triggers were increasing environmental problems such as smog and acid rain as a consequence of a growing world population and accelerating industrialisation. The report 'The limits to growth', published in 1972 by Meadows *et al.*,[21] forecasted a collapse of the industrial society unless conservation measures would be initiated, and led to an interest in describing industrial processes in terms of energy and raw materials consumption and emission into the environment. The oil crisis in the mid-1970s additionally accelerated this development. The first studies in this field focused on energy demand,[22] establishing balanced flow charts to describe the systems in this area. A number of further studies followed, especially on packaging,[23,24] taking into account environmental impacts from the raw materials extraction to the disposal of wastes. Discussions within the scientific community of the validity of the comparisons, undertaken in these reports, initiated a standardisation process. During the 1980s, the potential threat of ozone depletion as well as global warming appeared, leading to further activities. The evaluation methodology was extended to emissions into the compartments air and water.[25] Against this background, the term *Energy Analyses* was converted into terms such as *Eco-balance*, *Life Cycle Analysis* or *Life Cycle Assessment*. Since 1990, the LCA developments have been guided by the Society of Environmental Toxicology and Chemistry (SETAC). This multidisciplinary professional society serves as an umbrella organisation for practitioners in this field from academia, business, governmental and non-governmental organisations. Currently, six working groups coordinated by the SETAC LCA

Advisory Group are considering different facets of LCA.[26] In 1993, they developed the 'Code of Practice',[27] the first worldwide accepted technical framework for LCA. This was an important step towards the harmonisation of the method and has initiated and supported the standardisation process by ISO. Between 1997 and 2000, ISO produced the international series of standard defining the different stages of the LCA methodology (ISO 14040:1997,[28] ISO 14041:1998,[29] ISO 14042:2000[30] as well as ISO 14043:2000[31]). As mentioned above, these standards were replaced by two improved editions of life cycle assessment standards in 2006 (ISO 14040:2006[4] and ISO 14044:2006[5]).

The United Nations Environmental Programme (UNEP) represents a third international player in the field of LCA. In 1996, UNEP published an easy to read and user-friendly guide to LCA[32] and supported several international workshops since then, dealing with various topics in the context of LCA. The launch of the Life Cycle Initiative in 2002 within UNEP further emphasised its importance.[33]

Furthermore, the application of LCA is increasingly required by legislation, for example through the US Pollution Prevention Act of 1990,[34] the EU Directive on Integrated Pollution Prevention (IPPC)[35] and the Integrated Product Policy (IPP).[36] In the Pollution Prevention Act, a pollution prevention hierarchy has been established, where resource depletion reduction represents the option of first choice. This implicitly includes increased efficiency in the use of resources. The IPPC directive also represents a significant shift in the basis of environmental regulation. It requires that the Best Available Technique (BAT) must be chosen by considering the environment as a whole, including indirect releases, consumption of raw materials and waste disposal. Thus, IPPC embraces the life cycle both of the process (including construction and decommissioning) and of materials and energy (including resource use and waste).[37]

In recent times, the combinations of LCA with other assessment methods, for example, life cycle costing[38] and risk assessment[39] are gaining more and more importance.

7.2.3 Special requirements for an applicability of LCA in process development

Process development is a complex activity, including the implementation of new technologies, creation of new facilities, and retrofitting of existing processes, in which several iterative steps of data gathering, base-case development, and detailed design and optimisation are involved. Traditionally guided by economic, technical and safety criteria, it is now becoming increasingly obvious that environmental and enhanced social criteria should also be integrated to assess the issues of sustainability. As mentioned above, LCA is a suitable and established tool to evaluate the environmental part of sustainability. In contrast to a single evaluation of the process step under development, it allows for the consideration of upstream and downstream processes. Thus, the development of processes with optimised performance but with increased environmental impact outside of the system boundary is avoided.

According to Cano-Ruiz and McRae,[2] process design as a part of process development starts with problem framing. At this stage, the concept definition, scope of analysis, design objectives, constraints, evaluation and interruption criteria have to be determined. They argued that framing decisions are often made implicitly, whereas the critical importance of this step in determining the outcome of the design process is often overlooked. The problem framing phase is followed by the generation of alternatives and their analysis. Engineering analysis usually starts with mass and energy balances to make predictions about the expected performance of the system. Information about the inputs and outputs of the process, including the flow rates, compositions, pressure, temperature, and physical state of all material streams, as well as the energy consumption rate from various sources is gathered. These specifications can serve as an appropriate database for an assessment of the environmental impacts associated with the processing of those alternatives. During an evaluation step, the information is summarised as indicators of performance that can be used to assess whether the requirements specified during the objective formulation step are met. At last, the evaluation step ends with a ranking of alternatives according to their overall level of attractiveness.

The enhanced application of LCA in process development started during the 1990s, Burgess and Brennan[40] published an overview of these early works. At this time, Jödicke *et al.*[41] also applied the LCA methodology to support the development of environmentally sound processes for enantioselective reduction of ketoesters as pharmaceutical intermediates. They investigated four case studies concerning bio-catalysed and metal-catalysed processes and included the enantioselective reduction, work-up, waste treatment and recycling into the balance. The authors concluded that the environmental impact of the syntheses depends more on the process parameters than on the technology employed. In addition, (i) the concentration of the reactants and hence, the amount of solvent used, (ii) energy intensive steps such as distillations, and (iii) the catalyst losses, were pointed out as being mostly responsible for the environmental performance of chemical processes.

Examples of more recent work on this topic have been published, for example by Hellweg *et al.*[42] and Bauer and Maciel[43] using LCA for the comparison of different solvent alternatives for chemical processes, and investigating the production of propylene glycol by propylene oxide hydration, respectively.

In addition, a range of LCA studies have been published in the context of momentarily relevant issues in process development such as alternative energy systems,[44,45] supercritical solvents,[46] bio-fuels,[47] thermal drying processes for sewage sludge,[48] carbon dioxide recovery[49] and offshore produced drinking water.[50] In the following, a short overview of some selected studies is given.

Pehnt[44] suggested a dynamic approach towards the LCA of renewable energy technologies. He stressed that application of the prevailing LCA approach to future renewable energy systems does not distinguish between impacts which are 'imported' into the system due to the 'background system' (e.g., due to supply of materials or final energy for the production of the energy system), and does not point out the improvement potentials compared to the competitors. Also, he argued that future development will enable a further reduction of environmental impacts of renewable energy systems and that different factors are responsible for this development, such as improved efficiency, emissions characteristics, increased life time, and so on. As such, the application of renewable energy sources might modify not only the background system, but also further downstream aspects, such as consumer behaviour. Thus, he included these aspects into a dynamic LCA approach to support the development and fair assessment of future technologies.

Jungbluth *et al.*[45] investigated two emerging electricity systems based on renewable energy, photovoltaic and wind power. They considered twelve different grid-connected photovoltaic systems and four different wind turbine systems including the construction of the plants to find out the most environmentally benign alternative.

Gasafi *et al.*[46] performed LCA in combination with a dominance analysis in order to identify hot-spots in process chains in early phases of process design to efficiently improve the environmental performance. The authors illustrated their approach on the example of super-critical water gasification for the treatment of organic feedstock with high moisture content.

Kemppainen and Shonnard[47] compared the production of ethanol via fermentation-based processes starting with two lignocellulosic feedstocks: virgin timber resources or recycled newsprint from an urban area. For this purpose, they coupled a software-aided simulation of the fermentation process with impact assessment software.

Sadiq[50] investigated fourteen best available techniques for treatment of produced drinking water considering environmental, technological and economical drivers. They used a weighting scheme to integrate various criteria and policies so that a most appropriate alternative could be selected. On the basis of these results, seven new treatment strategies are developed and are further investigated.

In general, the freedom of design decreases during process development whereas both the knowledge and the costs of error elimination increase. Against this background, an implementation of environmental criteria into chemical process development as early as realisable is of great importance for the exploitation of its optimisation potential. For this purpose, Azapagic *et al.*[51] suggested the application of a sustainability assessment during all stages of process design. Starting with an identification of sustainability criteria and their qualitative evaluation during the project initiation phase, the extent of assessment increases during the preliminary design step, joining in a detailed LCA in case of environmental aspects at the detailed design phase. They illustrated this approach on a design case study of the vinyl chloride monomer process.

As discussed before on the example of Pehnt,[44] the application of LCA during process development commonly involves the challenge of comparing alternatives which are at different development stages. This is usually associated with a different level of knowledge about the alternative systems. Examples are the evaluation of different alternative solutions for a given design problem or the comparison of a newly developed, laboratory-scaled process with an industrially established production process. This challenge can be met by SLCA, based for example on generic data, or by calculation of case scenarios (see Section 7.3 for more detail). In all cases, the proof of reliability of the conclusions drawn from this contrast is an important part of the assessment. The method of SLCA is introduced below, followed by a case study of LCA application during process development.

7.2.4 Simplified life cycle assessment

The results obtained from the LCA methodology are well-founded and comprehensive, but its application is very time-consuming and requires an extensive database. However, data and time are often limited during the phase of process development and a comprehensive LCA study is therefore often not feasible. Nevertheless, the optimisation potential is highest and the costs to change the system are lowest at the beginning of the process design and development phase.[52,53] To meet this potential and to integrate ecological aspects and life cycle thinking into chemical process development, it may be useful in terms of cost and time reduction to consider the LCA as a set of flexible approaches rather than a rigid method.

This can be realised by a simplified or 'streamlined' LCA. According to SETAC-Europe,[54] SLCA is an application of the LCA methodology during a comprehensive screening assessment, which covers the whole life cycle, however on a superficial level. Thus, generic qualitative and quantitative data, standard modules for transportation or energy production can be used, followed by a simplified assessment which focuses on the most important environmental aspects with respect to potential environmental impacts, LCA stages and/or phases, thoroughly assessing the reliability of the results with the help of sensitivity and consistency checks.[5] At this, a SLCA consists of three steps which are iteratively interlinked: (i) screening, (ii) simplification and (iii) assessment of reliability. These three steps can be combined in an iterative procedure in several ways, for example in a de-simplification approach (starting with the screening of the main process sequence and increasing the level of detail and comprehensiveness step by step), a top-down approach (starting with a superficial identification of hot-spots within the whole life cycle, the level of detail is increased during every iteration step) or taking an existing comprehensive LCA as a starting point for an evaluation of similar case studies, for example with the help of screening indicators.

Following the SETAC-Europe report, screening indicators can be used in order to reduce the effort and detail in an ongoing LCA study. In this report, the Energy Demand, the Material Intensity per Service Unit (MIPS),[55] key substances (e.g., heavy metals or single impact categories such as global warming) as well as semi-quantitative screening indicator-methods (e.g., the ABC/XYZ-method[54]) are pointed out as possible screening indicators.

The ABC/XYZ-method has been suggested by Fleischer and Schmidt[52] in order to include environmental aspects into a screening and simplified LCA method already at the stage of product and process research and development (R&D). They proposed an interactive, iterative and integrative top-down approach to identify advanced materials. This approach named euroMat has been detailed by Fleischer *et al.*[56] The authors stress that the simplification of LCA is an approach to deal with data gaps, data asymmetries, and inconsistencies in life cycle inventories, and deduct that it is essential to apply LCA in the context of Design for Environment. Additionally, their approach was demonstrated on a comparison of aluminium and reinforced epoxy carbon fibres for lightweight containers in aerospace applications.

Other examples of the application of SLCA have been proposed for example by Weidenhaupt and Hungerbühler,[57] Hochschorner and Finnveden[58] and Hur *et al.*[59] However, despite the advantages in terms of time, data requirements and costs, a risk remains that not all key issues are identified with SLCA, which may result in incorrect decisions.

7.3 Application of LCA in process development – a case study

7.3.1 Goal and scope of the case study 1

The fine chemical industry has been characterised for decades by batch processes in laboratories as well as in production plants. Owing to the requirement of high flexibility (capacity, assortment and structure flexibility), facilities were often operated batch-wise rather than continuous.

Nowadays, the commercial interest in continuously running micro-processes clearly increases due to novel ways of process intensification offered by this engineering approach,[60–62] in addition to other reactor and process engineering advantages such as process safety or modularity.[61] Micro-process technology employs a system of reaction devices, mixers, heat exchangers and so on with internal structures on micro-meter scale in at least one spatial dimension. Compared to large-scale batch reactors, micro-structured devices are often orders of magnitude more efficient with regard to heat and mass transfer, due to their small channel sizes and high surface-to-volume ratios,[63] which may lead to lower cooling and heating power requirements. The excellent control of heat transfer can additionally result in improved yield, selectivity and product quality.[61] Especially fast and highly exothermic reactions can be performed in micro-reactors using defined residence times, thus leading to a reduction of unwanted side reactions.[64]

Owing to the characteristics of micro-structured devices, a considerable reduction of the environmental impact is commonly expected from the implementation of micro-reaction technology into industrial processes.[3,65] However, the potential advantages of micro-reaction technology mentioned above may also be accompanied by disadvantages such as an increased material and energy consumption during the manufacture of micro-reactors in comparison to the production of conventional equipment. This may especially be the case for micro-structured devices made from stainless steel. In addition, micro-reactors might have shorter life times caused by fouling (clogging) phenomena in the micro-channels than for example conventional stirred tank reactors.

The LCA methodology has been used by Kralisch and Kreisel[66,67] to investigate the pros and cons of micro-reaction technology at the stage of process development. The motivation for this study was the question whether ecological improvements can be expected for the chosen model reaction performed in the micro-structured device Cytos® produced by CPC Systems GmbH, Germany. Such improvements were the premise for the construction of a continuously operated pilot plant based on Cytos® micro-reactors.

Additionally, the influence of the life time of the micro-structured devices as well as the expenditure of the peripheral equipment was estimated in order to obtain insight into the ecological hot-spots of the system. A conventional macro-scaled semi-continuously operated batch process was chosen as a reference process. The comparison of both technological systems was performed by means of the two-step synthesis of *m*-anisaldehyde serving as model reaction.

The synthesis of *m*-anisaldehyde [3] is a very exothermic lithium-organic reaction (Scheme 7.1). In a macro-scaled batch reactor, this synthesis can only be carried out under extensive safety precautions and with high energetic effort for operating the cooling system.

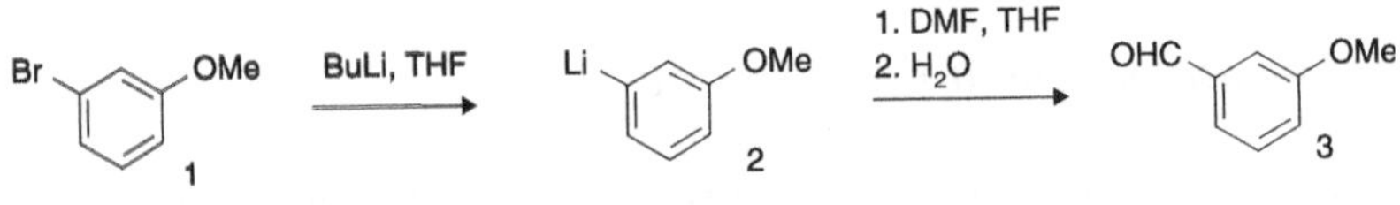

Scheme 7.1 Synthesis of *m*-anisaldehyde, model reaction of case study 1.

Contrary to this, the improved heat transfer in the micro-scale system allows for the reaction to be conducted isothermally at more moderate temperatures.

At stage I of the synthesis, *m*-bromoanisole [1] and *n*-butyllithium are converted *via* bromo/lithium exchange in order to obtain *m*-lithium anisole [2] and *n*-bromobutane. At stage II, the reaction mixture is treated with dimethylformamide, and then the reaction is quenched using 3 M hydrochloric acid. Tetrahydrofurane is used as solvent at both synthetic stages.

7.3.2 Life cycle inventory of the alternative technological systems

The reference (batch) experiment was conducted in a 10 L double-walled reactor. A temperature of 223 K was maintained using a thermostat. A yield of 60% *m*-anisaldehyde of the theory was achieved using this procedure. In continuous mode, the two-step synthesis was conducted at 273 K in two Cytos® Lab System modules connected in series. In this reaction mode, the yield amounted to 88% of the theory. When transferred to a continuous micro-scale set-up, the increased surface-to-volume ratio allows for an improved control of the heat supply and removal, and the reaction can be conducted isothermally, resulting in a decreased energy demand for cooling processes and an increase in yield. These phenomena can be attributed to the short residence time as well as the narrow residence time distribution of the instable lithium-containing intermediate [2] (a few seconds) before being submitted to stage II of the reaction. The short residence time of the intermediate (relative to its rate of decomposition) leads to a higher yield in comparison to the laboratory-scale batch process. In the latter instance, several hours for the dosage of the *n*-butyllithium are necessary under laboratory conditions, depending on the batch size.

The system boundary of the LCA study comprises the earlier life cycle stages of the chemicals used, the supply of energy and inert gas, the realisation of the model reaction, work-up and transports to the disposal of wastes (Figure 7.1). All calculations refer to a production of 10 kg *m*-anisaldehyde as functional unit. All processes within this system's boundaries were included into the balance. Also, the dependence of the additional

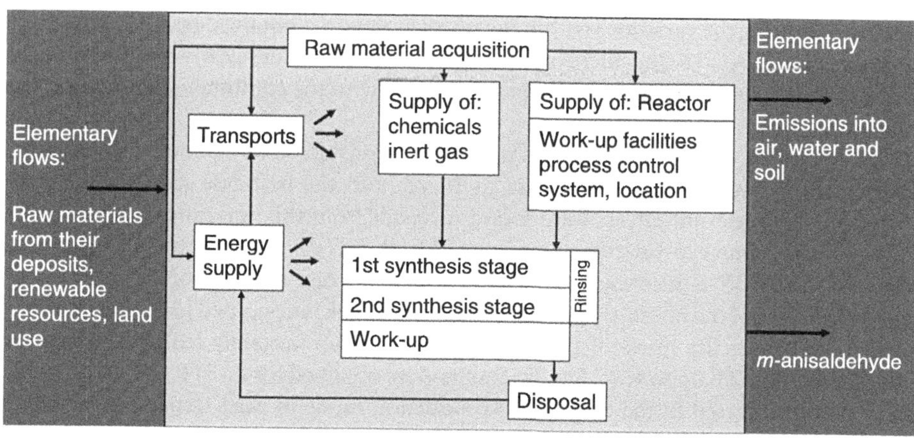

Figure 7.1 System boundary of the LCA – case study 1.

expenditure of materials and necessary energy on the choice of the reactor (stainless-steel micro-reactor *versus* borosilicate glass double-walled reactor) was investigated. This aspect was analysed with the help of different case scenarios, because there was no statistically proven knowledge about the life time of micro-structured stainless-steel reactors. The reasons for this lie in the novelty of such reaction devices in chemical process engineering, and in the strong dependence on the area of its application, chemical resistance and breakage safety. In the case of 'fouling' phenomena within the micro-channels, an immediate exchange of the micro-structured devices may be necessary. In order to elucidate the range of possible exchange cycles with respect to the resulting potentials of environmental impact, four scenarios were envisaged.

The 'worst case' scenario (Conti Sc wc) describes a life time of one week as a minimum prerequisite. In order to allow for a comparison of the two synthetic methods, the presumed 'best case' (Conti Sc3) is based on a life time equal to that of the double-walled glass reactor, which is ten years on average. Scenario 1 (Conti Sc1) and 2 (Conti Sc2) represent life time expectancies of three months and three years respectively, thus ranging between the worst- and the best-case scenario. A comprehensive estimation of the influence of the supply of the Cytos® reactors on the overall result is thus feasible.

The peripheral equipment such as control elements, thermostats, distillation equipment and so forth was included into the balance as well, but with the help of rough estimations based on their components or similar apparatuses. This has been done in order to provide a holistic view on its influence, but to reduce the effort of data acquisition. If a relevant influence would occur within the LCIA, the peripheral equipment would be investigated in more detail.

7.3.3 LCIA results

According to CML,[68] the following impact categories were considered in this specific LCA: ADP, GWP, ODP, POCP, AP, EP, HTP and the ETP sub-criteria Fresh Aquatic Eco-Toxicity Potential (FAETP), Marine Aquatic Eco-Toxicity Potential (MAETP) as well as the Terrestrial Eco-toxicity Potential (TETP). A summary of the LCIA results is presented in Figure 7.1, plotting the variations within the impact categories obtained by the change from the macro-scaled batch to the micro-reaction system. These results clearly indicate the possibility of significant ecological advantages associated with the continuous synthesis in the micro-reactor.

The life time of the micro-structured devices was considered in the four case scenarios (ranging from 1 week in the worst case to 10 years in the best-case scenario), and no environmentally relevant effects were found to result from this variation, except for the HTP and also some eco-toxicity potentials (FAETP, TETP). As an example, Figure 7.2 shows that the HTP is increased by 47% for the worst-case scenario of the continuous process (life time of the micro-structured devices: one week) as opposed to the batch mode. Realistically, longer life times of the Cytos® micro-reactor certainly led to a significant reduction of the HTP up to 46%. Similar results were obtained for FAETP and TETP, while MAETP was decreased in the continuous production mode in each scenario considered. This outcome was investigated in more detail by determining significant issues along the life cycle, taking the impact categories GWP and HTP as examples.

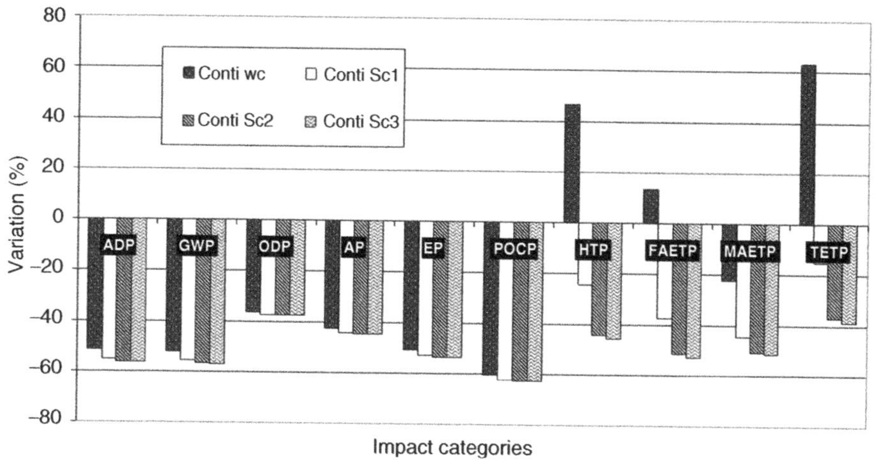

Figure 7.2 Variation of environmental impact potentials as a consequence of the change from macro-scaled batch (0%) to micro-reaction mode; $Y_{batch} = 60\%$, $Y_{conti} = 88\%$; four scenarios regarding the lifetime of the micro-structured devices (Conti wc: 1 week, Conti Sc1: 3 months, Conti Sc2: 3 years, Conti Sc3: 10 years).

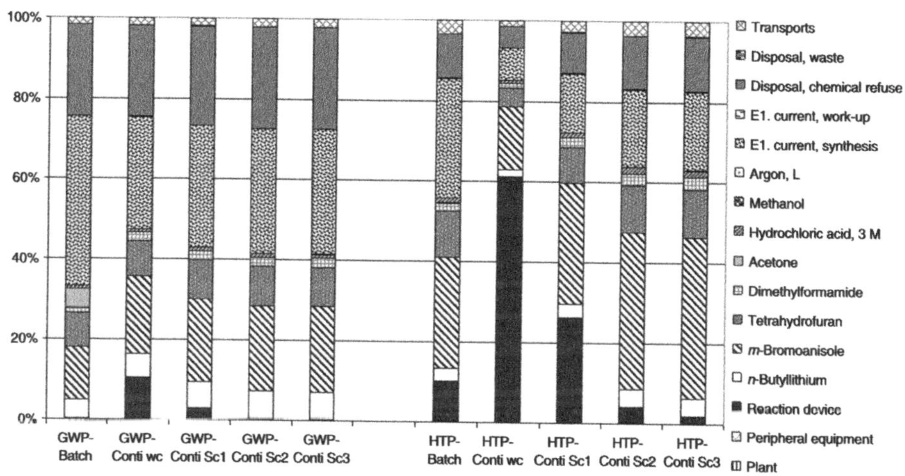

Figure 7.3 Identification of significant issues within GWP and HTP, $Y_{batch} = 60\%$, $Y_{conti} = 88\%$, FU = 10 kg *m*-anisaldehyde, four scenarios regarding the lifetime of the micro-structured devices (Conti wc: 1 week, Conti Sc1: 3 months, Conti Sc2: 3 years, Conti Sc3: 10 years).

7.3.4 Identification of significant issues

The determination of the significant issues helps to structure the results from the LCI or LCIA phases according to the goal and scope of the study.[5] Figure 7.3 demonstrates the reasons for the results obtained in the LCIA phase by means of two selected impact categories: GWP and HTP.

The results within the impact category GWP illustrate the great influence of both, the supply of chemicals (33% batch and 37% Conti wc, resp.) and the energy demand during synthesis (42% batch and 28% Conti wc, resp.) along the whole process chain. The disposal of the chemical refuse as well has a significant impact (23% both, batch and Conti wc). In the case of the worst-case scenario, the influence of the supply of the reaction device amounts to 10% (assumed life time of the micro-structured devices: 1 week) decreasing to 3% in Conti Sc1 (assumed life time of the micro-structured devices: 3 month).

The influence of this module strongly increases within the impact category HTP. Thus, the impact of the supply of the reaction device amounts to 10% in the batch mode versus 61% for the worst-case scenario of the continuous micro-reaction mode. This effect mainly results from the emission of chromium and nickel during the stainless-steel production representing an upstream process of the micro-reactor production. However, this can be counteracted by a longer life time of the micro-reactor (2% within the best-case scenario Conti Sc3, assumed life time of the micro-structured devices: 10 years). In both impact categories the supply of the peripheral equipment and of the storage place inside the plant exhibits almost no effect (percentage <0.2% in all cases).

7.3.5 Sensitivity analysis

On the laboratory scale, the transfer of the model reaction from macro-scaled batch mode (Y: 60%) to a continuous micro-scaled set-up (Y: 88%) resulted in a yield increase of 28%. This effect influences the results of the LCIA considerably. However, under industrially relevant reaction conditions, the problem of sufficient cooling of the batch system would be solved by a cryogenic cooling system. Under these conditions the same yield of 88% would be expected in batch mode. This important point for further decision-making was analysed by means of a sensitivity analysis. For this purpose, a yield of 88% was also assumed for the batch mode under laboratory conditions in order to answer the question whether the positive results obtained for the continuous micro-scaled process are due to the yield increase or rather the decrease in energy demand. The increase in energy demand connected to such a yield improvement was not included into the balance, since the parameters were not obtained under laboratory conditions, and thus the resulting energy demand could not be determined experimentally. Nevertheless, the results depicted in Figure 7.4 which represent a best-case scenario of the batch process, show that a positive trend regarding the continuous micro-reactor process is still detectable. On the basis of these results, a life time of more than 6 months for the micro-reactors was calculated, hence environmental benefits in all impact categories considered should occur compared to the macro-scaled batch mode.

7.3.6 Conclusions from case study 1

Within case study 1, the applicability and benefits of LCA during process development were demonstrated. Although not all parameter specifications of the alternative reaction technologies under investigation could be determined at this early stage, a range of helpful

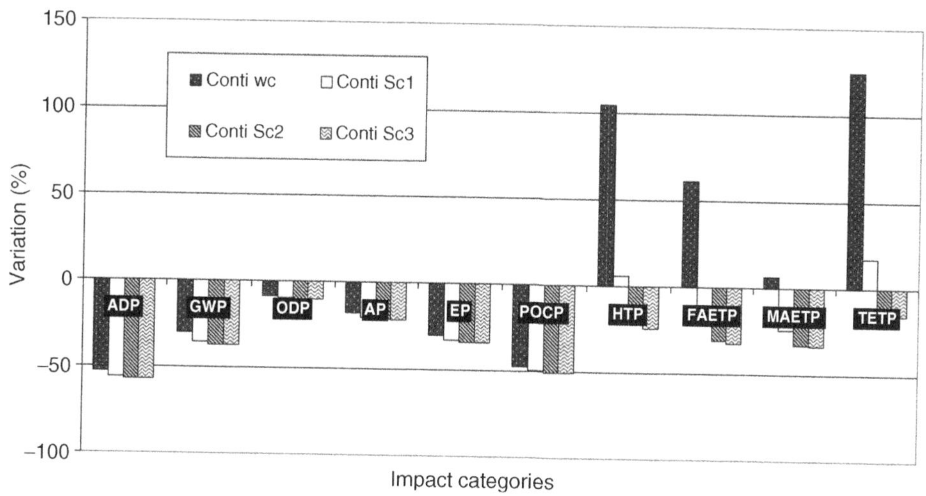

Figure 7.4 Variation of environmental impact potentials as consequence of the change from macro-scaled batch (0%) to micro-reaction mode; $Y_{\text{batch}} = 88\%$ (assumed), $Y_{\text{conti}} = 88\%$; four scenarios regarding the lifetime of the micro-structured devices (Conti wc: 1 week, Conti Sc1: 3 month, Conti Sc2: 3 years, Conti Sc3: 10 years).

information, insights into the interrelationships within these systems, as well as the influence of single life cycle stages on the overall results was gained. The calculation of case scenarios allowed for the investigation of a range of possible parameter values, and sensible assumptions and estimations based on generic data proved useful to calculate the proportion and relevance of single modules. With respect to this, the transition to an SLCA is seamless.

The results obtained within this LCA clearly indicated that significant ecological advantages can be gained from the transfer of the macro-scaled batch synthesis of *m*-anisaldehyde to continuous micro-reactor processing. This outcome is related (i) to the feasibility to conduct the reaction at a higher temperature (temperature difference: 50 K) resulting in a decrease of the energy demand during synthesis, and, (ii) to the reduction of solvent use and an increase in yield of the micro-reactor process. The results obtained by a sensitivity analysis pointed out that those environmental benefits would occur, even if the yields resulting from both considered set-ups were the same. Compared to these saving potentials, the additional energy and material demand during the fabrication step of the micro-structured devices only played a minor role within the majority of impact categories considered. Toxicity potentials resulting from the stainless-steel production became increasingly important only if a frequent exchange of the micro-reactors was required. Therefore, the authors conclude that special attention should be paid to the longevity of the micro-structured devices made of stainless steel.

Based on these positive results, a continuously operating Cytos® Pilot Plant was constructed by CPC Systems GmbH and is used for fine chemical production.[61] The results of the evaluation on production scale confirm those results obtained already during the process development stage, which have been discussed in detail elsewhere.[67]

7.4 Combination of LCA or SLCA and optimisation

Generally, two possible applications of LCA or SLCA for process development have to be differentiated. Firstly, they can be used to compare several alternatives, as discussed above. Secondly, they can be used to directly search for optimal configurations. If more than one objective is incorporated in the procedure, a multi-objective optimisation problem occurs, and therefore the problem of comparing several alternatives with respect to several objectives arises. This problem configuration is well-known in mathematics,[70] and can be solved using standardised algorithms for identifying *pareto-optimal*[71] solution candidates. A solution candidate is defined to be pareto-optimal if no objective can be improved without worsening the value of some other objective. In contrast, the so-called *dominated* alternatives can be differentiated, which are worse with respect to any other objective. This definition can be used to identify pareto-optimal alternatives which then constitute the basis for a total ranking, for example by using the AHP-process[72] or by using a partial ranking (outranking) procedure. Partial ranking procedures have the advantage that no compensation of good and bad degrees of performance regarding the objective function takes place. Therefore, no information is lost. Furthermore, indifferent, contradictory and incompletely assessed alternatives can be integrated by outranking procedures. Well-known examples for outranking methods are the ELECTRE method suggested by Roy[73] or the PROMETHEE method by Brans *et al.*[74] One advantage all multi-objective optimisation procedures have in common is that they do not require an *a priori* articulation of preferences. Thus, the whole set of pareto-optimal solutions can be explored by changing the preferences. The emphasis is to range the choices from a set of pareto-optimal solutions, rather than to define the preferences explicitly before analysing the alternative trade-offs. Thus, this approach shows explicitly what can be gained and lost by choosing each alternative.[75] This is necessary to understand the relationship between the levels of trade-off required.[76]

Kniel *et al.*[77] belonged to the first discussing the potential of combining LCA and multi-objective optimisation in a design tool for processes. They argued that this potential arises from the unique ability of LCA to link the environmental burdens of a process with its mass and energy flows. The approach was discussed on the example of a nitric acid plant, and the LCA was used to quantify and compare the environmental performance of a number of process design alternatives aimed at waste reduction. In addition, economic models for the alternatives were formulated and linked to the environmental models. For this purpose, the pressure in the absorption columns was chosen as a decision variable because it can be linked to the mass and energy flows and thus to the economic returns. The optimisation of the objective function was realised by an iterative calculation of pareto-optimal curves. The effect scores were multiplied with weighting factors for each process design alternative in order to allow for a ranking of these effect scores corresponding to different environmental burdens. A linear weighting and summation procedure has been chosen by the authors to reflect the aim of maximising the economic returns and minimising the environmental impact. The results indicated that operational changes may lead to significant environmental improvements at minimal economic costs.

Azapagic and Clift[7] also dealt with the coupling of multi-objective optimisation and LCA to facilitate decision-making in process development and optimisation. The system under investigation was simultaneously optimised on a number of environmental objective functions defined and quantified through the LCA approach.[75] Furthermore, the authors integrated economic performance into the optimisation process, thus enabling the choice of the Best Practicable Environmental Option and the Best Available Technique not Entailing Excessive Cost. This so-called decision aid tool Optimum LCA Performance methodology based on the results of the LCA study as starting point of the optimisation procedure refers to a predefined optimisation problem.

Stefanis *et al.*,[78] Alexander *et al.*,[76] Baitz *et al.*[79] and Sugiyama *et al.*[80] and so forth likewise presented the coupling of LCA and economic evaluation with a multi-objective evaluation to aid process design and development.

As an example, Baitz *et al.*[79] focused on different technologies and peripheral system conditions to reduce dust and heavy metal emissions from a refinery. They stressed that the knowledge of the sensitive life cycle parameters and a suitable database, and thus the possibility to quantify impacts, enables a sustainable decision-making in process design and process optimisation.

Sugiyama *et al.*[80] proposed a business model of simulation-oriented process design, including LCA, economic evaluation and multi-objective decision-making. The whole design procedure was described hierarchically with the focus on LCA-related activities. Further, they discussed the roles of decision-support activities such as sensitivity or uncertainty analyses. The design process of a chemical recycling process of beverage PET bottles served as case study.

The coupling of an SLCA approach, an economic analysis and multi-objective optimisation will be demonstrated in the following by means of case study 2.

7.5 SLCA and optimisation – a case study

7.5.1 *Intention of case study 2*

Ionic liquids gained a significant importance during the last decade. Due to their negligible vapour pressure and non-flammability, ionic liquids are considered as a substitute for commonly used volatile organic solvents. Furthermore, they feature specific solubility properties in mono, bi- and multi-phase reactions. Since ionic liquids are able to dissolve a wide range of organic, inorganic and polymeric materials, they have been investigated as solvents as well as auxiliaries in a great number of organic and organometallic syntheses, for example Heck reactions, hydrogenations, Diels–Alder reactions, catalytic oxidations and so on.[81–83] Additionally, ionic liquids are used as solvent for extractions,[84] biocatalytic reactions,[85] asymmetric catalysis[86] as well as electrolytes in electrochemistry.[87]

Until a few years ago, ionic liquids were uncritically referred to in the context of *Green Chemistry*.[88,89] Then, results on their partial toxicity, production effort and environmental impact have induced a more differentiated point of view. Nowadays, it is widely accepted that the *greenness* of ionic liquids heavily depends on their application, toxicological properties and on the environmental impact resulting from the production process. Especially the last aspect is presently not environmentally benign, and has not been optimised yet.[90]

Kralisch *et al.*[91] raised the question how synthesis pathways for ionic liquids can be optimised regarding ecological sustainability already at the R&D stage. In terms of a future applicability of these results, economic as well as ecological aspects should be included in such an optimisation strategy. For this reason they proposed a screening procedure based on the ECO Method, which can be used to search for optimal configurations in an iterative process. This method will be explained in more detail in the following.

7.5.2 The methodology

The ECO method was developed to aid environmental impact and cost optimisation of chemical synthesis pathways or processes suitable for the research and development (R&D) stage. In order to represent terms of ecological as well as economic sustainability, three objective functions which incorporate (i) energy demand (*EF*), (ii) risks concerning human health and the environment (*EHF*) and (iii) costs (*CF*), were defined. Their calculation follows the life cycle approach and is based on the data available already in R&D. Because the application of a comprehensive LCA is both, too complex and based on data which are partially not available at the R&D stage, the determination of the three objective functions is based on the SLCA approach[54] extended by economic issues. The key objectives are introduced below.

Energy factor. *EF* is defined as the energy demand (*E*), related to a product-based benefit for example the product molarity or the product mass. *E* incorporates the cumulative energy demand resulting from (i) the supply of the reactants, solvents and auxiliaries, (ii) the performance of the reaction, (iii) the energy demand necessary for the work-up, (iv) the application of the products and (v) the disposal of waste. The energy demand had been chosen to be one key objective, because it represents a screening indicator suitable for SLCA and reflects energy-related impact categories of LCIA such as ADP, GWP, POCP, ODP, AP and EP (see also Section 7.2.4).

Environmental and human health factor. *EHF* allows for a comparison of different chemical substances used, for example as reactants, solvents or auxiliaries regarding the resulting risks for human and environment during (i) their supply, (ii) product synthesis, (iii) product work-up, (iv) product application and (v) disposal. Toxicological risks are important aspects of an environmental benign synthesis pathway and thus, the identification of potential risks for human and ecosystems is essential during R&D. However, such aspects are difficult to estimate when newly developed synthetic pathways (with unknown disposal strategies) or novel chemicals (without defined impact factors) are evaluated. The EHS Method according to Koller *et al.*[92] is one suitable approach to facilitate the estimation of these risks already at an early stage of process design. That is why, the calculation of *EHF* is based on a summation of the Remaining Potential of Danger $(RPoD_i)_j$, calculated within the EHS Method and related to the molarity (or mass) of the product.

Cost factor. The calculation of *CF* is similar to current approaches of LCC analyses, again tailored to the evaluation of chemical synthesis strategies. *CF* includes (i) the costs of the supply of reactants, solvents and auxiliaries; (ii) costs resulting from synthesis, (iii) work-up, (iv) application and (v) disposal. Again, this effort is related to the molarity (or mass) of the product.

These objective functions have to be minimised during the optimisation process to maximise the benefit in terms of ecological and economic sustainability. They are calculated on the basis of experimental results obtained for process parameter variations during the ongoing process development. In order to limit the time and costs for accessing experimental data, a screening on the basis of selected parameter values with a high degree of information is executed. If a comparatively high optimisation potential is detected for a specific parameter, the number of tests will be incremented at this point. The decision-making process is guided by an outranking of pareto-optimal solution candidates referring these three key objectives.

7.5.2.1 The evaluation and optimisation procedure

The evaluation is carried out in close collaboration with an expert in the particular field of R&D or by the researcher himself. At the beginning, a pre-selection of optimisation parameters and a range of parameter variations is made, which at this point rarely represents a starting configuration, and can be varied by considering actual results at any time.

After the starting configuration has been specified, the experiments required to characterise the system are performed. Having the experimental data resulting from a parameter variation on hand, the calculation of the sustainability factors can be executed. Finally, the values of all factors are collected in a performance matrix. This performance matrix constitutes the basis of optimisation. Its compilation is an iterative process and it may change significantly during the optimisation process. The ranking of alternatives is executed using the partial ranking procedure PROMETHEE.[74] This decision support tool allows for the partial ranking of the parameter alternatives under investigation according to the three key objectives *EF*, *EHF* and *CF*. Freely selectable weighting factors are assigned to each of the objectives. By this means, scenarios that reflect different decision preferences can be considered. Thus, unfavourable (dominated) alternatives can be excluded in ongoing R&D work to restrict the experimental effort. With the help of this procedure the expert gains insight into the interrelationship between single determining factors of the decision problem.

In order to limit the extent of the database required for both the evaluation as well as the calculation effort, the ECO Method utilises a local optimisation of single process steps (e.g., a synthesis step including the work-up procedure). A selection of preferred parameter values for each single step is resulted, leading into new insights, which can be used to further optimise and/or identify new areas of process/product development. Kralisch *et al.*[91] stress that if this procedure is executed along the process chain, a local optimisation of the entire process regarding ecological and economic sustainability can be approximated. The application of the ECO Method is demonstrated below on the example of an ionic liquids synthesis.

7.5.3 Demonstration of the ECO method on the example of ionic liquids R&D

The synthesis of ionic liquids is usually carried out in a two-step pathway. The first step, the alkylation of a *N*-, *P*- or *S*-containing organic compound, for example *N*-methylimidazole, pyridine, alkyl phosphonates or sulphonates, is followed by an anion exchange. Both steps, particularly the alkylation step, involve energy and time consuming synthesis and work-up procedures. Especially the work-up (extraction) often requires a high input of organic solvents, resulting in energy consuming distillation steps.

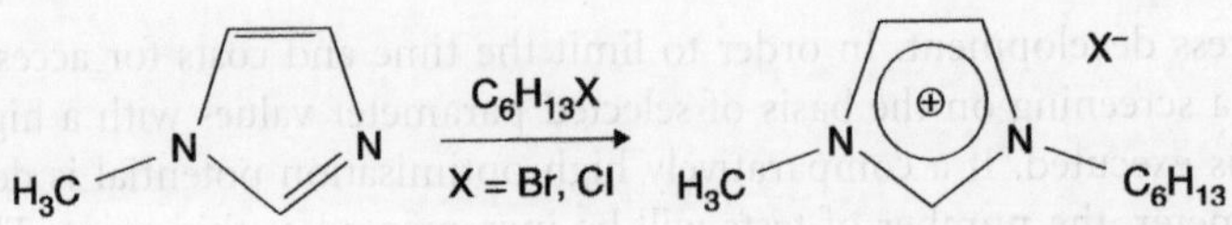

Scheme 7.2 Preparation of *N*-methylimidazolium-based ionic liquids – model reaction of case study 2.

To pursue the development of environmentally benign synthesis routes for ionic liquids, the alkylation step (Menschutkin reaction) was investigated by the authors in detail. The preparation of the ionic liquid 1-hexyl-3-methylimidazolium chloride ($[C_6MIM]Cl$) was taken as a representative experiment (Scheme 7.2). The process parameters: temperature ($T = 70$–100°C), solvent (ethanol, xylene, cyclohexane, *n*-heptane, solvent free), concentration of the *N*-base ($c = 1.6$–6.7 M), molar ratio ($n{:}n = 1{:}0.5$–$1{:}4$) and reaction time ($t = 10$–144 h) were investigated. In addition, the *N*-base was altered in order to proof the transferability of the reaction parameters.

In the case of the preparation of pure $[C_6MIM]Cl$ (purity > 99%), the evaluation of the actual experimental results and the outranking of the different solution candidates resulted in the following optimal synthesis prescription: $T = 100$°C, $t = 30$ h, $n_{MIM}/n_{C_6H_{13}Cl} = 1{:}1$, $c_{N-base} = 4.6\ \mathrm{mol\,L^{-1}}$, solvent free. This parameter configuration has been found to represent the best trade-off between *EF*, *EHF* and *CF*.

In relation to the starting point of the optimisation procedure ($T = 80$°C, $t = 30$ h, $n_{MIM}/n_{C_6H_{13}Cl} = 1{:}1.2$, $c_{N-base} = 3\ \mathrm{mol\,L^{-1}}$, solvent: *n*-heptane), this synthesis protocol implies a reduction of *EF* by 78%, of *EHF* by 98% and of *CF* by 87%.

7.5.4 Conclusions from case study 2

The ECO method can be seen as an instrument for decision support and optimisation in the context of an application oriented chemical R&D. It is not suitable to provide concrete values describing an environmental impact or specific costs, but allows for an unbiased development of environmentally benign and cost-optimised synthesis pathways, work-up procedures, processes or functionalities of chemical compounds by providing detailed information about dependencies between the different parameters under investigation. The decision-making process remains with the expert of the specific R&D field, because the selection of significant parameter values and the weighting of the objectives require special expertise. The implementation of this method in ionic liquids R&D demonstrated the high optimisation potential within common synthesis strategies for these compounds.

7.6 Conclusion

The environmental impacts caused by using a specific technology or resulting from specific process conditions are mostly predefined within the process development phase. Therefore, it is of great importance to integrate the analysis of environmental aspects already in this phase of the decision-making process. A broad scope of environmentally evaluated methods, ranging from easily calculable single-point metrics to complex life cycle approaches, has been established in this context. However, if the assessment of environmental issues is limited to the process under investigation itself, a lack of relevant data concerning preliminary and post-processes occurs. Often, consequences of a modification only become apparent at a later process step. Otherwise, the environmental burden resulting from the production of a reactant may dominate the overall balance of a chemical process. Due to this, the LCA methodology is a powerful tool to aid the development of processes and technologies, which are environmentally benign from a holistic point of view.

To meet the limitations of data and time during the early stages of process development, the application of a Simplified Life Cycle Assessment is a suitable approach for screening and hot-spot detection along the entire process chain. The results obtained from this should be scrutinised regarding validity and correctness by iteratively increasing the detail while the knowledge increases with ongoing development progress so that a comprehensive LCA is performed at last, quantifying the environmental impact potentials related to this process in a responsible way.

One criticism of decision guidance during process development by LCA is that it supports retrofitting systems rather than fundamental conceptual shifts. However, this can be avoided by case scenario calculation including future changes, for example in legislation, raw material availability, disposal strategy and so on, and thus by providing an insight into the interrelations and the influence of constraints of the systems under development. For this purpose, LCA should rather be considered as a flexible and dynamic toolbox. Then, the advantages of the LCA method (holistic, comprehensive, internationally recognised etc.) can be favourably exploited in process development without restrictions by overstated guidelines.

Further guidance in sustainable process development can be achieved by coupling both, the LCA and the SLCA methodology, with multi-objective optimisation procedures. Thus, an extended range of alternative parameter configurations becomes systematically accessible. In addition, the optimisation of the entire process chain regarding different, maybe even contradictory, key objectives becomes feasible. Due to the powerful potential of coupling evaluation methods and computer-aided optimisation procedures, future process development will be characterised by an increasing implementation of such approaches.

References

1. Grossmann, I.E. (2004) Challenges in the New Millennium: Product Discovery and Design, Enterprise and Supply Chain Optimisation, Global Life Cycle Assessment. *Computers and Chemical Engineering*, **29**(1), 29–39.
2. Cano-Ruiz, J.A. & McRae, G.J. (1998) Environmentally Conscious Chemical Process Design. *Annual Review of Energy and the Environment*, **23**(1), 499–536.
3. Jenck, J.F., Agterberg, F., Droescher, M.J. (2004) Products and Processes for a Sustainable Chemical Industry: A Review of Achievements and Prospects. *Green Chemistry*, **6**(11), 544–556.
4. ISO 14040:2006 (2006) Environmental Management – Life Cycle Assessment – Principles and Framework. European Commitee for Standardisation, Brussels, Belgium.
5. ISO 14044:2006 (2006) Environmental Management – Life Cycle Assessment – Requirements and Guidelines. European Commitee for Standardisation, Brussels, Belgium.
6. Life Cycle Assessment – An Operational Guideline to the ISO Standards (2001) Ministry of Housing, Spatial Planning and the Environment and Centre of Environmental Science – Leiden University, The Netherlands.

7. Azapagic, A. & Clift, R. (1999) The Application of Life Cycle Assessment to Process Optimisation. *Computers and Chemical Engineering*, **23**(10), 1509–1526.
8. Guinée, J.B. (ed.), Gorree, M., Heijungs, R. *et al.* (2001) Life Cycle Assessment – An Operational Guideline to the ISO Standards – Part 2a. Ministry of Housing, Spatial Planning and the Environment and Centre of Environmental Science – Leiden University, The Netherlands.
9. The eco-indicator 99 (2001) A Damage Oriented Method for Life Cycle Impact Assessment. PRé Consultant.
10. Habersatter, K. (1990) *Ökobilanz von Packstoffen* [132]. BUWAL. Schriftenreihe Umwelt, Bern, Switzerland.
11. Ahbe, S., Braunschweig, A., Müller-Wenk, R. (1990) *Methodik für Ökobilanzen auf der Basis ökologischer Optimierung* [133]. BUWAL. Schriftenreihe Umwelt, Bern, Switzerland.
12. Stehen, B. & Ryding, S.O. (1995) The EPS Enviroaccounting Method. Swedish Environmental Research Institute, Federation of Swedish Industries, Gothenborg, Sweden.
13. Umberto (2006) v.5.5. ifu Institut für Umweltinformatik; ifeu Institut für Energie- und Umweltforschung, Germany.
14. GABI (2007) v.4. PE International, Germany.
15. SimaPro (2006) v.7. PRé Consultants, The Netherlands.
16. Ecoinvent database by Frischknecht *et al.* (2006) v.1.3. Swiss Centre for Life Cycle Inventories, Switzerland.
17. http://lca.plasticseurope.org/index.htm (2007).
18. http://www.probas.umweltbundesamt.de/php/index.php (2007).
19. McDonough, W. & Braungart, M. (2002) *Cradle to Cradle: Remaking the Way We Make Things.* Rodale Books, US.
20. McDonough, W., Braungart, M., Anastas, P.T., Zimmerman, J.B. (2003) Applying the Principle of Green Engineering to Cradle-to-Cradle Design. *Environmental Science and Technology*, **37**(23), 434A–441A.
21. Meadows, D.H., Meadows, D.L., Randers, J., Behrens, W.W. (1972) The Limits to Growth. *A report for the Club of Rome's Project on the predicament of mankind.* Universe Books, New York.
22. Sundström, G. (1973) Investigation of the Requirements from Raw Materials to Garbage Treatment for Four Swedish Beer Packaging Alternatives. Report for Rigello Pak AM.
23. US EPA (1974) Resource and Environmental Profile Analysis of Nine Beverage Container Alternatives. US EPA (US Environmental Protection Agency); Report No.: EPA/530/SW-91c.
24. Boustead, I. & Hancock, G.F. (1975) T*he Glass Industry: A Report to the Glass Manufacturers Federation*, Ellis Horwood, London.
25. Boustead, I. (1999) Eco-Profiles for Plastics and Related Intermediates – Methodology. The European Centre for Plastics in the Environment of the Association of Plastics Manufacturers in Europe (APME), Brussels, Belgium.
26. http://www setac org (2007).
27. Consoli, F. & Allen, D. (1993) *Guidelines for Life-Cycle Assessment: A "Code of Practice".* Society of Environmental Toxicology and Chemistry (SETAC), Sesimbra, Portugal.
28. ISO 14040:1997 (1997) Environmental Management – Life Cycle Assessment – Principles and Framework. European Commitee for Standardisation, Brussels, Belgium.
29. ISO 14041:1998 (1998) Environmental Management – Life Cycle Assessment – Goal and Scope Definition and Life Cycle Inventory Analysis. European Commitee for Standardisation, Brussels, Belgium.
30. ISO 14042:2000 (2000) Environmental Management – Life Cycle Assessment – Life Cycle Impact Assessment. European Commitee for Standardisation, Brussels, Belgium.
31. ISO 14043:2000 (2000) Environmental Management – Life Cycle Assessment – Life Cycle Interpretation. European Commitee for Standardisation, Brussels, Belgium.
32. UNEP report (1996) Life Cycle Assessment: What It is and What to Do about It.

33. Solgaard, A. & de Leeuw, B. (2002) UNEP/SETAC Life Cycle Initiative: Promoting the Life-Cycle Approach. *International Journal of Life Cycle Assessment*, 7, 1999–2002.
34. Allen, D.T. & Rosselot, K.S. (1997) *Pollution Prevention for Chemical Processes.* Wiley, New York.
35. Council directive 96/61/EC concerning integrated pollution prevention and control (1996) European Communities. *Official Journal of the European Communities*, **L257**(10/10/1996).
36. Integrated Product Policy: Building on Environmental Life-Cycle Thinking (2003) European Communities, Report No. 302, Brussels, Belgium.
37. Nicholas, M.J. (1998) Legislation's Going Holistic. *Chemical Engineering*, **4**, 33–34.
38. Rebitzer, G. & Hunkeler, D. (2003) Life Cycle Costing in LCM: Ambitions, Opportunities, and Limitations – Discussing a Framework. *International Journal of Life Cycle Assessment*, **8**(5), 253–256.
39. Herrchen, M. & Klein, W. (2000) Use of the Life-Cycle Assessment (LCA) Toolbox for an Environmental Evaluation of Production Processes. *Pure Applied Chemistry*, **72**(7), 1247–1252.
40. Burgess, A.A. & Brennan, D.J. (2001) Application of Life Cycle Assessment to Chemical Processes. *Chemical Engineering Science*, **56**(8), 2589–2604.
41. Jödicke, G., Zenklusen, O., Weidenhaupt, A., Hungerbühler, K. (1999) Developing Environmentally-Sound Processes in the Chemical Industry: a Case Study on Pharmaceutical Intermediates. *Journal of Cleaner Production*, **7**, 159–166.
42. Hellweg, S., Fischer, U., Scheringer, M., Hungerbühler, K. (2004) Environmental Assessment of Chemicals: Methods and Application to a Case Study of Organic Solvents. *Green Chemistry*, **6**(8), 418–427.
43. Bauer, P.E. & Maciel, R. (2004) Incorporation of Environmental Impact Criteria in the Design and Operation of Chemical Processes. *Brazilian Journal of Chemical Engineering*, **21**(3), 405–414.
44. Pehnt, M. (2006) Dynamic Life Cycle Assessment (LCA) of Renewable Energy Technologies. *Renewable Energy*, **31**(1), 55–71.
45. Jungbluth, N., Bauer, C., Dones, R., Frischknecht, R. (2005) Life Cycle Assessment for Emerging Technologies: Case Studies for Photovoltaic and Wind Power. *International Journal of Life Cycle Assessment*, **10**(1), 24–34.
46. Gasafi, E., Meyer, L., Schebek, L. (2004) Using Life-Cycle Assessment in Process Design: Supercritical Water Gasification of Organic Feedstocks. *Journal of Industrial Ecology*, **7**(3–4), 75–91.
47. Kemppainen, A.J. & Shonnard, D.R. (2005) Comparative Life-Cycle Assessments for Biomass-to-Ethanol Production from Different Regional Feedstocks. *Biotechnology Progress*, **21**(4), 1075–1084.
48. Peregrina, C.A., Lecomte, D., Arlabosse, P., Rudolph, V. (2006) Life Cycle Assessment (LCA) Applied to the Design of an Innovative Drying Process for Sewage Sludge. *Process Safety and Environment Protection*, **84**(B4), 270–279.
49. Khoo, H.H. & Tan, R.B.H. (2006) Life Cycle Evaluation of CO_2 Recovery and Mineral Sequestration Alternatives. *Environment Progress*, **25**(3), 208–217.
50. Sadiq, R., Khan, F.I., Veitch, B. (2005) Evaluating Offshore Technologies for Produced Water Management Using Greenpro-1 – A Risk-Based Life Cycle Analysis for Green and Clean Process Selection and Design. *Computers and Chemical Engineering*, **29**(5), 1023–1039.
51. Azapagic, A., Millington, A., Collett, A. (2006) A Methodology for Integrating Sustainability Considerations into Process Design. *Chemical Engineering Research and Design*, **84**(A6), 439–452.
52. Fleischer, G. & Schmidt, W.P. (1997) Iterative Screening LCA in an Eco-Design Tool. *International Journal of Life Cycle Assessment*, **2**(1), 20–24.
53. Hoffmann, V.H., Hungerbühler, K., McRae, G.J. (2001) Multiobjective Screening and Evaluation of Chemical Process Technologies. *Industrial Engineering and Chemical Research*, **40**(21), 4513–4524.

54. Christiansen, K. (ed.) (1997) *Simplifying LCA: Just a Cut?* Final Report of the SETAC-Europe LCA Screening and Streamlining Working Group. SETAC-Europe, Brussels, Belgium, ISBN 90-5607-006-1.
55. Schmidt-Bleek, F. (1993) *Wieviel Umwelt braucht der Mensch? MIPS – das Maß für ökologisches Wirtschaften.* Birkhäuser, Berlin, Germany.
56. Fleischer, G., Gerner, K., Kunst, H., Lichtenvort, K., Rebitzer, G. (2001) A Semi-Quantitative Method for the Impact Assessment of Emissions within a Simplified Life Cycle Assessment. *International Journal of Life Cycle Assessment,* **6**(3), 149–156.
57. Weidenhaupt, A. & Hungerbühler, K. (1997) Integrated Product Design in Chemical Industry. A Plea for Adequate Life-Cycle Screening Indicators. *Chimia,* **51**(5), 217–221.
58. Hochschorner, E. & Finnveden, G. (2003) Evaluation of Two Simplified Life Cycle Assessment Methods. *International Journal of Life Cycle Assessment,* **8**(3), 119–128.
59. Hur, T., Lee, J., Ryu, H., Kwon, E. (2005) Simplified LCA and Matrix Methods in Identifying the Environmental Aspects of a Product System. *Journal of Environment Management,* **75**(3), 229–237.
60. Murphy, E.R., Martinelli, J.R., Zaborenko, N., Buchwald, S.L., Jensen, K.F. (2007) Accelerating Reactions with Micro-Reactors at Elevated Temperatures and Pressures: Profiling Aminocarbonylation Reactions. *Angewandte Chemie,* **119**(10), 1764–1767.
61. Hessel, V., Loeb, P., Loewe, H. (2005) Development of Microstructured Reactors to Enable Organic Synthesis Rather Than Subduing Chemistry. *Current Organic Chemistry,* **9**(8), 765–787.
62. Wehle, D. (2005) Micro-Process Technology Dynamically Enables Process Intensification. *Chemical and Industrial Digest,* **18**(12), 51–55.
63. Jähnisch, K., Hessel, V., Löwe, H., Baerns, M. (2004) Chemie in Mikrostrukturreaktoren. *Angewandte Chemie,* **116**(4), 410–451.
64. Schwalbe, T., Autze, V., Hohmann, M., Stirner, W. (2004) Novel Innovation Systems for a Cellular Approach to Continuous Process Chemistry from Discovery to Market. *Organic Process Research and Development,* **8**(3), 440–454.
65. Tonkovich, A.L., Kuhlmann, D., Rogers, A. *et al.* (2005) Microchannel Technology Scale-up to Commercial Capacity. *Chemical Engineering Research and Design,* **83**(A6), 634–639.
66. Kralisch, D. & Kreisel, G. (2005) Bewertung der ökologischen Potenziale der Mikroverfahrenstechnik. *Chemie-Ingenieur-Technik,* **77**(6), 62–69.
67. Kralisch, D. & Kreisel, G. (2007) Assessment of the Ecological Potential of Microreaction Technology. *Chemical Engineering Science,* **62**(4), 1094–1100.
68. Guinée, J.B. (ed.), Gorree, M., Heijungs, R. *et al.* (2001) Life Cycle Assessment – An Operational Guideline to the ISO Standards – Part 2b. Ministry of Housing, Spatial Planning and the Environment and Centre of Environmental Science – Leiden University, The Netherlands.
69. http://www synthacon biz, 2007.
70. Horn, J. (1997) Multicriteria Decision Making. In: Bäck, T., Fogel, D.B., Ichalewicz, Z. (eds) *Handbook of Evolutionary Computation.* Institute of Physics Publishing, Bristol, UK.
71. Pareto, V. (1896) *Cours d'economie politique.* F. Rouge, Lausanne.
72. Saaty, T.L. (1980) *The Analytic Hierarchy Process.* McGraw-Hill, New York.
73. Roy, B. (1991) The Outranking Approach and the Foundations of ELECTRE Methods. *Theory Decision,* **31**, 49–73.
74. Brans, J.P., Vincke, P., Mareschal, B. (1986) How to Select and How to Rank Projects: The PROMETHEE Method. *European Journal of Operational Research,* **24**, 228–238.
75. Azapagic, A. (1999) Life Cycle Assessment and Its Application to Process Selection, Design and Optimisation. *Chemical Engineering Journal,* **73**(1), 1–21.
76. Alexander, B., Barton, G., Petrie, J., Romagnoli, J. (2000) Process Synthesis and Optimisation Tools for Environmental Design: Methodology and Structure. *Computers in Chemical Engineering,* **24**(2–7), 1195–1200.

77. Kniel, G.E., Delmarco, K., Petrie, J.G. (1996) Life Cycle Assessment Applied to Process Design: Environmental and Economic Analysis and Optimisation of a Nitric Acid Plant. *Environment Progress*, **15**(4), 221–228.
78. Stefanis, S.K., Livingston, A.G., Pistikopoulos, E.N. (1997) Environmental Impact Considerations in the Optimal Design and Scheduling of Batch Processes. *Computers in Chemical Engineering*, **21**(10), 1073–1094.
79. Baitz, M., Wolf, M.A., Faltenbacher, M. (2001) Life-Cycle Related Technical-Environmental Multi-Parameter Trade-Off Optimisation of Processes. *Journal of Advances in Science*, **13**(3), 199–202.
80. Sugiyama, H., Hirao, M., Mendivil, R., Fischer, U., Hungerbühler, K. (2006) A Hierarchical Activity Model of Chemical Process Design Based on Life Cycle Assessment. *Process Safety and Environmental Protection*, **84**(B1), 63–74.
81. Seddon, K.R. & Stark, A. (2002) Selective Catalytic Oxidation of Benzyl Alcohol and Alkylbenzenes in Ionic Liquids. *Green Chemistry*, **4**(2), 119–123.
82. Sheldon, R.A. (2001) Catalytic Reactions in Ionic Liquids. *Chemical Communications*, **23**, 2399–2407.
83. Welton, T. (1999) Room-Temperature Ionic Liquids. Solvents for Synthesis and Catalysis. *Chemical Review*, **99**(8), 2071–2083.
84. Visser, A.E., Swatloski, R.P., Reichert, W.M. *et al.* (2001) Task-Specific Ionic Liquids for the Extraction of Metal Ions from Aqueous Solutions. *Chemical Communications*, **1**, 135–136.
85. Sheldon, R.A., Lau, R.M., Sorgedrager, M.J., van Rantwijk, F., Seddon, K.R. (2002) Biocatalysis in Ionic Liquids. *Green Chemistry*, **4**(2), 147–151.
86. Gruttadauria, M., Riela, S., Lo Meo, P., D'Anna, F., Noto, R. (2004) Supported Ionic Liquid Asymmetric Catalysis: A New Method for Chiral Catalysts Recycling. the Case of Proline-Catalysed Aldol Reaction. *Tetrahedon Letters*, **45**(32), 6113–6116.
87. Galinski, M., Lewandowski, A., Stepniak, I. (2006) Ionic Liquids as Electrolytes. *Electrochimica Acta*, **51**(26), 5567–5580.
88. Earle, M.J. & Seddon, K.R. (2002) Ionic Liquids: Green Solvents for the Future. *Clean Solvents*, **819**, 10–25.
89. Wasserscheid, P. & Keim, W. (2000) Ionic Liquids – New 'Solutions' for Transition Metal Catalysis. *Angewandte Chemie International Edition*, **39**(21), 3773–3789.
90. Kralisch, D., Stark, A., Körsten, S., Kreisel, G., Ondruschka, B. (2005) Energetic, Environmental and Economic Balances: Spice up Your Ionic Liquid Research Efficiency! *Green Chemistry*, **7**(5), 301–309.
91. Kralisch, D., Reinhardt, D., Kreisel, G. (2007) Implementing Objectives of Sustainability into Ionic Liquids Research and Development. *Green Chemistry*, **9**(12), 1308–1318, DOI: 10.1039/b708721g.
92. Koller, G. (2000) Identification and Assessment of Relevant Environmental, Health and Safety Aspects during Early Phases of Process Development. *Technische Wissenschaften Eth Zürich, Zürich.*

Chapter 8
Tools and Strategies for Greening Chemical Inventories in Small Businesses

Lauren Heine and Teresa McGrath

8.1 Introduction and background

Small and medium-sized enterprises (SMEs) often have the advantage over larger businesses of being flexible and able to implement change quickly. Small businesses, however, carry a disproportionate burden of complying with regulations. A study by the United States Small Business Administration, showed that the cost of complying to federal environmental regulations is 364 per cent higher per employee in small firms (<20 employees) than in large firms (500+ employees).[1] Companies that stay ahead of chemical regulatory and policy pressures by proactively implementing green chemistry into their products and processes may avoid some of these costs. However, small businesses are tasked with the challenge of implementing green chemistry with limited budget and staff resources. In this chapter we present strategies, tools and metrics that can be used to help companies, particularly SMEs, focus their resources and prioritize areas within their chemical inventories to strategically improve and benchmark the overall hazard profile of the chemicals they purchase and use.

The 12 Principles of Green Chemistry do an excellent job of laying out the principles and practices that define green chemistry for practicing chemists. However, in order to successfully realize green chemistry in products and processes, it is necessary to develop metrics. As the adage goes, 'what gets measured, gets managed'. Focusing on metrics for chemical hazard facilitates informed chemical and material selection and promotes green chemistry. But, it is important to remember that hazard assessment fits within a bigger picture of sustainable product design and development and sustainable material flows.

A sustainable product has been defined as one that:

- Meets market requirements
- Has positive social effects (for individuals and communities)
- Is safe for human and ecological health
- Is sourced from renewable or repeatedly recycled materials
- Is sourced from renewable energy
- Is designed for safe, productive return to nature or industry
- Is recovered and recycled at highest quality after use.

This is based on what McDonough and Braungart call 'Cradle-to-Cradle' design.[2] Green chemistry can support the design of chemicals, materials and products that are safe and healthy for humans and the environment during their full life cycle with a focus on design for safe, productive return to nature or industry. For example, cleaning products are typically discharged to water or air and should therefore be designed to fully degrade to innocuous by-products once they are released to the environment. As such, metaphorically, they should be designed as biological nutrients destined for biological metabolisms such as wastewater treatment. In contrast, polymers or textiles may be designed as technical nutrients to be recovered and recycled for high value recovery in industrial or technical metabolisms. In this chapter we focus what 'is safe for human and ecological health'. The goal is to improve the overall hazard profile of chemical inventories, while keeping in mind that this is just one piece of the big picture of sustainable material flows and 'Cradle-to-Cradle' design.

Since the 12 Principles of Green Chemistry were derived in part by observing best practices, metrics for some of them already exist. For example, metrics for catalytic selectivity, activity and stability and the use of renewable feedstock for energy or reagents, and so on can be combined under a framework of green chemical processes and products (Principles 6, 7 and 9). Additional metrics have been developed primarily for assessing green chemistry practices such as Atom Economy and E-Factor. Atom Economy is defined as the number of atoms contained in reactants that end up in the desired product. It is calculated by dividing the molecular weight (MW) of the product by the sum of the MW of the reactants and does not take into account auxiliaries or actual reaction yields.[3] The E-factor was developed to take into account both reagents and auxiliaries and is defined as 'the mass ratio of waste to desired product'.[4] E-factor is calculated by dividing the sum of the mass of all reactants, solvents and auxiliaries by the mass of the desired product, taking into account per cent yield.

Developing metrics for hazard is more complex because there are numerous chemical hazards and combinations of hazards, and the severity of hazards or hazard combinations is somewhat subjective. Hazard as defined by the Organisation for Economic Co-operation and Development (OECD), as the 'inherent property of an agent or situation having the potential to cause adverse effects when an organism, system or (sub) population is exposed to that agent'.[5] In the field of green chemistry, the term hazard is used to include toxicity but it also includes broader threats to human and environmental health such as depletion of non-renewable resources, stratospheric ozone depletion, tropospheric ozone generation, contribution to climate change and more. Typically, chemical hazards are assessed as part of a risk assessment where risk is defined as the probability of an adverse effect due to exposure to a hazard. A fundamental premise behind green chemistry is that the most effective way to reduce risk from chemicals is to use chemicals with reduced hazard. Comparative hazard assessment is a way to inform decision making and to promote green chemistry. By understanding the hazards of a material or chemical, one is better prepared to consider hazard in product design and development and presumably more likely to select chemicals with more benign properties when other factors are equivalent.

Several of the Principles of Green Chemistry call for the use or generation of chemicals with reduced hazard and improved health and safety attributes. For example, Principle 3, *Less Hazardous Chemical Syntheses*, states, 'Wherever practicable, synthetic methods should be designed to use and generate substances that possess little or no toxicity to human health and the environment'. To apply this principle one must define what is meant by toxicity and how it should be measured. Principle 4, *Designing Safer Chemicals*, states that 'Chemical products should be designed to effect their desired function while minimizing their toxicity'. Again, metrics are useful to determine whether or not toxicity has been successfully minimized. Principle 5, *Safer Solvents and Auxiliaries* states that 'the use of auxiliary substances (e.g., solvents, separation agents, etc.) should be made unnecessary wherever possible and innocuous when used'. To determine whether or not auxiliaries are innocuous means that their hazards must be assessed and evaluated. Principle 10, *Design for Degradation* states that, 'Chemical products should be designed so that at the end of their function they break down into innocuous degradation products and do not persist in the environment'. This depends on better understanding of the type and rate of degradation (i.e., photolysis, hydrolysis, biodegradation, etc.) which determines the degradation products formed. Principle 12, *Inherently Safer Chemistry for Accident Prevention*, states that 'Substances and the form of a substance used in a chemical process should be chosen to minimize the potential for chemical accidents, including releases, explosions, and fires'. Metrics are needed for physical hazards to determine whether or not green chemistry practices are reducing the potential for accidents.

Green Chemistry is defined as 'the design of products and processes that reduce or eliminate the use or generation of hazardous substances'.[6] While green chemistry primarily focuses on the synthesis of new chemicals and materials, it also applies to formulated products. Formulated products are created by combining chemicals and materials in a product to perform a particular function. Oddly, the university curriculum offers little education in the art and science of product formulation. Yet, formulated specialty chemicals and consumer products account for 30–35% of the chemical products sold.[7] Formulated products include lubricants, paints, coatings, pigments, dyes, cosmetics, packaging materials, polymers, personal care products, cleaning products and more. Even pharmaceutical products are active ingredients compounded with other agents to help with stability and effective drug delivery. Depending on how you define 'formulated', one could argue that most manufactured products are formulated products.

In this chapter, we focus on strategies, tools and metrics that help 'chemical choosers' to green their chemical product inventories and to benchmark progress. The term 'chemical chooser' applies to those who purchase chemicals for use in the development of formulated products or those who purchase formulated chemical products for use in their use in activities such as maintenance, repair and operations. Most product manufacturers are chemical and material choosers in contrast to chemical or raw material manufacturers who process, synthesize and supply chemicals as raw materials.

Given the scope and impact of formulated products, tools are needed to help chemical choosers – whether product developers or purchasers – to identify chemicals with desirable characteristics from the hazard perspective and to determine how the use of a particular chemical will affect the overall hazard profile or 'greenness' of a product. We focus in particular on publicly available tools and resources that can support small businesses who may not have extensive environmental health and safety resources but who seek to green their chemical products and inventories.

8.2 The path to green chemical inventories

A decision to integrate green chemistry into products, processes and inventories is a commitment to organizational change and to continual improvement. Management systems such as the International Organization for Standardization (ISO) Quality Management Systems (ISO 9000 series) and Environmental Management Systems (ISO 14000 series) are also useful models for greening a chemical inventory. A management system approach (loosely defined) involves the following steps:

1 Awareness and motivation
2 Inventory
3 Assessment and evaluation
4 Setting and reaching goals
5 Reviewing and reporting
6 Repeating.

Strategies, tools and metrics support the process of change toward the adoption of green chemical inventories at each step.

8.2.1 Awareness and motivation

Before an organization sets down the path to green chemistry they must become aware of the value of green chemistry and be motivated to implement change. A company should consider why it is embarking on the path to green chemistry. Potential motivations and drivers include:

- Desire for decreased liability
- Desire for increased sales/profits/market share
- Response to national and international chemical bans and phase-out initiatives
- Response to environmentally preferable purchasing initiatives
- Desire for product recognition/certification
- Response to increased public awareness
- Desire to differentiate one's product in the marketplace
- Desire to meet internal corporate objectives
- Desire to innovate and to develop new products.

Through chemicals bans, phase-outs and initiatives, like the 'restriction of the use of certain hazardous substances in electrical and electronic equipment' (RoHS Directive)[8] and the 'waste electrical and electronic equipment' (WEEE Directive),[9] companies are pushed to find green chemistry solutions to traditional chemistry. The US has a growing number of state and federal initiatives including Environmentally Preferable Purchasing (EPP) initiatives such as Federal Executive Order 13101,[10] and a growing number of state and city environmentally preferable purchasing programs such as those in Minnesota,[11] Massachusetts,[12] Michigan,[13] Seattle, Washington[14] and San Francisco, California.[15] Policy is also developing in support of green chemistry research in the US. This includes the pending legislation, the *Green Chemistry Research and Development Act of 2008*,[16] the *California Green Chemistry Initiative*,[17] the *Executive Directive 2006-6 for Michigan State: "Promotion of Green Chemistry for Sustainable Economic Development and Protection of Public Health"*[18] and a report *Green Chemistry in California: A Framework for Leadership in Chemicals Policy and Innovation*, which proposes a 'modern, comprehensive chemicals policy' in California.[19]

Green chemistry is not just policy driven. Marketing to the conscious consumer has cultivated a series of product recognition, certification and eco-labelling programs. These include a growing number of US Green Building Council's Leadership in Energy and Environmental Design (LEED) programs,[20] which address new and existing buildings, commercial interiors and even schools, retail and neighbourhood development. Eco-labelling programs such as EcoLogo[21] and Green Seal[22] certify a variety of products and services that are better for human health and the environment. The US Environmental Protection Agency's Design for the Environment (DfE) Program promotes green chemistry by reviewing and recognizing chemical products and formulations that are most favorable with respect to environmental and human health when compared to conventional products.[23]

Another driver for green chemistry is the growing availability of *effective* alternative ingredients (i.e., surfactants, chelating agents, solvents, etc.). In the industrial and institutional cleaning industry, at one time, 'green' products were known for the absence of chemicals of concern and also for the absence of good performance. Today, with the emergence of effective alternative ingredients, the expectation that green products are inferior appears to be replaced with the expectation of effective and innovative new products.

Activist and public health organizations continue to raise awareness of health and environmental issues related to chemicals through campaigns and emerging science such as Chemical Body Burden[24]; The Campaign for Safe Cosmetics,[25] and endocrine disruption research. Forward thinking companies recognize the increase in public awareness and trends and feel continued pressure to find greener alternatives. We often hear manufacturers say, I know what I cannot use, but what CAN I use? They seek tools and strategies for assessing and evaluating chemicals and agreement on what is 'green' so that they can measure and benchmark their positions.

Most companies prefer to be in a proactive position – ahead of the demand curve and knowing that their efforts to find greener alternatives will not be wasted – rather than in the more vulnerable position of reacting to chemical by chemical bans based on regulations or policies. For example, while the polybrominated diphenyl ether (poly-BDE) flame retardants octa-BDE and penta-BDE were voluntarily phased out by the major US manufacturer at the end of 2004, IBM prohibited octa-BDE, penta-BDE and deca-BDE ethers above 0.1% in any homogeneous material. IBM prohibits the intentional addition of any PBDE in any homogenous material and does not exempt deca-BDE ether from this restriction.[26] Similarly, on 1 November 2005, Hewlett Packard (HP) announced a goal to eliminate the brominated flame retardant tetrabromobisphenol A from external case parts of all new HP brand products introduced after 31 December 2006, again, coming before legislation, particularly in the Unites States.[27]

8.2.1.1 Metrics for the awareness and motivation step

Metrics are needed to demonstrate advancement in organizational awareness and motivation toward implementing green chemistry. The type of metrics depends in part on how the metrics will be used. Possible applications include:

- Supporting an internal process of continual improvement
- Public reporting to management and stockholders
- Reporting to customers
- Public reporting to the broader community through Sustainability Reporting Guidelines such as the Global Reporting Initiative.[28]

Two of the most effective activities related to awareness and motivation are corporate policies, and education and training. A corporate policy sends a signal to employees and articulates a vision for the company. Education and training help employees to align with the corporate policy. Corporate policies, education and training are leading indicators that will determine in part both your success and your credibility. They are tangible actions that can be measured in the sense that stakeholders, including stockholders and investors are increasingly looking at toxics liability in their evaluation of a company. The Investor Environmental Health Network (IEHN) is a collaborative partnership of institutional investors that is advised by non-governmental organizations concerned about the market and health risks associated with corporate toxics policies. It serves as an informational resource and secretariat for investors working to reduce portfolio risk related to toxics. These investors recognize the strategic market opportunities that are emerging as governments, businesses, and consumers demand safer, less toxic products and believe that companies should keep investors fully informed of these risks and rewards. The principal elements of desirable policies and practices have been suggested in a benchmarking framework published by Dr. Richard Liroff, the Director of IEHN.[29] A commitment as reflected in a corporate policy and education and training for employees is considered to be a strong indicator of a company's commitment to green chemistry by external observers.

8.2.2 Inventory (strategies, tools and metrics)

Chemical and material supply chains can be long and it can be difficult to know the exact composition of the chemicals in your raw materials and products. Nevertheless, examining the human and environmental health profiles of the chemicals in your design palette and your products is worth the effort. Companies that are not aware of the chemical constituents of the products they sell are more vulnerable to liability should toxicity become apparent. In an extreme recent example (February 2006), a four-year-old boy in Minneapolis, Minnesota swallowed a 'silver' charm that came on a bracelet with a box of Reebok children's shoes. The charm was 99% lead and killed the child within a few days.[30] The Consumer Product Safety Commission recommends that jewelry contain no more than 0.06% lead. Did the buyers of this 'silver' charm know its composition? What management infrastructures could have prevented such toxins from being used in products? As society grows more knowledgeable about chemicals of concern, companies that are unaware of the presence of chemical hazards in the products and even packaging materials that they make and sell are vulnerable to liability.

Strategies for gaining more knowledge and control over one's chemical inventory include:

- Review of Material Safety Data Sheets (MSDSs)
- Full ingredient disclosure (to a set concentration level, that is, 1000 ppm or 100 ppm)
- Use of Restricted Substances Lists (RSLs, X-lists)
- Selection of preferred materials.

8.2.2.1 Examining Material Safety Data Sheets (MSDSs)

Examining Materials Safety Data Sheets (MSDSs) for chemical components and chemical hazards in product formulations is far less than ideal, but pragmatic. The Occupational

Safety and Health Administration (OSHA) requires all manufacturers and importers to obtain or develop a Material Safety Data Sheet for every chemical they produce or import. In addition to general information, MSDSs must list all ingredients at 1% or greater which have been determined to be 'health hazards'. All carcinogens (present at 0.1% or greater) must be listed if defined by International Agency for Research on Cancer (IARC) or the National Toxicology Program (NTP) as a carcinogen or potential carcinogen or regulated by OSHA as a carcinogen. Chemicals present at $<1\%$ need only to be reported if the release would exceed an OSHA Permissible Exposure Limit (PEL) or American Conference of Governmental Industrial Hygienists (ACGIH) Threshold Limit Value (TLV). The chemicals must be accompanied by the health hazards and appropriate OSHA or ACGIH limits. All ingredients that are considered a 'physical hazard' when present in the mixture must also be listed.[31]

The advantages to using MSDSs for chemical inventories are that MSDSs are publicly available and required by law. There is no extra cost or labour needed to supply them. The limitations to using MSDSs as a source of inventory data are that currently, in the United States, a standardized reporting format is not mandatory. MSDSs are not verified or audited and MSDS information can be incomplete and/or inaccurate. In addition, OSHA's definition of hazard does not include the broader scope of hazard used in green chemistry. Therefore, chemicals that are potential hazards from the green chemistry perspective but are not defined by OSHA as hazards, are not required to be identified on a MSDS.

8.2.2.2 Obtaining full ingredient disclosure

Obtaining full ingredient disclosure from suppliers is ideal but can be more challenging. Ideally, all chemicals intentionally added as constituents or present at a threshold of 100 parts per million (100 ppm or 0.01%) or higher should be identified by Chemical Abstract Service number (CAS number).* If it is not possible to identify all constituents to 100 ppm, then identifying all constituents intentionally added or present at a threshold of 1000 ppm (0.1%) or higher is a good start. Full ingredient disclosure allows for comprehensive hazard assessment of all chemicals in products and processes.

A growing number of household and commercial product suppliers are willing to provide full ingredient disclosure to third parties or to their customers under the protection of non-disclosure agreements (NDAs). The third party can review the formulation, perform a hazard assessment and report the results to the chooser or to the public without publicly revealing the exact formulation of the product which would be considered confidential business information. Examples of programs through which chemical formulations are fully disclosed and assessed to a third party in order to provide human and environmental health and safety data include the US Environmental Protection Agency's DfE Furniture Flame Retardancy Partnership and the CleanGredients^{SM} database projects described in more detail later in this chapter. In the case of the DfE Partnership, the US EPA reviews flame-retardant formulations and evaluates all constituents against a set of hazard endpoints used by the US EPA New Chemicals Program. The results of the assessment for each endpoint are publicly reported and summarized in an easy to read matrix (Figure 8.3). What is NOT reported is the per cent composition of the individual components and the exact identity

*This is not applicable to food and drugs which is beyond the scope of this discussion.

of certain chemicals that are considered proprietary.[32] This does not take away from the chemical choosers who seek the details of the hazard assessment for each chemical for a comprehensive set of hazard endpoints. In the CleanGredients^{SM} database, cleaning product ingredients may be identified by either CAS number and/or product name along with data for key environmental attributes. Data are validated and reported by a third party, in this case NSF International.[33]

8.2.2.3 Creating restricted substances lists (RSLs or X-Lists)

Another strategy for gaining better knowledge and subsequently control over one's inventory is to create restricted substances lists (RSLs or X-Lists) to identify chemicals that are prohibited or limited in certain product classes. At a minimum, RSLs may reflect chemicals that are regulated nationally or internationally. A more conservative, in the sense of precautionary, approach is to also include chemicals that are not regulated but that reflect public health concerns. We describe strategies for identifying chemicals of concern in the next step, Assessment and Evaluation. By establishing RSLs and requiring suppliers to sign legally binding agreements or affidavits confirming that their products meet the RSL requirements, a company can have some confidence about what is NOT present in the chemical product formulations. RSLs help to drive change down through the supply chain. While the use of RSLs does not reflect a comprehensive approach to greening chemical inventories, it does help to eliminate some of the 'bad actors'. RSLs send a signal to suppliers and set in motion the exploration of more favourable alternatives. An RSL program can be reinforced with random testing to ensure compliance by suppliers. This is a quasi-regulatory approach that sidesteps full ingredient disclosure. The Business for Social Responsibility (BSR) organization provides an example of an RSL that identifies chemicals of concern in the apparel industry.[34]

8.2.2.4 Identification and solicitation of preferred chemicals or materials

In contrast to specifying to suppliers what chemicals or materials are restricted, it is useful to specify exactly what chemicals and materials are desired. Once a material or chemical is well characterized, and it is considered benign with respect to human and environmental health, it can be added to a preferred or 'positive' list (i.e., P-list). For example, a textile manufacturer may source certified organic cotton, or polyester made with antimony-free catalysts, to develop a product line based on these fibres. Or a cleaning product formulator may seek bio-based solvents or rapidly biodegradable surfactants consistent with their product development objectives.

8.2.2.5 Metrics for the inventory step

The measures of success for the inventory step include tracking current and incoming chemical products and recording their constituents in an inventory tracking system. A high bar of achievement is demonstrated by obtaining full ingredient disclosure to a set concentration level (i.e., 1000 or 100 ppm). Additional tracking of products screened against RSLs or made with substances from preferred chemical lists can also support organizations in gaining better knowledge and control over their chemical inventories.

8.2.3 Assessment and evaluation (strategies, tools and metrics)

Once the inventory of chemicals in raw materials or products is complete, the next step on the path to integrating green chemical inventories is to assess and evaluate the associated hazards. Using the following rather simplistic distinctions helps to assess and evaluate hazards. Each of these distinctions brings with it a different set of strategies, tools and metrics.

- Chemicals of concern, that is 'known bads'
- Better alternatives
- Positively defined chemicals, that is 'known goods'.

8.2.3.1 *Identifying chemicals of concern*

Many individuals and organizations start down the path to green chemistry by first becoming aware of chemicals of concern in the products that they make or use. This awareness may come via regulations, internal priorities or external pressure from concerned stakeholders. Chemicals of concern satisfy two requirements:

1 there is sufficient knowledge of the chemical's hazardous properties, and
2 those properties are considered highly problematic for human and/or environmental health.

For example, we understand dioxins and DDT to be highly problematic because they are persistent, bioaccumulative, and toxic compounds (PBTs).[35] Chemicals of concern are often targeted for elimination by state, federal and/or international regulations and policies, and are usually identified on regulatory or policy lists. But, they may also include chemicals that are of public concern for particular product classes. For example, nonylphenol ethoxylate surfactants are not regulated in the United States, but they are prohibited in environmentally preferable product recognition programs due to concerns about certain degradation products which can be endocrine disrupting and more aquatically toxic and persistent than the parent compound.[36] Manufacturers who value green chemistry will seek to avoid chemicals that are associated with human or environmental health issues regardless of regulatory status. The process of identifying chemicals of concern can support an existing RSL.

In general, there are two pragmatic approaches to identifying chemicals of concern. One is to identify chemicals that are 'listed' on regulatory, policy or other lists based on hazardous properties of concern and the other is to identify chemicals with specific attributes of concern as determined by test methods or other criteria. By law, organizations must comply with chemical regulations, but an increasing number of companies are seeking to move beyond compliance to reduce or eliminate the use of unregulated chemicals that are identified as hazards. Selecting chemicals of concern is a subjective process that depends in part on the particular product class and the company's values. It also depends on the perceived importance of specific hazards and the perceived value or authority of certain lists. For example, electronics manufacturers who sell primarily in the US may seek to eliminate chemicals of concern targeted by international regulations such as the Restriction of Hazardous Substances (RoHS) Directive (2002/95/EC); manufacturers of cleaning products may prioritize chemicals that are water and air pollutants, or skin, eye or respiratory irritants

Box 8.1 Examples of chemical hazard categories

Carcinogens
Mutagens
Developmental toxins
Reproductive toxins
Endocrine disrupting substances
PBTs (in addition to those listed by governmental organizations)
- vPvBs (very persistent, very bioaccumulating) substances
- vPvTs or vBvTs (very persistent and toxic or very bioaccumulative and toxic) substances

Physical hazards
- Flammable
- Explosive
- Corrosive, etc.

Substances with very high acute mammalian toxicity
Substances with very high acute or chronic aquatic toxicity
Neurotoxins
Skin or respiratory sensitizers
Asthmagens
Skin or eye irritants
Cardiovascular or blood toxicants
Internal organ toxicants (e.g., liver, kidney)
Volatility
Substances with strong greenhouse gas potential
Chemical classes and substances of concern based on public health and public concerns
- Polyvinyl chloride (pvc)
- Brominated flame retardants
- Bisphenol A
- Phthalates

and sensitizers; while manufacturers of toys for young children may choose to eliminate chemicals that are suspected endocrine disrupting substances.

In addition to regulated chemicals, examples of hazard categories that may be selected for identifying chemicals of concern include but are not limited to Box 8.1.

Appendix 1 contains a number of references and links to chemical lists and resources to help identify chemicals of concern. In addition, it may be useful to review the criteria used to classify substances via EU's Dangerous Substance Directive (67/548/EEC),[37] Environment Canada's Ecological Categorization of Substances on the Domestic Substances List,[38] or Health Canada's Categorization of Substances on the Domestic Substances List[39] when establishing your organization's criteria for selecting chemicals of concern.

Once the hazard categories are identified, organizations may use a list-based or a criteria-based approach to identifying chemicals of concern. Using a list-based approach, an organization must first identify which regulatory, policy or hazard lists to consider. For example, carcinogens are identified via numerous national and international governmental and non-governmental organizations and they are categorized based on the weight of evidence and likelihood of carcinogenicity. Some carcinogens are classified as 'known' while others are classified as 'reasonably anticipated to be human carcinogens'[40] or even 'potential' carcinogens.[41] Some carcinogen lists are compiled from other sources (e.g., California

Proposition 65) while others are based on research and reports performed and/or assessed by the authoritative body (e.g., National Toxicology Program of the National Institutes of Health). Examples of carcinogenicity source lists (with a US focus) are:

- California Proposition 65
- International Agency for Research on Cancer (IARC)
- National Toxicology Program (NTP)
- Scorecard List of recognized carcinogens
- Scorecard List of suspected carcinogens
- American Conference of Governmental Industrial Hygienists Carcinogens List (ACGIH)
- EPA Integrated Risk Information System (IRIS)
- National Institute of Occupational Health and Safety Carcinogens List (NIOSH)
- EU Consolidated List of C/M/R/-Substances
- Occupational Safety and Health Administration
- US EPA Water Disinfection By-Products with Carcinogenicity Estimates.

Organizations will need to determine which lists and sublists to include by considering the credibility of the list source and the extent to which they seek to be inclusive of chemicals with different levels of potential concern.

Chemicals of concern can also be defined based on hazard criteria that apply to chemicals regardless of whether or not they are on regulatory or policy lists. Below we describe several tools that can help businesses to identify chemicals of concern through modelling programs and by tracking regulatory lists around the world.

The PBT Profiler. In addition to state, national and international PBT lists, there is a user-friendly software tool, called the PBT Profiler that will model the persistence, bioaccumulation potential and aquatic toxicity of organic chemicals based on chemical CAS number or molecular structure.[42] The PBT Profiler was designed to be an easy to use, widely available, no-cost tool to screen chemicals lacking experimental data in order to help identify pollution prevention opportunities. The methodology the PBT Profiler uses was developed by government, academic, and private-sector researchers over the past 20 years and represents some of the best techniques currently available for estimating persistence, bioaccumulation potential and aquatic toxicity. There are some limitations to the PBT Profiler, however, that are described on the website. It is best suited for neutral organic chemicals and is not well suited for inorganics, polymers or surfactants.[43]

There are a number of commercially available database tools that support regulatory compliance and the identification of chemicals of concern based on regulatory or policy lists.

Ariel™ WebInsight. Ariel™ WebInsight is a subscription-based online compliance management tool with an easy-to-use interface and robust search, query, reporting and analysis features. The tool provides access to current, accurate, comprehensive global regulatory content containing more than 800 regulatory lists covering more than 75 countries; full-text repositories of legislation, international transportation data, as well as chemical property and hazard data. This data aids companies in making critical business decisions and determining the impact of regulatory changes on the products they manufacture. Quick access to this information improves resource effectiveness in performing routine tasks throughout the lifecycle of a product (formulation, registration, handling, storage, distribution and disposal).[44]

ChemAdvisor and the List Of LIsts. ChemAdvisor and the LOLI (List Of LIsts) Database contains over 2700 regulatory lists from around the world, including environmental, health and safety, international, inventories, and reference material necessary for Material Safety Data Sheet preparation.[45]

Green Product Selector™. Dolphin Software, Inc. develops software tools that support MSDS management and regulatory compliance but has recently developed applications that also support movement toward green chemical inventories.[46] The Green Product Selector™ (GPS) and related software tools are based on approximately 60 regulatory and policy lists, selected for their relevance to hazard as defined in green chemistry. The system is highly configurable so that additional lists and criteria can be included for more in-depth screening evaluation. The GPS software evaluates chemical product constituents by CAS number against the hazard lists. A flexible algorithm weights the different lists and scores chemicals and the associated products in which they are found based on their concentration and presence on certain lists. For example, a high score is ascribed to chemicals that are known carcinogens as defined by the National Toxicology Program. Product cost information can also be included so that the system will allow users to compare products in over 700 use categories by cost, environmental hazard and human health hazard using the list screening infrastructure. This information can be graphically displayed to help companies identify and select safer, more affordable chemical products within a use class (i.e., comparing glass cleaners to glass cleaners and adhesives to adhesives) while eliminating the most hazardous and expensive ones. Comparing products within a use class based on hazard only is reasonable because such products have similar exposure pathways. Figure 8.1 illustrates a sample output from the Green Product Selector™. Additional views allow for comparison of combined health and environmental hazard versus cost and access to MSDS, relevant hazard lists and non-proprietary formulation information.

The Green Product Selector™ helps users to screen out products of concern, to reduce their total chemical product inventory, to set purchasing screens, and to select products with the least cost and best relative human and environmental health profiles based on the selected hazard criteria. A laboratory version of the Green Product Selector (LGPS) is under development. This product will allow chemists to create new molecules, reactions and syntheses and to continually move towards greener methods utilizing all 12 Green Chemistry Principles. Additionally, it will allow different reaction and synthetic pathways to be compared, to facilitate awareness of the greenest pathways. The software is being developed to allow for customization so that each chemist can prioritize the importance of each principle for a particular reaction or synthesis.

As noted earlier, the obvious limitations of a system based on information reported on MSDSs and hazard lists are that MSDS information can be incomplete and inaccurate and screening hazard lists does not reflect a complete hazard assessment. However, the GPS platform is designed so that additional formulation information – including full ingredient disclosure – can be added to product MSDS information. And fields for chemical

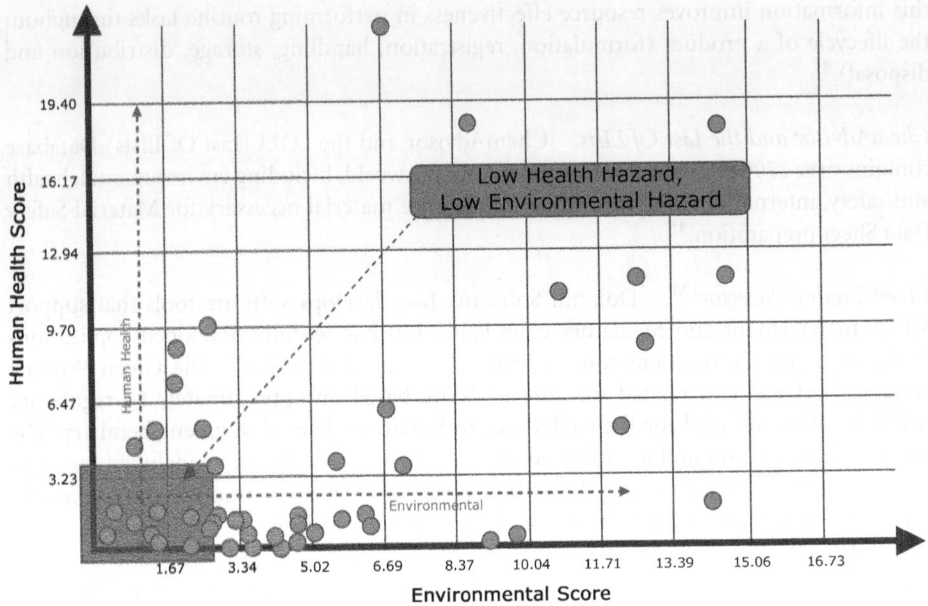

Figure 8.1 The Green Product Selector[TM]. Reproduced with permission of Mark L. Wysong, Dolphin Software, Inc.

and product level data may be added so that hazard list screening can be supplemented with actual test data. For example, chemical choosers may want to know if the product is irritating or sensitizing and/or the exact values for acute mammalian toxicity, acute aquatic toxicity and/or biodegradability.

8.2.3.2 Comparing alternatives

Whether voluntarily or not, as companies phase out chemicals of concern, they are subsequently faced with the challenging task of finding alternative chemicals or processes to serve the same desired functions. Simply because a chemical is not a known hazard does not make it benign. To avoid going from the frying pan to the fire so to speak, it is necessary to assess and compare alternatives. In this section we describe several tools and approaches for assessing the human and environmental health and safety aspects of chemicals. They include:

- US EPA DfE Alternatives Assessment protocol
- The Green Screen for Safer Chemicals
- SC Johnson & Son, Inc. GreenList® Criteria
- US EPA DfE Formulator Program
- CleanGredients[SM].

The US EPA's DfE Program. The EPA's DfE Program,[47] housed in the EPA's Office of Pollution Prevention and Toxics, was created in 1992 and is one of the EPA's premier partnership programs, working with industry sectors to compare and improve the human health and environmental risks, as well as the performance and cost of existing and alternative products, processes and practices. The DfE Program promotes pollution prevention and other risk reduction activities in industrial sectors. To accomplish this mission:

- DfE forms partnerships with industry and other interested parties to develop information on environmental and human health impacts, performance, and cost of cleaner technologies and approaches.
- DfE disseminates information to help businesses design and redesign cost-effective products and processes that are cleaner and safer for workers and the public.

Through partnerships EPA supports organizations in gathering and assessing information that they may not be able to gather independently – whether due to cost or because the information is considered proprietary. Chemical manufacturers share proprietary formulations with DfE through these voluntary partnerships because of the way the hazard information is handled and presented. DfE receives full ingredient disclosure and makes the results of the hazard assessments publicly accessible but protects proprietary formulation information by not revealing exact percentages of constituents and masking the identity of certain chemicals.

A good example of the many successful DfE Partnerships is the Furniture Flame Retardancy Partnerhip. Pentabromodiphenylether (PentaBDE) was the primary flame retardant used in low density, flexible polyurethane furniture foam. Due to concerns over its use and the fact that the chemical was found widespread in the environment and in human tissue and breast milk, PentaBDE was voluntarily phased out of production by US manufacturers in January 2004. The industry needed alternatives in order to meet furniture flame retardancy requirements, but did not have the human and environmental health and safety information needed in order to compare the alternatives. DfE worked with the furniture manufacturers, foam manufacturers, and flame-retardant chemical suppliers along with governmental and environmental groups to evaluate possible alternatives.

Fourteen formulations of chemical alternatives were submitted to EPA under confidentiality and they were assessed based on numerous human health and ecotoxicity endpoints in addition to bioaccumulation potential and environmental persistence. They were also screened for potential exposure to workers, users and the aquatic environment. Where data gaps existed, EPA experts used models and chemical analogs to estimate the hazard for a particular endpoint. The literature and test data reviews were published in the final report, 'Environmentally Preferable Options for Furniture Fire Safety: Low Density Furniture Foam'.[32] In addition, each hazard endpoint was ranked with a concern level (High, Moderate or Low) based on the criteria used by the EPA's New Chemicals Program to rate the concern level of new chemicals submitted under the Toxic Substance Control Act (TSCA). As seen in Figure 8.2, where the hazard endpoint rankings are bold, the value is based on experimental data. Where the hazard endpoints are presented in italic font, the value is estimated based on models or chemical analogs. In this way, detailed hazard information was summarized and presented in a clear and concise format.

The process did not rank the overall risk of the flame retardants; but rather it provided the industry with the information necessary to make informed decisions when choosing appropriate alternatives for their particular use.

L = Low hazard concern
M[a] = Moderate hazard concern
H = High hazard concern
L, *M*[a], or *H* = Endpoint assigned using estimated values and professional judgment (Structure Activity Relationships)

N = No
Y = Yes
P = Yes for pure chemical

*Ongoing studies may result in a change in this endpoint
▲Persistent degradation products expected[b]

Company	Chemical	% in Formulation[c]	Human Health Effects							Ecotoxicity		Environmental		Potential Routes of Exposure							Reactive or Additive?
														Worker			General Population				
			Cancer Hazard	Skin Sensitizer	Reproductive	Developmental	Neurological	Systemic	Genotoxicity	Acute	Chronic	Persistence	Bioaccumulation	Inhalation	Dermal	Ingestion	Inhalation	Dermal	Ingestion	Aquatic	
Supresta	AB053																				
	Tris(1,3-dichloro-2-propyl)Phosphate CAS # 13674-87-8		M	L	M	M	L	M	M	M	M	M	*L*	N	Y	Y	N	Y	Y	Y	Additive
Supresta	AC003																				
	Proprietary I Organic phosphate ester	92–99%	*L*	L	L	L	*L*	*M*	L	H	H	H	L	P	Y	Y	N	Y	Y	Y	Additive
	Triphenyl Phosphate CAS # 115-86-6	1–8%	*L*	L	L	L	L	M	L	H	*H*	L	L	Y	Y	Y	Y	Y	Y	Y	Additive
Supresta	AC073																				
	Triphenyl Phosphate CAS # 115-86-6	38–48%	*L*	L	L	L	L	M	L	H	*H*	L	L	Y	Y	Y	Y	Y	Y	Y	Additive
	Proprietary J Aryl phosphate	40–46%	*L*	*L*	L	*L*	L	M	M*	*L*	*H*	L	*L*	Y	Y	Y	Y	Y	Y	Y	Additive
	Proprietary K Aryl phosphate	12–18%	*L*	*L*	*L*	*L*	*L*	*M*	*L*	*L*	*L*	*L*	*L*	P	Y	Y	N	Y	N	N	Additive
	Proprietary L Aryl phosphate	1–38%	*L*	*L*	*L*	*L*	*L*	*M*	*L*	*L*	*L*	*L*	*L*	P	Y	Y	N	Y	N	N	Additive
Supresta	Fyrol FR-2																				
	Tris(1,3-dichloro-2-propyl)phosphate CAS # 13674-87-8	99%	M	L	M	M	L	M	M	M	M	M	*L*	N	Y	Y	N	Y	Y	Y	Additive

[a] The moderate designation captures a broad range of concerns for hazard.
[b] More information on degradation products can be found in Sections 4.1.1 and 5.1.
[c] Chemical concentrations are listed in descending order; only chemicals with concentrations greater than 1% in the formulation were evaluated.

Figure 8.2 Excerpt from the US EPA Furniture Flame Retardants Hazard Assessment Matrix. Reproduced with permission of Clive Davies, US Environmental Protection Agency.

The Green Screen for Safer Chemicals. The Green Screen for Safer Chemicals (The Green Screen) is a publicly available chemical hazard assessment and evaluation method developed by Clean Production Action that supports alternatives assessment through the identification of chemicals that are safer for human health and the environment.[48] It builds on and advances the DfE chemical assessment approach and information presentation format. By expanding the number and scope of hazard endpoints and integrating them into a decision support logic based on judgement about the significance of hazard endpoints – both individually and in combination with other endpoints, the Green Screen answers a need for a publicly accessible, scientifically based hazard assessment method and for agreement on what defines green. It is designed to inform decision making by businesses, governments, and individuals concerned with the risks posed by chemicals, to advance the development of green chemistry and to be a building block on the path to sustainable product design and sustainable material flows. The development of the Green Screen method involved three major steps:

1 Establish the hazard endpoints that will be used to evaluate a chemical.
2 Define the levels of concern – high, moderate, and low – for each hazard.
3 Specify the hazard criteria for each of the four benchmarks.

All of the hazard and benchmark criteria developed for the Green Screen are presented in the report, along with information on government and other precedents for classification that were used to help establish the thresholds. The list of hazard categories and threshold values used to define levels of concern in the Green Screen are presented in Table 8.1.

The Green Screen defines four benchmarks on the path to safer chemicals:

Benchmark 1 characterizes chemical products as 'Avoid – Chemical of High Concern';
Benchmark 2 characterizes chemical products as 'Use but Search for Safer Substitutes';
Benchmark 3 characterizes chemicals as 'Use but Still Opportunity for Improvement'; and
Benchmark 4 characterizes chemicals as 'Prefer – Safer Chemical'.

Each benchmark consists of a set of hazard criteria that encompass a combination of hazards and threshold values that a chemical along with its known and predicted breakdown products (environmental degradation products and metabolites) must pass. To progress from Benchmark 1 to Benchmark 2, the chemical (along with its breakdown products and metabolites) must pass all the criteria specified under Benchmark 1. Similarly to progress from Benchmark 2 to Benchmark 3 and from Benchmark 3 to Benchmark 4, the chemical (along with its breakdown products and metabolites) must pass all criteria specified under each respective benchmark. The criteria become increasingly more demanding for environmental and human health and safety for each benchmark, with Benchmark 4 representing the safest chemical. Figure 8.3 details the hazard criteria a chemical (along with its known or predicted degradation products and metabolites) needs to pass for each benchmark.

Table 8.1 Threshold values for each organic chemical hazard endpoint included in the Green Screen v1.0.

Hazard	Very high (v)	High (H)	Moderate (M)	Low (L)
Environmental fate				
Persistence – P (half-life in days)[a]	• Soil or sediment >180 days; or • Water >60 days	• Soil, sediment >60 to 180 days; • Water >40 to 60 days; or • Potential for long-range environmental transport	• Soil, sediment 30 to 60 days; or • Water 7 to 40 days	• Soil, sediment <30 days; • Water <7 days; or • Ready bio-degradability
Bioaccumulation potential – B[a]	• BCF/BAF >5000; or • Absent such data, log K_{ow} > 5	• BCF/BAF >1000 to 5000; • Absent such data, log K_{ow} > 4.5–5; or • Weight of evidence demonstrates bioaccumulation in humans or wildlife	• BCF/BAF 500 to 1000; • Absent such data, log K_{ow} > 4–4.5; or • Suggestive evidence of bioaccumulation in humans or wildlife	• BCF/BAF <500; or • Absent such data, log K_{ow} < 4
Ecotoxicity				
Acute aquatic toxicity[a]		• $LC_{50}/EC_{50}/IC_{50}$ <1 mg/l; or • GHS Category 1	• $LC_{50}/EC_{50}/IC_{50}$ 1–100 mg/l; or • GHS Category 2 or 3	• $LC_{50}/EC_{50}/IC_{50}$ > 100 mg/l
Chronic aquatic toxicity[a]		• NOEC <0.1 mg/l; or • GHS Category 1	• NOEC 0.1–10 mg/l; or • GHS Category 2, 3 or 4	• NOEC >10 mg/l
Human health				
Carcinogenicity[b]		• Evidence of adverse effects in humans; • Weight of evidence demonstrates potential for adverse effects in humans; • NTP known or reasonably anticipated to be human carcinogen; • OSHA carcinogen; • California Prop 65; • IARC Group 1 or 2A; • EU Category 1 or 2; or • GHS Category 1A or 1B	• Suggestive animal studies; • Analog data; • Chemical class known to produce toxicity; • IARC Group 2B; • EU Category 3; or • GHS Category 2	• No basis for concern identified or • IARC Group 3 or 4

Mutagenicity/Genotoxicity[b]	• Evidence of adverse effects in humans; • Weight of evidence demonstrates potential for adverse effects in humans; • EU Category 1 or 2; or • GHS Category 1A or 1B	• Suggestive animal studies; • Analog data; • Chemical class known to produce toxicity; • EU Category 3; or • GHS Category 2	No basis for concern identified
Reproductive toxicity[b]	• Evidence of adverse effects in humans; • Weight of evidence demonstrates potential for adverse effects in humans; • NTP Center for the Evaluation of Risks to Human Reproduction; • California Prop 65; • EU Category 1 or 2; or • GHS Category 1A or 1B	• Suggestive animal studies; • Analog data; • Chemical class known to produce toxicity; • EU Category 3; or • GHS Category 2	No basis for concern identified
Developmental toxicity[b]	• Evidence of adverse effects in humans; • Weight of evidence demonstrates potential for adverse effects in humans; • NTP Center for the Evaluation of Risks to Human Reproduction; or • California Prop 65	• Suggestive animal studies; • Analog data; or • Chemical class known to produce toxicity	No basis for concern identified
Endocrine disruption[b]	• Evidence of adverse effects in humans; or • Weight of evidence demonstrates that mechanisms of action lead to adverse effects	• Suggestive animal studies; • Analog data; • Chemical class known to produce toxicity; • EU Draft List – Category 1 or 2; or • Japanese list	No basis for concern identified

Table 8.1 Continued.

Hazard	Very high (v)	High (H)	Moderate (M)	Low (L)
Neurotoxicity[b]		• Evidence of adverse effects in humans; or • Weight of evidence demonstrates potential for adverse effects in humans	• Suggestive animal studies; • Analog data; or • Chemical class known to produce toxicity	No basis for concern identified
Acute toxicity (oral, dermal or inhalation)		• LD_{50} <50 mg/kg bodyweight (oral); • LD_{50} <200 mg/kg bodyweight (dermal); • LC_{50} <500 ppm (gas); • LC_{50} <2.0 mg/l (vapour); • LC_{50} <0.5 mg/l (dust or mist); • US EPA Extremely Hazardous Substance List; or • GHS Category 1 or 2	• LD_{50} 50–2000 mg/kg bodyweight (oral); • LD_{50} 200–2000 mg/kg bodyweight (dermal); • LC_{50} 500–5000 ppm (gas); • LC_{50} 2–20 mg/l (vapour); • LC_{50} 0.5–5 mg/l (dust or mist); or • GHS Category 3 or 4	No basis for concern identified
Corrosion/irritation of the skin or eye		• Evidence of irreversible effects in studies of human populations; • Weight of evidence of irreversible effects in animal studies; or • GHS Category 1 (skin or eye)	• Evidence of reversible effects in humans or animals; • GHS Category 2 or 3 – skin irritation; or • GHS Category 2A or 2B – eye	No basis for concern identified
Sensitization of the skin or respiratory system		• Evidence of adverse effects in humans; • Weight of evidence demonstrates potential for adverse effects in humans; • GHS Category 1 – (skin or respiratory); or • Positive responses in predictive Human Repeat Insult Patch Tests (HRIPT) (skin)	• Suggestive animal studies; • Analog data; or • Chemical class known to produce toxicity	No basis for concern identified

Immune system effects	• Evidence of adverse effects in humans; or • Weight of evidence demonstrates potential for adverse effects in humans	• Suggestive animal studies; • Analog data; or • Chemical class known to produce toxicity	No basis for concern identified
Systemic toxicity/organ effects (via single or repeated exposure)	• Evidence of adverse effects in humans; • Weight of evidence demonstrates potential for adverse effects in humans; • GHS Category 1 – organ/systemic toxicity following single or repeated exposure	• Suggestive animal studies; • Analog data; • Chemical class known to produce toxicity; • GHS Category 2 or 3 single exposure; or • Category 2 repeated exposure	No basis for concern identified
Physical/chemical properties			
Explosive	• GHS Category: Unstable Explosives or Divisions 1.1, 1.2 or 1.3	• GHS Category: Divisions 1.4, 1.5	No basis for concern identified
Flammable	• GHS Category 1 – Flammable gases; • GHS Category 1 – Flammable aerosols; or • GHS Category 1 or 2 – Flammable liquids	• GHS Category 2 – Flammable Gases; • GHS Category 2 – Flammable Aerosols; or • GHS Category 3 or 4 – Flammable Liquids	No basis for concern identified

[a] Experimental data are preferred. Absent experimental data, values based on structure activity relationships are sufficient.
[b] Priority human health effect.

Abbreviations: BAF = bioaccumulation factor; BCF = bioconcentration factor; EC_{50} = median effective concentration; EU = European Union; GHS = Globally Harmonized System of Classification and Labelling of Chemicals; IARC = International Agency for Research on Cancer; IC_{50} = mean inhibitory concentration; LC_{50} = median lethal concentration: the concentration at which 50% of test animals diec after exposure; LD_{50} = median lethal dose: the dose at which 50% of test animals died during exposure; log K_{ow} = log octanol water partition coefficient; NOEC = no observec effect concentration; NTP = National Toxicology Program; OSHA = Occupation Safety and Health Administration.

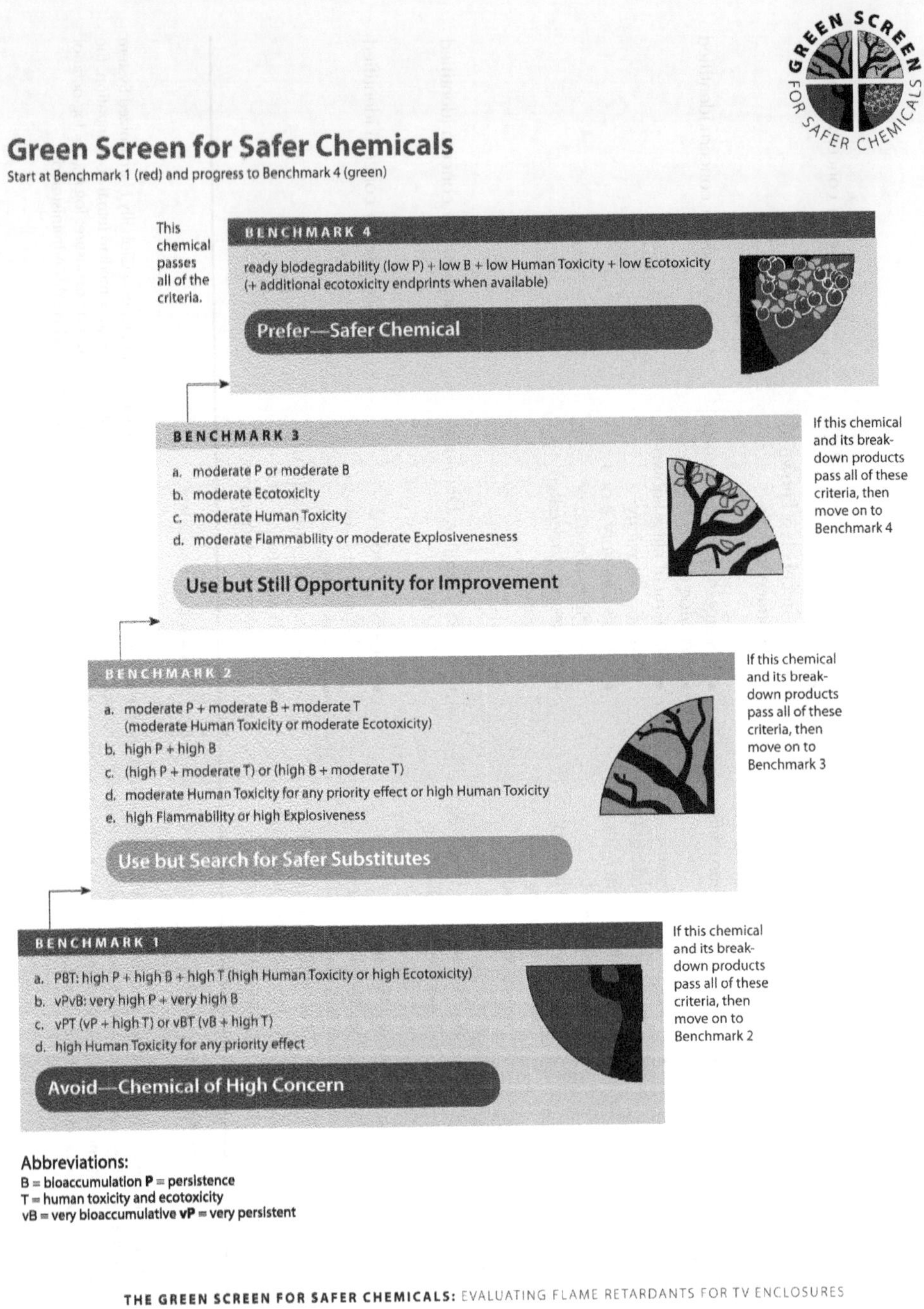

Figure 8.3 The Green Screen for Safer Chemicals Benchmarks. Reproduced with permission of Clean Production Action.

Benchmark 1: Avoid – High Concern. The criteria for Benchmark 1 are:

1(a) PBT: high P + high B + high T[†] (high human toxicity or high ecotoxicity); or
1(b) vPvB: very high P + very high B; or
1(c) vPT (vP + high T) or vBT (vB + high T); or
1(d) high human toxicity for any priority effect.[‡]

The chemicals that stop at Benchmark 1 encompass the hazard criteria that are leading governments to restrict the use of chemicals: high/very high persistence (P/vP), high/very high bioaccumulation (B/vB) and/or high toxicity (T). The European Union's new REACH legislation, for example, targets chemicals that are PBTs, vPvBs or highly toxic to humans (carcinogenic, mutagenic, reproductive toxicant, or endocrine disruptor). Chemicals that stop at Benchmark 1 are those for which any release into the environment and exposure to humans is viewed as problematic and not amenable to management through pollution control measures.

Benchmark 2: Use *but* Search for Substitutes. (See criteria for Benchmark 2 in Figure 8.3.)

Benchmark 2 continues the emphasis on persistence, bioaccumulation, and toxicity, but at lower threshold values. In addition, Benchmark 2 includes flammability and explosiveness. It is anticipated that many chemicals will not move past Benchmark 2 because of the broad scope of hazards and challenging threshold values included in the Green Screen.

Benchmark 3: Use *but* Still Opportunity for Improvement. (See criteria for Benchmark 3 in Figure 8.3.)

The hazard criteria for Benchmark 3 define chemicals that are on the cusp of being benign; they have some hazard characteristics of modest concern, but no characteristics of high concern.

Benchmark 4: Prefer – Safer Chemical. Only organic chemicals with low inherent toxicity to humans and wildlife, that do not bioaccumulate, and rapidly and completely degrade to benign degradation products or metabolites reach Benchmark 4. These are chemicals that would meet the principles of green chemistry that relate to hazard.

The Green Screen includes a chemical's breakdown products, that is, metabolites and degradation products in a hazard assessment because they may be more hazardous than the parent compound. The final benchmark for a parent chemical is the lowest benchmark achieved by either it or its breakdown products. For example, if a parent chemical achieved Benchmark 2, but its breakdown product achieved Benchmark 1, the final benchmark for the parent chemical is Benchmark 1. Thus the degradation product or metabolite of a chemical is considered equivalent to the parent compound with respect to its benchmark unless it can be demonstrated that the breakdown product is insignificant (i.e., transient, not actually formed, etc.).

The Green Screen is a much needed method for comparing and benchmarking chemicals based on hazard. Green Screen assessments can be incorporated into broader alternatives

[†] Toxicity – 'T' = human toxicity and ecotoxicity.

[‡] The 'priority effects' are carcinogenicity, mutagenicity, reproductive or developmental toxicity, endocrine disruption and neurotoxicity. 'Human toxicity' is broader than priority effects, including: acute toxicity, systemic toxicity (organ effects), immune system effects and skin/eye/respiratory damage as well as the priority effects. And 'toxicity' as 'T' includes both human toxicity and ecotoxicity.

assessment frameworks that include additional sustainability criteria and non-chemical alternatives.

SC Johnson's GreenList™. SC Johnson & Son, Inc. (SC Johnson) won a Presidential Green Chemistry Award in 2006 for their GreenList™ raw material evaluation system. The patented system[49] was developed to help SC Johnson reduce the environment and human health impact of their products beyond regulatory requirements, to set internal goals and to track their progress. GreenList™ is a shining example of how hazard assessment can be used as a metric for product and inventory evaluation.

GreenList™ was developed by SC Johnson by first dividing their chemical inventory into chemical classes, which include over 90% of their raw materials:

- Surfactants
- Solvents
- Propellants
- Insecticides
- Resins
- Chelants
- Preservatives
- Waxes
- Fragrances
- Inorganic acids and bases
- Non-woven fabrics
- Packaging.

For each chemical class, a set of 4–7 key attributes were identified that can be used to differentiate a chemical from others in the same class. These may include biodegradability, aquatic toxicity, sensitization potential, renewable resource derived, and so forth. Each attribute was selected to be:

- Meaningful
- Discriminating
- Readily available
- Or able to address other significant concerns.

Criteria were then developed for each of the 4–7 key attributes in order to rate a chemical as Best (3), Better (2), Acceptable (1) or as a Restricted Use Material (RUM) (0). For example, for the attribute, acute oral toxicity, an $LD_{50} > 2000$ mg/kg will earn a chemical a 3, while an LD_{50} of 500–2000 mg/kg a 2, and $LD_{50} < 500$ mg/kg a 1. The scores of the different attributes are then averaged to give an overall rating for a chemical within a chemical class. Table 8.2 illustrates the key attributes and scoring system for chemicals used as chelating agents.

The GreenList™ system also includes a kind of 'wild card'. When a chemical is found on a regulatory list or banned in a country, it is considered to have 'Other Significant Concerns' (OSCs). OSCs are applied to the overall rating of a chemical – not its individual attributes – to reduce the overall score by one point. For example chemicals have OSCs if they are:

- Classified as a PBT/POP
- Listed as an endocrine disrupter

Table 8.2 GreenList criteria for chelating agents.

	Class 1: Acceptable	Class 2: Better	Class 3: Best
Aquatic toxicity (LC/EC_{50})	<1 mg/l	≥1 mg/l (<3 species tested)	≥1 mg/l (>3 species tested)
Ultimate biodegradability (OECD 301)	<60% within 28 days	>60% within 28 days	>60% within 10 days window
EU classification	N_1 R50, N_1 R50–53, N_1 R51–53, R52–53, R52 or R53 and/or not both readily biodegradable and aquatic toxicity >1 mg/l	Readily biodegradable and aquatic toxicity >1 mg/l	Readily biodegradable and aquatic toxicity >100 mg/l
Supplier	Acceptable supplier rating	Environmental management system in place and >25% green chemistry preferred	ISO 14001 and 75% green chemistry preferred

- Listed as a carcinogen or on Proposition 65
- Listed as a reproductive toxin or on Proposition 65
- Banned in one or more countries
- Banned unofficially by trade associations.

Using this system, SC Johnson scores both raw materials and formulated products, tracks the score of all raw material purchases and measures the overall percentage of materials that fall into each rating category, that is considered 'Best' or 'RUM'. The scores are used to set company goals, as well as to reward the formulators and chemists who use the system to improve the environmental and human health profiles of individual products and product lines.

The GreenList™ system is available to interested users via a free licensing agreement with SC Johnson.[50]

8.2.3.3 Positively defining chemicals 'known goods'

As organizations become more proactive in promoting green chemistry in their products and processes, they move from identifying chemicals of concern; to assessing and evaluating alternatives; to positively defining the desired attributes of the chemicals they seek. 'Defining what "good" looks like' is necessary for advancing green chemical inventories. We have observed two broad approaches to defining what 'green/good' looks like for chemicals. The first approach defines green/safe/healthy chemicals as those that are *absent known hazard.* The second approach defines chemicals that are '*best in class*'.

Defining positive as the absence of hazard. Using Clean Production Action's Green Screen, the ideal chemicals are defined as those that have low hazard for all relevant hazard endpoints as exemplified by Benchmark 4. While it is anticipated that the majority of chemicals will not currently reach Benchmark 4, it is still important to have a vision of what good (and better) looks like in order to guide the development of new molecules and the selection of chemical alternatives.

Best in class approach. The best in class approach is based on a continuum in order to compare chemicals or materials within a particular functional class such as surfactants or chelating agents used for cleaning products.

Both approaches are useful and they are also complementary because it is important to know where a chemical that may be best in its class falls out with respect to hazard. For example, a surfactant that is best in its class will be rapidly biodegradable, but most surfactants have some aquatic toxicity because they are surface active. However, surfactants as a class are typically close to the 'green' end of the hazard spectrum because they tend to have low hazard ratings for most other endpoints. It is also possible to have chemicals that are best in their class but that are still problematic. For example, some dioxin congeners are less toxic than others but one would not presume that a dioxin congener that is best in its class is 'green'. Concurrent use of the best in class approach with the absence of hazard approach is also important because it drives continual advancement within a class toward the ideal green chemistry. Once innovation occurs and a chemical or product is developed that meets the same or better performance criteria with lower hazard, what was once considered best in class shifts.

Below we describe three examples of frameworks that use the 'Best in Class' approach including another US EPA DfE Program called the Formulator Program, the CleanGredients^{SM} database and SC Johnson's GreenList^{TM}.

DfE Formulator Program. The DfE Formulator Program is a Partnership program that encourages and assists formulators in designing products with more positive environmental and health profiles than conventional products. DfE can provide formulators with information on chemical characteristics and toxicities of raw materials and additives. DfE partners enjoy Agency recognition, including use of the DfE logo on products that are deemed as currently best in their class.

The DfE review team screens every ingredient in the product for potential human health and environmental effects based on currently available information, predictive models and expert judgement. The formulator is informed when hazards are identified and recommendations for alternatives are made. The formulator may then reformulate the product and submit the reformulation for review. If the DfE review team confirms that the product contains only those ingredients that pose the least concern among chemicals in their class, the formulator is awarded recognition and permitted to use the DfE logo on their product for three years. DfE uses a continuum approach, pushing toward the ideal and raising the bar after each innovation. For example, once effective floor finishes were made without zinc, this expectation was applied to all floor finishes reviewed from that time forward. More information about the DfE review criteria, Partners and recognized products is available on the DfE website.[51,52]

CleanGredients^{SM}. CleanGredients^{SM} is an online database of institutional and industrial (I&I) cleaning ingredients to support green product formulation.[53] CleanGredients^{SM} aligns broad environmental and human health goals with the cleaning product industry's business objectives to support the formulation of products with human and environmental health benefits, whether to meet corporate internal objectives, regulations, voluntary product recognition programs, or national and international eco-labels. CleanGredients^{SM} is designed to:

- help formulators identify ingredients that have potential environmental and human health and safety benefits, and
- provide opportunity for manufacturers and producers of cleaning ingredients to showcase their ingredients with potential environmental and human health and safety benefits.

CleanGredients^{SM} is a searchable database that allows users to search for candidate chemicals by physical properties and functionality. Search results are displayed by product trade name and/or CAS number along with verified human and environmental health and safety information, thus allowing formulators to consider health and safety implications at the early stages of the product design process. The first module for CleanGredients^{SM} comprises surfactants and solvents for hard surface and carpet cleaners, laundry and hand dish soap. Modules for fragrances and chelating agents are under development. Additional ingredient classes such as preservatives and colorants will be added as the attributes and criteria are established.

In order to list surfactants on the CleanGredients^{SM}, full ingredient formulation information along with data for aquatic toxicity and biodegradability are submitted to a qualified third party reviewer (currently, NSF International) for review. At a minimum, surfactants listed in CleanGredients^{SM} must be ultimately biodegradable with no degradation products of concern; similar to the requirements of the EU Detergent Directive. NSF's reviews are carried out under confidentiality, providing verification of claims for the key ingredient attributes for surfactants without compromising proprietary formulations.

NSF also evaluates surfactant products against the DfE Screen for Surfactants (the DfE Screen), a set of criteria defined by the US. EPA DfE Program that designates surfactants with especially positive environmental characteristics defined in Table 8.3.[54] The requirements for listing and the DfE Screen represent two levels by which surfactants (and in the future, other ingredients classes) can be evaluated. CleanGredients^{SM} uses the 'best in class' approach by identifying the most desirable criteria for chemicals within an ingredient class.

The CleanGredients^{SM} development model engages a multi-stakeholder coalition of participants from government entities, environmental organizations, industry associations, cleaning product formulators and distributors, and chemical manufacturers and suppliers, working to shift the direction of industrial and institutional (I&I) cleaning product

Table 8.3 The DfE screen for surfactants.

Aquatic toxicity (L/E/IC50 value)	Ideal DfE Candidate
≤1 mg/l	May be acceptable if biodegradation occurs within a 10-day window without products of concern
>1 ppm and ≤10 ppm	Biodegradation occurs within a 10-day window without products of concern
>10 ppm	Biodegradation occurs within 28 days without products of concern

development toward products that are benign with respect to human and ecological health and safety. Initial funding and support for CleanGredients^{SM} was provided by the US EPA DfE Program, which continues to collaborate closely during the development process. CleanGredients^{SM} supports product recognition by waiving chemical review costs for formulators who use ingredients that meet the DfE Screen. CleanGredients^{SM} also supports ecolabelling efforts. Ecolabel certifiers cannot recommend chemicals for use in reformulation to clients due to the potential for conflict of interest, but they can refer to CleanGredients^{SM} as a resource for greener chemicals that will help to meet ecolabel criteria.[55]

SC Johnson & Son, Inc. GreenList^{TM}. As described previously, the GreenList^{TM} metrics allow for the comparison and ranking of chemical alternatives within an ingredient class (i.e., solvents, chelating agents, etc.) based on a set of key criteria. Scores for formulated products are then calculated based on a weighted average of the scores of the individual ingredients. Ingredients that scored a 3 represent what is considered 'Best in Class'. As displayed in Table 8.2, chelating agents fall into Category 3 or 'Best' if the chelating agents are not labelled with a risk phrase in the European Union, are rapidly biodegradable with low aquatic toxicity and if they are supplied by a supplier with a high supplier rating based on having an environmental management system and a commitment to green chemistry as measured by '% green chemistry preferred'.

The Green Screen for Safer Chemicals, the DfE Alternatives Assessment methodology, the DfE Formulator Program, CleanGredients^{SM} and GreenList^{TM} all provide publicly accessible tools and metrics to evaluate chemical hazards. They identify key hazard endpoints, provide thresholds within each hazard endpoint and provide a system for ranking and communicating hazard values based on discrete and transparent thresholds. They also illustrate contrasting approaches in defining what 'good' looks like – whether defined as the absence of hazard across all endpoints or as best in class based on functional use. It is useful to have both frameworks to ensure that the bar for best in class is held high in the context of overall hazard.

8.2.4 Implementation: Setting and reaching goals

As noted earlier, steps on the path to green chemical inventories include promoting awareness and motivation within an organization, performing an inventory of chemicals in products and processes, and assessment and evaluation of the chemicals used. The next step is implementation whereby the organization establishes goals and targets along with supporting metrics.

In Section 8.2.3, we simplistically characterized chemicals as (i) chemicals of concern, (ii) better alternatives and (iii) positively defined 'known goods'. We described some tools and metrics that are useful for each level of characterization. The distinctions themselves implicitly suggest different strategies for implementation. For example, once chemicals of concern are identified, it makes sense to focus on restricting them from use. Where one is evaluating alternatives, it makes sense to assess and evaluate chemical options by collecting data and using frameworks such as the Green Screen for Safer Chemicals or the GreenList^{TM} and to select safer alternatives. And where certain chemicals or materials have been positively defined, that is fully characterized and determined to be positive with respect to performance, human and environmental health and safety, it makes sense to find ways to integrate such chemicals and materials into products and processes. Depending on resources, the nature of one's current inventory, and the results of the prior steps, companies may engage in one or more of these different implementation strategies simultaneously. In this section, we provide examples of companies and programs that support implementation.

8.2.4.1 Eliminate chemicals of concern

Eliminating chemicals of concern often starts as a reactive strategy but it is an important first step on the path to implementing greener chemical inventories. In this section, we discuss two initiatives that focus on eliminating chemicals of concern. While we use an example from a very large company, their approach can be applied by small and large businesses alike.

Wal-Mart. Wal-Mart has recently established preferred chemical principles for the chemical products they sell. In addition, they have established a Chemical Intensive Products (CIP) Network whose stated mission is to 'provide to our customers affordable and effective products where all chemical ingredients are preferred for Mother, Child and the Environment delivered in the most efficient and effective way'.[56] In practice, this means that when an ingredient in a product or the product itself is considered to be capable of causing harm to human health and the environment, Wal-Mart will act to find better alternatives.

Wal-Mart's CIP Network's Preferred Chemical Principles states that they will favour those products that do not contain the following:

Chemicals harmful to human health

- Carcinogens (C) – can cause cancer
- Mutagens (M) – can damage genetic material
- Reproductive toxicants (R) – may affect reproduction or the unborn.

Chemicals harmful to the environment

- Persistent (P) – do not break down in the environment
- Bioaccumulative (B) – builds up in the food chain
- Toxic (T) – causes death or damage to organisms in the environment.

Based on these principles, Wal-Mart has proposed to screen all of their chemically intensive products (CIPs) for ingredients that have characteristics of CMRs or PBTs. Based on the results of the screening, formulated chemical products will be evaluated and the information will be used to guide purchasers in decision making. Wal-Mart is currently considering software platforms that will allow them to automate the product screening process. They are working with their suppliers and environmental advocacy organizations to optimize the criteria for screening. In addition, Wal-Mart will work with suppliers to obtain products that do not contain either of the two pesticides: propoxur and permethrin (both used in household insect control products) or a surfactant commonly used in cleaning products: nonylphenol ethoxylates (NPEs).[57] This recently launched strategy reflects a desire to identify and eliminate chemicals of concern that will have far-reaching effects throughout their enormous supply chain.

Safer Detergent Stewardship Initiative (SDSI). An example of a program that focuses on eliminating chemicals of concern is the US EPA DfE Program's SDSI.

SDSI is being developed to recognize companies, facilities, and others who voluntarily phase out or commit to phasing out the manufacture or use of NPE surfactants. NPEs are used in detergents in cleaning and other products. Both NPEs and their breakdown products, such as nonylphenol, can harm aquatic life. The Safer Detergents Initiative complements EPA's Aquatic Life Ambient Water Quality Criteria for Nonylphenol which are designed to protect aquatic life in both fresh and saltwater and can form the basis for state and tribal water quality standards.

SDSI will engage and recognize two types of participants – those who have phased-out the manufacture or use of NPE surfactants, and those who commit to a phase-out and take action to implement it. Participants may include suppliers, formulators, retailers/distributors, end-users, and others who actively encourage the use of safer surfactants.[58]

Eliminating chemicals of concern can lead a company to become more proactive by creating demand for 'greener' alternatives and by positively defining chemicals. While it may appear at first to be restrictive in terms of disallowing chemicals, in practice it tends to drive innovation and lead to the development of new chemicals and products – part of what makes green chemistry so attractive to people of widely divergent views.

8.2.4.2 Evaluate alternatives

The need to evaluate alternatives often arises once the decision to eliminate chemicals of concern is made. Sometimes a chemical of concern can be eliminated through product design such as through the use of a material that is inherently flame retardant or a fabric barrier without the need for added flame-retardant chemicals. But when the only currently feasible option appears to be a chemical substitution, then the alternatives must be evaluated and assessed to ensure that they are in fact 'greener' than the chemical they are intended to replace. We provide as examples of implementing alternatives assessment, the Green Screen for Safer Chemicals and the DfE Flame Retardants in Printed Circuit Boards Partnership.

The Green Screen for Safer Chemicals. The Green Screen for Safer Chemicals was successfully applied to evaluate three flame retardants that currently meet performance criteria for use in the external plastic housing of televisions (TVs). The three flame retardants evaluated are:

- decabromodiphenyl ether (decaBDE)
- resorcinol bis(diphenylphosphate) (RDP), and
- bisphenol A diphosphate (BAPP or BPADP).

The method for applying the Green Screen involved

- identifying viable candidates for flame-retarding TV enclosures
- assessing the hazards of the three flame retardants and their associated breakdown products
- applying the Green Screen benchmarks to the results of the hazard assessments.

The results of the application concluded that of the three flame retardants, RDP was the only flame retardant to pass all criteria under Benchmark 1 of the Green Screen. Based on this assessment, it was determined that RDP, at Benchmark 2, was the preferred alternative of the three flame-retardant options.

The importance of the Green Screen is that it considers a comprehensive set of hazards and provides a useful framework for comparing and evaluating chemical alternatives. By demonstrating that alternatives that meet performance requirements are indeed preferable with respect to human and environmental health and safety, decision makers from government, industry, purchasing entities and environmental organizations can have more confidence in selecting alternatives to chemicals of concern.

Flame retardants in printed circuit boards partnership.[59] Circuit boards are commonly used in electronics in consumer and industrial products, including computers and cell phones. In order to ensure fire safety, manufacturers commonly produce circuit boards with flame-retardant chemicals. While serving an important performance function, some flame-retardant chemicals can be harmful if released into the environment. To better understand the issues and the full range of options for flame-retarding circuit boards, DfE is engaging with the electronics industry and other stakeholders in a partnership.

Most circuit boards are FR-4 boards that meet standards for fire safety by the use of brominated epoxy resins in which the reactive flame-retardant tetrabromobisphenol A (TBBPA) forms part of the polymeric backbone of the resin. Alternative flame-retardant materials are used in only 3–5 per cent of the FR-4 boards, but additional alternative flame-retardant materials are also under development. Little information exists concerning the potential environmental and human health impacts of the materials which are being developed as alternatives to those used today that are based on brominated epoxy resins.

The DfE partnership will develop information to improve the understanding of the environmental and human health impacts of new and current materials that can be used to meet the fire safety requirements for circuit boards. Project partners include electronics manufacturers, component and board manufacturers, chemical companies, trade associations, environmental groups, universities and governments. The project will consider environmental and human health impacts that can occur throughout the life cycle of a material, from development and manufacture, through product use and end of life of the material or product. Stakeholders have expressed a particular interest in understanding the combustion products that could be formed during certain end of life scenarios. The list of active partners includes:

- Acer in cooperation with ITRI (Industrial Technology Research Institute)
- Albemarle
- BSEF (The Bromine Science and Environmental Forum)
- Ciba Specialty Chemicals
- Chemtura
- Clariant
- Clean Production Action
- Dell
- The Dow Chemical Company
- EIA (Electronic Industries Alliance)
- Fujitsu Siemens Computers
- Greenpeace

- HDP User Group International, Inc. (High Density Packaging)
- HFFREC (The Halogen-Free Flame Retardants Electronics Consortium)
- Hitachi Chemical Company America, Ltd.
- HP (Hewlett-Packard Company)
- IBM
- ICL Industrial Products
- IPC - Association Connecting Electronics Industries
- Intel
- iNEMI (International Electronics Manufacturing Initiative)
- ITEQ Corporation
- Lenovo
- Matsushita Electric Works
- Nan Ya Plastics
- NEMA (National Electrical Manufacturers Association)
- Nokia
- Panasonic
- PRé Consultants
- Purdue University
- Supresta
- The Swedish Institute for Fibre and Polymer Research (IFP Research) in cooperation with KemI
- Tyco Printed Circuit Group, LP.

Evaluating chemicals is challenging because decision makers are often faced with both a lack of information and with lack of consensus on what constitutes 'green'. Both of these deficiencies are addressed through the US EPA DfE Partnerships and the Green Screen for Safer Chemicals.

8.2.4.3 Using positively defined chemicals ('known goods')

Once a chemical or material is positively defined, that is it is fully characterized and determined to be preferable as used with respect to human and environmental health and safety, organizations may establish a platform from which to develop products and processes. We provide as examples of positively defining chemicals and chemical products the CleanGredients^{SM} database and the Coastwide Laboratories' Sustainable Earth® Green Chemistry (SEGC) 114 Standard for green chemical product development.

CleanGredients^{SM}.[53] As noted earlier, surfactant preparations listed in CleanGredients^{SM} are, at a minimum, ultimately biodegradable with no known degradation products of concern.[54] In order to be listed on CleanGredients^{SM}, suppliers must submit full ingredient formulation information along with data for aquatic toxicity and biodegradability to a third party (currently NSF International) for review. In addition to validating data, NSF also evaluates surfactant products against the DfE Screen for Surfactants criteria (Table 8.2) defined by the US. EPA DfE Program that designates surfactants with especially positive environmental characteristics.[60] NSF's reviews are carried out under confidentiality, providing verification of claims for the key ingredient attributes for surfactants without compromising proprietary formulations. CleanGredients^{SM} promotes green chemistry by positively defining chemicals for use in cleaning product formulations. Chemicals are positively defined in two ways: (1) they are well characterized, that is, claims about key human and/or environmental health and safety attributes are supported with appropriate test data, and (2) the attributes that they exhibit are determined to be preferable with respect to human and environmental health and safety. The attributes and criteria for listing solvents and for identifying those that meet the DfE Screen are currently under development.

Positively characterized chemical ingredients provide chemical manufacturers, product formulators and purchasers with chemical product candidates for use in green product formulation. By using ingredients that are listed in CleanGredients^{SM} and/or that meet the DfE Screen, formulators can have more confidence that their products have verifiable green chemistry characteristics and they are more likely to be successful with third party product recognition and certification programs.

Coastwide Laboratories/Corporate Express and SEGC 114. Coastwide Laboratories, now owned by Corporate Express uses green chemistry to develop products that meet rigorous performance, environmental, and human health criteria. They developed a product development standard called the Sustainable Earth® Green Chemistry Standard 114 (SEGC 114), to establish positive criteria for product efficacy as well as human and environmental health benefits.[61] Their strategy involves:

1 fully assessing all ingredients to understand their potential human health, environmental health, and lifecycle impacts,
2 formulating products to meet SEGC 114.

This strategy has resulted in entirely new formulations with remarkable benefits. Sustainable Earth® (SE) cleaning products combine reagents determined to be safer for human and environmental health with a positively characterized hybrid surfactant system containing a stabilized oxidizing compound. This system eliminates conventional ingredients such as alkyl glycol ethers, alkali builders, alkylphenol ethoxylates, EDTA and ethanolamine.

SE products have increased functionality and performance, use fewer, more benign ingredients and reduce waste and emissions. Current SE products include many types of cleaners, as well as an odour eliminator, floor finish, wax stripper and dust mop treatment. The success of the SE products contributed much to the company's success. The company has committed to formulating ALL of their formulated products to meet the SEGC 114 standard by 2009. Having a framework to positively define chemical products and through which to develop new products helps the company to gain human and environmental health and safety advantages, demonstrates a commitment to green chemistry and helps to achieve product recognition in the marketplace.

8.2.5 Reviewing and reporting

Reviewing and reporting is part of the management system approach to greening chemical inventories and is critical to assessing progress. We limit our discussion here to two examples, the Global Reporting Initiative and the work of the Investors Environmental Health Network to benchmark corporate efforts to lower product toxicity and to promote safer alternatives.

8.2.5.1 Global Reporting Initiative

The Global Reporting Initiative (GRI) is an international network of thousands of individuals from business, civil society, labour, and professional institutions who collaborate to create the content of the Reporting Framework.[62] The Global Reporting Initiative's (GRI) vision is that reporting on economic, environmental and social performance by all organizations becomes as routine and comparable as financial reporting. GRI accomplishes this vision by developing, continually improving and building capacity around the use of its Sustainability Reporting Framework. The Sustainability Reporting Guidelines consist of:

- Reporting principles,
- Reporting guidance, and
- Standard disclosures (including Performance indicators).

The core Guidelines are in their third generation (G3) and were released in October 2006. They provide the foundation upon which all other reporting guidance is based and outline core content for reporting that is broadly relevant to all organizations regardless of size, sector or locations. Protocols are the 'recipe' behind each indicator in the Guidelines and include definitions for key terms, compilation methodologies, scope and technical references. Sector supplements complement the core Guidelines by capturing unique sets of sustainability issues faced by different sectors such as mining, automotive, banking public agencies and others.

The GRI provides a useful framework for sustainability reporting and continues to evolve. While the GRI currently focuses on a broad set of sustainability metrics, it could be adapted to include green chemistry metrics.

8.2.5.2 The Investors Environmental Health Network

The work of the Investors Environmental Health Network is targeted toward investors and senior executives and focuses on corporate management of product detoxification. From an article published in the journal, *Corporate Environmental Strategy: International Journal for Sustainable Business*, Richard Liroff wrote:

> Companies face increasing inquiries about their knowledge and management of toxic chemicals in their products. Inquiries are fueled by reports of rising levels of contaminants in human blood and breast milk, scientific findings about links between chemical exposures and human health, activist campaigns against cancer-causing ingredients in cosmetics and hazardous chemicals in electronics products and shareholder resolutions urging changes in corporate management of toxic chemicals. Companies that do not understand toxic hazards in their products and who do not take steps to reduce or eliminate them face the risk of disruption to their supply chains, exclusion from markets, damage to their reputation, foregone profits, and toxic tort litigation. On the other hand, such concerns present a remarkable opportunity for innovation and entrepreneurship that can contribute to competitive advantage, reduced operating costs, increased profits, and enhanced shareholder value.[63]

Liroff offers a benchmarking tool to assess progress in corporate management of product detoxification. The tool can be used internally by senior corporate management teams or externally by investors and investment analysts to screen investments, assess 'best in class' environmental performance, manage portfolio risk and launch shareholder actions. The framework describes activities that can be done to demonstrate:

- Corporate commitment to safer alternatives policy
- Investor and public accountability
- Data development
- Internal capacity and 'greening' the supply chain.

He offers examples of how prominent companies in the electronics, retailing and consumer products sectors have adopted elements of the framework.

8.2.6 Repeat

Engaging on the path to green chemical inventories is part of an ongoing process of continual improvement that results in continued advancement toward aspirational goals consistent with corporate policies. The final step, repeat, becomes the first step in a reiteration that leads to further improvements. By continuing to chip away at restricting chemicals of concern, selecting safer alternatives based on alternatives assessment and evaluation, and identifying and deploying preferred chemical platforms as they become available, organizations drive innovation and improve the human and environmental health and safety profiles of their chemical products and projects – benefiting people, the environment and their bottom line. The steps outlined in this chapter practiced in concert with the strategies, tools and metrics can be applied to current inventories and to all chemicals considered for future purchase and use. Businesses can track their successes and use the results to develop marketing materials and public reports while continuing to advance on the path toward green chemical inventories.

Conclusion

In this chapter we describe strategies, tools and metrics that are currently and publicly available for advancing green chemical inventories in products and processes. The science of green chemistry will continue to advance as will the frameworks, strategies, tools and metrics that support understanding, implementation and reporting.

While advancing green chemical inventories is critical to a sustainable future, it is important to remember the big picture when designing products and processes – it is not just about hazard. Once the hazard assessment and evaluations are complete, and as part of the implementation stage, it is important to step back and consider the big picture when developing products. As noted earlier, a sustainable product – one that meets the criteria for Cradle to Cradle' Design:

- Meets market requirements
- Has positive social effects (for individuals and communities)
- Is safe for human and ecological health
- Is sourced from renewable or repeatedly recycled materials
- Is sourced from renewable energy

- Is designed for safe, productive return to nature or industry
- Is recovered and recycled at highest quality after use.

Strategies for sustainable product design should include the design of products as biological nutrients within biological metabolisms or technical nutrients within technical metabolisms.

Assessing and evaluating chemical hazards in order to seek the greenest options provides the foundation from which lifecycle impacts and exposure can be considered. By first selecting chemical candidates with the least inherent hazard, companies can make more informed choices by considering the potential for exposure and lifecycle impacts.

Appendix 1: List of Resources for Hazard and Toxicological Screening and Research

A Resources to identify chemicals of concern
B Chemical toxicity
C Exposure assessment tools
D Hazard and risk assessment tools
E Safer chemistry design tools

A. Resources to Identify Chemicals of Concern

CA Proposition 65

www.oehha.ca.gov/prop65.html

'Prop 65' is a list of chemicals that have been confirmed by the state of California to be carcinogens and or reproductive toxins. Chemicals are listed by name and by CAS #

National Toxicology Program Report on Carcinogens

ntp.niehs.nih.gov/ntp/roc/toc11.html

NTP maintains a list of chemicals that are carcinogens, likely carcinogens or probable carcinogens.

Toxics Release Inventory Resources (TRI)

www.epa.gov/tri/index.htm

TRI is a publicly available EPA database that contains information on toxic chemical releases and other waste management activities reported annually by certain covered industry groups as well as federal facilities.

Clean Air Act: Hazardous Air Pollutant List

www.epa.gov/ttn/atw/188polls.html; www.epa.gov/ttn/atw/pollutants/atwsmod.html

US Congress amended the Federal Clean Air Act in 1990 to address a large number of air pollutants that are known to cause or may reasonably be anticipated to cause adverse effects to human health or adverse environmental effects. 188 specific pollutants and chemical groups were initially identified as hazardous air pollutants (HAPs), and the list has been modified over time.

SARA/EPCRA 313 List (TRI)

www.epa.gov/tri/chemical/index.htm

Section 313 of the Emergency Planning and Community Right to Know Act (EPCRA) of 1986 was enacted to facilitate emergency planning, to minimize the effects of potential toxic chemical accidents, and to provide the public with information on releases of toxic chemicals in their communities. The current list contains 581 individually listed chemicals and 30 chemical categories.

Clean Water Act Priority Pollutants List

oaspub.epa.gov/wqsdatabase/wqsi_epa_criteria.rep_parameter

Section 307 of the CWA defines a list of 126 priority pollutants for which the US EPA must establish ambient water-quality criteria and effluent limitations.

Comprehensive Environmental Response, Compensation, and Liability Act (CERCLA)

www.epa.gov/ceppo/pubs/title3.pdf

List of over 800 chemicals from, Section 102 of CERCLA, Clean Water Act list of hazardous substances and priority pollutants (Section 211(b)(2)(a) or 307(a)), Any hazardous waste as defined under Section 3001 of Resource Conservation and Recovery Act; Clean Air Act list of hazardous air pollutants (HAPs) (Section 112); Toxic Substances Control Act list of imminent hazards (Section 7).

Canadian Environmental Protection Act, 1999 (CEPA)

www.ec.gc.ca/CEPARegistry/subs_list/Toxicupdate.cfm

Under CEPA substances that are determined to be 'toxic' are recommended for addition to the List of Toxic Substances (Schedule 1) of the Act.

EU Risk Phrase

ec.europa.eu/environment/dansub/consolidated_en.htm

Council Directive 67/548/EEC of 27 June 1967 on the approximation of laws, regulations and administrative provisions relating to the classification, packaging and labelling of dangerous substances. Annex I of the directive assigns Risk Phrases to chemical substances.

European Commission's Community Strategy for Endocrine Disrupters

http://ec.europa.eu/environment/endocrine/strategy/substances_en.htm#priority_list

The European Commission (EC) has sponsored four reports that evaluate a set of 553 substances selected by experts and stakeholders for assessment for endocrine disruption. Each report addresses a subset of the total set based on priorities such as whether the chemicals are persistent, bioaccumulating or High Production Volume chemicals and/or whether or not there is already regulatory control of the chemical.

WHO Water Quality Guidelines

www.who.int/water_sanitation_health/dwq/gdwq3/en/

International standards for selected chemical, microbiology and other parameters.

The Danish List of Undesirable Substances (LOUS)

glwww.mst.dk/homepage/default.asp?Sub=http://glwww.mst.dk/udgiv/publications/2004/87-7614-477-1/html/kolofon_eng.htm

The Danish List of Undesirable Substances is a list of chemicals of concern that the government believes should be avoided to the extent feasible in commerce. Using a systematic analysis, substances are selected automatically if they meet some clear and defined criteria, for example, problematic classifications, because they are under suspicion for being PBT/vPvB (Persistent, Bioaccumulative, Toxic/very Persistent, very Bioaccumulative) or endocrine-disrupting.

International Agency for Research on Cancer (IARC)

monographs.iarc.fr/index.php

IARC Monographs are the result of interdisciplinary working groups of expert scientists who review published studies and evaluate the weight of the evidence that an agent can increase the risk of cancer. Since 1971, more than 900 agents have been evaluated, of which approximately 400 have been identified as carcinogenic or potentially carcinogenic to humans.

National Institute for Occupational Safety and Health (NIOSH) Carcinogen List

www.cdc.gov/niosh/npotocca.html

NIOSH maintains a list of substances considered to be potential occupational carcinogens.

Occupational Safety and Health Administration (OSHA)

www.osha.gov

OSHA maintains a list of potential carcinogens. In addition, OSHA sets enforceable permissible exposure limits (PELS) to protect workers against the health effects of exposure to hazardous substances.

EPA Water Disinfection By-Products with Carcinogenicity Estimates (DBPCAN)

www.epa.gov/ncct/dsstox/sdf_dbpcan.html

The DBPCAN database contains predicted estimates of carcinogenic potential for 209 chemicals detected in finished drinking water samples having undergone water disinfection treatment.

PBT Profiler

www.pbtprofiler.net/

EPA has developed an evaluation tool, the PBT Profiler, which predicts PBT potential of chemicals. The PBT Profiler estimates environmental persistence (P), bioconcentration potential (B), and aquatic toxicity (T) of discrete chemicals based on their molecular structure. It is Internet-based and there is no cost for use.

Health & Safety Executive (HSE) Direct

www.hse.gov.uk/legislation/services.htm

The Health and Safety Commission is responsible for health and safety regulation in Great Britain. HSE Direct is a subscription service providing full-text access to the full range of HSE guidance publications as well as the consolidated and annotated text of health and safety legislation and recent legislative changes. It includes a 'Stop- Press' feature which informs the user of very recent legislative changes. HSE direct provides full-text versions of HSE Legal Services material (Approved Codes of Practice, 'L' series, Codes of Practice and HSR series); HSE forms; European Directives; Safety, Health and Environment cases; and summaries of British Standards. Users can choose from either 'day ticket' access or regular subscription options.

CA Toxic Air Contaminants List

www.arb.ca.gov/toxics/id/taclist.htm

California maintains a list of Toxic Air Contaminants (TACs) and a program for adding additional TACs.

B. Chemical Toxicity Data

Ecological Structure Activity Relationships (ECOSAR)

http://www.epa.gov/oppt/newchems/tools/21ecosar.htm

ECOSAR is a personal computer software program that is used to estimate the toxicity of chemicals used in industry and discharged into water. The program predicts the toxicity of industrial chemicals to aquatic organisms such as fish, invertebrates and algae by using Structure Activity Relationships (SARs). The program estimates a chemical's acute (short-term) toxicity and, when available, chronic (long-term or delayed) toxicity.

Ecotox database

http://cfpub.epa.gov/ecotox/

The Ecotox database provides single chemical toxicity information for aquatic and terrestial life. This is a useful tool for evaluating the impact of chemicals on the environment.

EPA Triage Database

http://www.epa.gov/8e_triag/

Triage is a searchable database of scientific studies on the health and environmental effects of toxic chemicals related to Section 8(e) of TSCA.

Integrated Risk Information System (IRIS) Database

http:/www.epa.gov/iris/

IRIS is a database of human health effects that may result from exposure to various substances found in the environment. IRIS was initially developed for EPA staff in response to a growing demand for consistent information on chemical substances for use in risk assessments, decision-making and regulatory activities.

Toxic Substances Control Act Test Submissions (TSCATS)

http://www.rtknet.org/tsc/

TSCATS is an online index to unpublished, nonconfidential studies covering chemical testing results and adverse effects of chemicals on health and ecological systems. The studies are submitted by US industry to EPA under several sections of the Toxic Substance Control Act (TSCA). There are four types of documents in the database: Section 4 chemical testing results, Section 8(d) health and safety studies, Section 8(e) substantial risk of injury to health or the environment notices, and voluntary documents submitted to EPA known as a For Your Information (FYI) notice.

High Production Volume Information System (HPVIS)

www.epa.gov/hpvis/index.html

HPVIS provides complete and easy access to technical health and environmental effect information on chemicals that are manufactured or imported to US in volumes greater than 1MM lbs per year. Information in this database are submitted through HPV Challenge Program. HPVIS allows users to search for summary information, test plans, and new data on high production volume chemicals as they are developed.

Chemical Fact Sheets

www.epa.gov/chemfact/

The EPA's Office of Pollution Prevention and Toxics developed Chemical Fact sheets to summarize information on a particular chemical including exposure, environment and human health hazard, environmental fate, regulatory information, and whom to contact for additional information.

TOXNET

toxnet.nlm.nih.gov/

TOXNET is a series of databases on chemical toxicity hosted by the National Institutes of Medicine, which allows multiple searching options. The databases include: ChemIDplus, HSDB, Toxline, CCRIS, DART, GENETOX, IRIS, ITER, LactMed, Multi-Database, TRI, Haz-Map, Household Products and TOXMAP.

Enhanced ToxSeek Meta-Search Engine and Clustering Tool

toxseek.nlm.nih.gov

ToxSeek is an NLM metasearch engine and clustering tool that enables the simultaneous searching of many different toxicology and environmental health information databases and web sites. This tool includes 59 databases including the TOXNET Search tool, as well as information sources from NLM, NIH, US Government, International and other sources.

Library of Chemical Information

www.cfsan.fda.gov/~dms/chemist.html

The Library of Chemical Information is maintained by the US Food and Drug Administration's Center for Food Safety and Applied Nutrition and is an excellent database for multiple classes of chemicals including food additives, cosmetics, colour additives, pesticides and other chemicals.

Fragrance or Flavor Components Database

rifm.org/nd/Login.cfm

The Research Institute on Fragrance Materials maintains the most comprehensive worldwide source database on fragrance/flavour components including acute aquatic toxicity, biodegradation data, human health issues, for example, carcinogenesis, sensitization. A password is required to access the database.

The Scorecard Database

www.scorecard.org/

The Scorecard Database provides information on chemical releases, risk prioritization of substances and other relevant information for chemicals and facilities.

International Chemical Safety Cards (ICSCs)

http://www.cdc.gov/niosh/ipcs/icstart.html

ICSCs are made available by the National Institute for Occupational Safety and Health. They summarize safety information in 8 languages and indexed in a variety of ways.

CleanGredients

www.cleangredients.org/

CleanGredients^{TM} is an online database of cleaning product ingredients 'A one-stop-shop for green formulation'. The database contains physical chemical data, MSDSs, technical datasheets, and environmental and human health hazard data on raw materials used to formulate cleaning products.

Cefas Building, Assessing and Standardising Information on the Atlantic Coasts (BASIC) Toxicity Database

www.cefas.co.uk/basic/toxdata.htm

The BASIC toxicity database contains information on the aquatic toxicity of a number of hazardous substances. In many cases, the information is given as some sort of 'safe' level such as UK Environmental Quality Standards (EQSs) or the national/international equivalent. For substances for which no such levels have been set, a brief literature review was performed in order to produce an environmental hazard/risk assessment.

PubMed® MEDLINE®

www.pubmed.gov

MEDLINE contains bibliographic citations and author abstracts from more than 5000 biomedical journals published in the United States and 80 other countries. The database contains over 15 million citations dating back to the mid-1950s.

Chemical Backgrounders

www.nsc.org/library/chemical/index.htm

Chemical Backgrounders is maintained by the National Safety Council and contains chemical descriptors including properties, health effects, exposure and regulatory information. (I can't access is membership required?).

Managerial Technologies Corp (MTC) Safety Library

www.mtclibraries.com/lib/resource_locate.php?id=sl26949

The safety professional's online library, covering topics related to safety, safety management, ergonomics, fleet and environment. The Safety Library contains over 40 environmental databases and over 70 databases of chemical profiles. Subscription is required.

Chemistry Links for Chemists

www.liv.ac.uk/Chemistry/Links/refdatabases.html

This website contains a list of 105 different databases from US and around the world.

Environmental Health and Safety Freeware

www.ehsfreeware.com/cheminfo.htm

This website contains a list of almost 100 different EH&S databases that are all freeware plus several commercial databases.

Ariel™ WebInsight

www.3ecompany.com/ariel_webinsight.com

Ariel™ WebInsight is a subscription-based online compliance management tool with an easy-to-use interface and robust search, query, reporting and analysis features. The tool provides access to current, accurate, comprehensive global regulatory content containing more than 700 searchable regulatory lists covering more than 75 countries; full-text repositories of legislation, international transportation data, as well as chemical property and hazard data.

Infochems database

www.infochems.com/main/default.asp

Commercial service that provides chemical information for US and Global regulations. Contains chemical, physical and toxicity information. The database includes generic chemical names and commercial chemical names.

American Conference of Governmental Industrial Hygienists (ACGIH)

www.acgih.org

ACGIH maintains annual editions of the TLVs® and BEIs® which are used worldwide as a guide for evaluation and control of workplace exposures to chemical substances and physical agents. Threshold Limit Value (TLV®) occupational exposure guidelines are recommended for more than 700 chemical substances and physical agents. There are more than 50 Biological Exposure Indices (BEIs®) that cover more than 80 chemical substances.

The Canadian Centre for Occupational Health and Safety (CCOHS's Databases)

ccinfoweb.ccohs.ca/

CCOHS's Web Information, a paid service, provides simple, one-step searching across many database collections, including:

- *MSDS* is a comprehensive database of more than 310 000 Material Safety Data Sheets, obtained directly from 2000 North American manufacturers and suppliers.
- *MSDS Management Service* is designed to help you manage your occupational health and safety responsibilities relating to MSDSs and WHMIS requirements.
- *CHEMINFO* provides chemical health and safety information for more than 1300 important workplace chemicals.
- *CHEMpendium*™ regularly updated information on the transport of hazardous materials, chemical toxicity, industrial chemicals and environmental contaminants, workplace safety, regulatory compliance, WHMIS and Right-to-Know, Canada's DLS and NDSL, Emergency response.
- *RTECS*® helps you find critical toxicological information with citations for over 160 000 chemical substances from more than 2500 sources. This database is also available in French and Spanish.

- *OSH References* allows you to search over 300 000 summaries. This collection of bibliographic databases provides you with convenient access to international sources of OSH-related information including OSHline®, NIOSHTIC®, NIOSHTIC-2, HSELINE, CISILO, and Canadiana.
- *Canadian enviroOSH Legislation plus Standards* is a comprehensive, easy to search compilation of the full text of the Canadian health, safety and environmental legislation, critical guidelines and codes of practice from all jurisdictions and regular updates, with amendments highlighted.

CrossFire Beilstein

www.mdl.com/products/knowledge/crossfire_beilstein/

The CrossFire Beilstein database is the world's largest compilation of chemical facts. This database indexes three primary data domains: substances, reactions and literature. The substance domain stores structural information with all associated facts and literature references, including chemical, physical and bioactivity data. The reaction domain details the preparation of substances, enabling scientists to investigate specific reaction pathways with reaction search queries. The literature domain includes citations, titles and abstracts, which are hyperlinked to the substance and reaction domain entries. It contains over 320 million experimental data, over 10 million reactions and data indexed from over 175 journals.

European Chemical Substances Information System (ESIS)

ecb.jrc.it/esis/

ESIS is an IT system which provides you with information on chemicals including:

- EINECS (European Inventory of Existing Commercial Chemical Substances)
- ELINCS (European List of Notified Chemical Substances)
- NLP (No-Longer Polymers)
- HPVCs (High Production Volume Chemicals) and LPVCs (Low Production Volume Chemicals), including EU Producers/Importers lists
- C&L (Classification and Labelling), Risk and Safety Phrases, Danger and so on
- IUCLID (International Uniform ChemicaL Information Database) Chemical Data Sheets, IUCLID Export Files, OECD-IUCLID Export Files, EUSES Export Files
- Priority Lists, Risk Assessment process and tracking system in relation to Council Regulation (EEC) 793/93 also known as Existing Substances Regulation (ESR).

eChemPortal

webnet3.oecd.org/eChemPortal/Home.aspx

The eChemPortal is an effort of the Organisation for Economic Co-operation and Development (OECD) in collaboration with the European Commission, the United States, Canada, Japan, the International Council of Chemical Associations, the Business and Industry Advisory Committee, the World Health Organization's International Program on Chemical Safety, the United Nations Environment Programme on Chemicals and environmental non-governmental organizations.

eChemPortal offers free public access to information on properties and effects of chemical substances. It is an integrated system that allows users to simultaneously search multiple databases prepared for government chemical and review programs around the world.

The current version of eChemPortal offers the possibility to retrieve information by searching on chemical names or CAS Registry numbers. The second phase will incorporate additional search options to retrieve and compile specific hazard or other effects data (e.g., toxicity endpoints) from the participating databases.

At this time, the following databases participate in eChemPortal™* European Chemical Substances Information System (ESIS, European Commission), *CHRIP (Japan's Information on Biodegradation and Bioconcentration of the Existing Chemical Substances in the Chemical Risk information platform), *OECD HPV Database (OECD), *Screening Information Datasets for High Volume Production Chemicals (UNEP Chemicals), *HPVIS (US Environmental Protection Agency), *INCHEM (IPCS).

C. Exposure Assessment Tools

Environmental Fate Database (EFDB)

www.syrres.com/eSc/efdb.htm

EFDB has been developed in support of US EPA. It is comprised of DATALOG and BIOLOG, which contain environmental fate, microbial toxicity and biodegradation data.

EPA's Environmental Fate Database for actives

cfpub.epa.gov/pfate/home.cfm

This database includes information on the environmental fate of pesticide actives.

EPA's OPPT Exposure Assessment Tools and Models

www.epa.gov/opptintr/exposure/index.htm

The Office of Pollution Prevention and Toxics (OPPT) has developed a series of methods, databases, and predictive models to help in evaluating what happens to chemicals when they are used and released into the environment. These tools are intended to be used by scientists and engineers familiar with exposure assessment principles.

Greatest Potential for Human Exposure report

http://www.hc-sc.gc.ca/ewh-semt/pubs/contaminants/existsub/exposure/index_e.html

Health Canada Proposal for Priority Setting for Existing Substances on the Domestic Substances List under the Canadian Environmental Protection Act, 1999: Greatest Potential for Human Exposure. This report describes a proposed priority setting process of existing substances in Canada. A stakeholder meeting was convened to discuss the Complex Exposure Model (comET).

D. Hazard and Risk Assessment Tools

International Uniform Chemical Information Database (IUCLID5)

ecb.jrc.it/iuclid5/

IUCLID5 is a software tool for entering and storing information on chemicals, as well as for preparing and submitting dossiers to fulfil legislation requirements. For the EU Chemicals Agency and Member States, it is the central data repository for all dossiers submitted, the basis for evaluating the risks of substances and requiring new information and the basis for restricting and authorizing the use of chemicals to manage risks. IUCLID data will comply with REACH, OECD, EU HPV Chemicals Program, US HPV Challenge Program, Japan Challenge Program and EU Biocides.

Massachusetts Toxics Use Reduction Institute Pollution Prevention Options Analysis Tool (p2oasys)

www.turi.org/content/content/view/full/1125/

The Institute has developed P2OaSys to help companies determine whether the Toxic Use Reduction (TUR) options they are considering may have unforeseen negative environmental, worker or public health impacts. P2OASys allows companies to assess the potential environmental, worker, and public health impacts of alternative technologies aimed at reducing toxics use. The goal is more comprehensive and systematic thinking about the potential hazards posed by current and alternative processes identified during the TUR planning process.

Report on the Advisory List for Self-Classification of Dangerous Substances

www.mst.dk/homepage/default.asp?Sub=http://www.mst.dk/udgiv/publications/2004/87-7614-477-1/html/default_eng.htm

The Danish EPA has developed an advisory list for self-classification of dangerous substances including 20 624 substances. The substances have been identified by means of QSAR models (Quantitative Structure-Activity Relationship) as having acute oral toxicity, sensitization, mutagenicity, carcinogenicity, and/or danger to the aquatic environment.

E. Safer Chemistry Design Tools

Sustainable Futures Program (SF)

www.epa.gov/oppt/newchems/pubs/sustainablefutures.htm

The SF program is an approach that encourages pollution prevention in new chemical development through the transfer of OPPT's chemical risk screening methodologies.

The Green Chemistry Expert System (GCES)

www.epa.gov/opptintr/greenchemistry/tools.html

GCES allows users to build a green chemical process, design a green chemical, or survey the field of green chemistry. The system is equally useful for new and existing chemicals and their synthetic processes. The GCES features are contained in five modules:

1 Synthetic Methodology Assessment for Reduction Techniques (SMART)
2 Green Synthetic Reactions
3 Designing Safer Chemicals
4 Green Solvents/Reaction Conditions
5 Green Chemistry References.

The Green Screen for Safer Chemicals

www.cleanproduction.org/Green.Greenscreen.php

The Green Screen for Safer Chemicals is a method for assessing and benchmarking chemical alternatives based on their hazard characteristics developed by Clean Production Action. The Green Screen builds on new chemicals assessment protocols developed by the USEPA and adapted for comparing alternatives by the USEPA Design for the Environment Program. It identifies chemical characteristics of high concern and characteristics of increasingly greener chemicals.

Alternatives Assessment Methodologies

www.turi.org

The Massachusetts Toxics Use Reduction Institute has developed an Alternatives Assessment Process Guidance for its analysis of substitutes to five chemicals in Massachusetts.

Swedish Chemicals Inspectorate PRIO Program

www.kemi.se/templates/PRIOEngframes_4144.aspx

PRIO is a web-based tool intended to be used to preventively reduce risks to human health and the environment from chemicals. The aim of PRIO is to facilitate in the assessment of health and environmental risks of chemicals so that people who work as environmental managers, purchasers and product developers can identify the need for risk reduction. To achieve this PRIO provides a guide for decision-making that can be used in setting risk reduction priorities.

Control of Substances Hazardous to Health Regulations (COSHH)

www.coshh-essentials.org.uk/

Developed by the UK Health and Safety Executive, COSHH Essentials provides advice on controlling the use of chemicals for a range of common tasks, for example, mixing or drying.

Safer Solvent Alternatives

www.irta.us

The Institute for Research and Technical Assistance (IRTA) identifies, develops, tests and demonstrates safer alternatives in cleaning applications, dry cleaning, paint stripping, coatings, adhesives and lubricants.

References

1. Crane, W.M. (2005) *The Impact of Regulatory Costs on Small Firms.* Lafayette College, Easton, PA for the United States Small Business Administration Office of Advocacy. Under contract SBHQ-03-M-0522.
2. McDonough, W. & Braungart, M. (2002) *Cradle to Cradle: Remaking the Way We Make Things.* North Point Press, New York.
3. Trost, B.M. (1991) The Atom Economy – A Search for Synthetic Efficiency. *Science,* **254**(5037), 1471–1477.
4. Sheldon, R.A. (2000) Atom Utilization, E Factors and the Catalytic Solution. *C.R. Acad. Sci. Paris, Serie, IIc, Chimie/Chemistry,* **3**, 541–551.
5. Organisation for Economic Co-Operation and Development (OECD) (2003) *Descriptions of Selected Key Generic Terms Used in Chemical Hazard/Risk Assessment.* OECD, Paris.
6. Anastas, P.T. & Warner, J.C.(1998) *Green Chemistry: Theory and Practice.* Oxford University Press, New York, p. 30. By permission of Oxford University Press.
7. American Chemistry Council (2002) *Guide to the Business of Chemistry.* American Chemistry Council, Arlington, VA.
8. Directive 2002/95/EC of the European Parliament and of the Council of 27 January 2003 on the restriction of the use of certain hazardous substances in electrical and electronic equipment (2003) *Official Journal of the European Union,* L37/19–37/23.
9. Directive 2002/96/EC of the European Parliament and of the Council of 27 January 2003 on waste electrical and electronic equipment (WEEE) (2003) *Official Journal of the European Union,* L37 13/02/2003, 24–39.
10. Clinton, W.J. (1998) Executive Order 13101Greening the Government through Waste Prevention, Recycling, and Federal Acquisition. *Federal Register,* **63**(179), 49643–49651.
11. Office of Revisor of Statutes, State of Minnesota (2006) Purchase of Recycled, Repairable, and Durable Materials. 16B.121 Minnesota Statues 2006.
12. State of Massachusetts (2007) Recycled and Environmentally Preferable Products and Services Guide: Massachusetts Statewide Contracts: Operational Services Division. Volume XXV.
13. State of Michigan (2007) Environmental Purchasing [Online – cited 24 April 2007] Available from: URL: http://www.michigan.gov/deq/0,1607,7-135-3585_30068_4174–,00.html
14. City of Seattle (2007) Sustainable Purchasing. [Online – cited 24 April 2007] Available from: URL: http://www.ci.seattle.wa.us/environment/purchasing.htm
15. City of San Francisco (2007) City and County of San Francisco Municipal Code: Environmental Code. Ordinance 70-07, File Number 061432, approved 2 April 2007. [Online – cited 24 April 2007] Available from: URL: http://www.municode.com/Resources/gateway.asp?pid=14134&sid=5
16. GovTrack.us. S. 2669 – 110th Congress (2008) Green Chemistry Research and Development Act of 2008, GovTrack.us (database of federal legislation) [Online – cited 29 May 2008]. Available from: URL: http://www.govtrack.us/congress/bill.xpd?bill=s110-2669.
17. http://www.dtsc.ca.gov/PollutionPrevention/GreenChemistryInitiative/index.cfm
18. Granholm, J.M. (2006) Promotion of Green Chemistry for Sustainable Economic Development and Protection of Public Health. *Executive Directive 2006-6.* State of Michigan.
19. Wilson, M.P. (2006) *Green Chemistry in California: A Framework for Leadership in Chemicals Policy and Innovation.* Prepared for California Senate Environmental Quality Committee and the

California Committee on Environmental Safety and Toxic Materials. Regents of the University of California, Berkeley, CA.
20. US Green Building Council (2005) LEED-EB. Green Building Rating System for Existing Buildings. Upgrades, Operations and Maintenance. Version 2.
21. www.environmentalchoice.com
22. www.greenseal.com
23. www.epa.gov/dfe/pubs/projects/formulat/index.htm
24. www.chemicalbodyburden.org/
25. www.safecosmetics.org/
26. Horn, D. & Mahanna, J. (2007) IBM Engineering Specification 46G3772: Baseline Environmental Requirements for Supplier Deliverables to IBM. [Online – cited 24 April 2007] Available from: URL: http://www.ibm.com/ibm/environment/products/46g3772_011007.pdf
27. Hewlett-Packard Development Company, L.P. (2005) HP to Eliminate Brominated Flame Retardants from External Case Parts of All New HP Brand Products. [Online – cited 24 April 2007] Available from: URL: http://www.hp.com/hpinfo/newsroom/press/2005/051101a.html
28. http://www.globalreporting.org/Home
29. Liroff, R. (2005). Benchmarking Corporate Management of Safer Chemicals in Consumer Products – A Tool for Investors and Senior Executives. Corporate Environmental Strategy: *International Journal for Sustainable Business.* [Online – accessed 24 April 2007] Available from: URL: http://www.iehn.org/files/CES%20Benchmarking%20article.pdf.
30. Reebok, Ltd. (2006) Reebok Recalls Bracelet Linked to Child's Lead Poisoning Death. [Online – cited 24 April 2007] Available from URL: http://www.reebok.com/Static/useng/Initiatives/Recall/Recall_0323.htm
31. OSHA Hazard Communication Standard. Subpart Z, Toxic and Hazardous Substances 29 CFR 1910.1200 section g. Material Safety Data Sheets.
32. Eastern Research Group and Syracuse Research Corporation (2005) Furniture Flame Retardancy Partnership: Environmental Profiles of Chemical Flame-Retardant Additives for Low-Density Polyurethane Foam. United States Environmental Protection Agency, EPA 742-R-05-002A.
33. http://www.nsf.org/business/laboratory_services/cleangredient_lab/CleanGredient.asp.
34. The Business for Social Responsibility (2002) Restricted Substances in Apparel Products: Implementation Resources. [Online – accessed 1 May 2007] Available from URL: http://www.bsr.org/CSRResources/Environment/RSLImplementationResources.pdf.
35. US EPA (2007) Fact Sheet L Multimedia Strategy for Priority Persistent, Bioaccumulative, and Toxic (PBT) Chemicals. [Online – accessed 1 May 2007] Available from URL: http://www.epa.gov/pbt/pubs/fact.htm
36. www.epa.gov/dfe (Formulator Program, SDSI Program); www.greenseal.org; www.ecologo.org
37. http://ec.europa.eu/environment/dansub/main67_548/index_en.htm
38. http://www.ec.gc.ca/substances/ese/eng/dsl/cat_index.cfm
39. http://www.hc-sc.gc.ca/ewh-semt/contaminants/existsub/categor/index_e.html
40. http://ntp.niehs.nih.gov/index.cfm?objectid=32BA9724-F1F6-975E-7FCE50709CB4C932
41. http://www.osha.gov/pls/oshaweb/owadisp.show_document?p_table=STANDARDS&p_id=11373
42. http://www.pbtprofiler.net/
43. http://www.pbtprofiler.net/about.asp
44. https://webinsight.arielresearch.com/Home.aspx
45. http://www.chemadvisor.com/
46. www.dolphinmsds.com and www.dolphinsafesource.com
47. http://www.epa.gov/dfe
48. Rossi, M. & Heine, L. (2007) The Green Screen for Safer Chemicals. Version 1.0. [Online – accessed 1 May 2007] Available from URL: http://www.cleanproduction.org/Green.Greenscreen.php

49. US patent No 6,973,362.
50. Fast Track Media LLC. (2007) SC Johnson Makes Greenlist Available Royalty-Free. [Online – accessed 1 May 2007] Available from URL: http://www.environmentalleader.com/2007/02/22/sc-johnson-makes-greenlist-available-royalty-free/
51. http://www.epa.gov/dfe/pubs/formulat/formulator-review1.pdf
52. http://www.epa.gov/dfe/pubs/projects/formulat/formpart.htm
53. http://www.cleangredients.org
54. http://www.cleangredients.org/about/ingredient-attributes/surfactants/dfe-screen
55. Terrachoice (2007) Database Streamlines Ingredient Selection. [Online – accessed 1 May 2007] Available from URL http://www.environmentalchoice.com/images/press/2007/PR_02_07.pdf
56. From presentation from the 'Molecule to Molecule' meeting in October 2006.
57. Walmart (2006) Walmart Stores, Inc. Launches Innovative Program to Inspire Use of Preferred Substances in Chemical Intensive Products. [Online – accessed 1 May 2007] Available from URL: http://www.walmartfacts.com/articles/4556.aspx
58. http://www.epa.gov/dfe/pubs/projects/formulat/sdsi.htm
59. http://www.epa.gov/dfe/pubs/projects/pcb/index.htm
60. http://www.cleangredients.org/about/ingredient-attributes/surfactants/dfe-screen
61. Coastwide Laboratories (2005) Standard Specification for Certification of Environmentally Preferable and Sustainable Institutional and Industrial (I&I) Cleaning Products [Online – accessed 1 May 2007] Available online from URL: http://www.coastwidelabs.com/Technical%20Articles/SEGC%20114-05%20August%202005.pdf
62. http://www.globalreporting.org/Home
63. Liroff, R. (2005) Benchmarking Corporate Management of Safer Chemicals in Consumer Products – A Tool for Investors and Senior Executives. Corporate Environmental Strategy: *International Journal for Sustainable Business.* [Online – accessed 1 May 2007] Available from URL: http://www.iehn.org/files/CES%20Benchmarking%20article.pdf

Index

Printed and bound in the UK by
CPI Antony Rowe, Eastbourne

Printed and bound by CPI Group (UK) Ltd, Croydon, CR0 4YY
15/09/2026
14962393-0004